DNA Fingerprinting

An Introduction

Breakthroughs in Molecular Biology
DNA Fingerprinting is the second volume to appear in this exciting new series of high quality, affordable books on the fields of molecular biology and immunology. This series is dedicated to the rapid publication of the latest breakthroughs and cutting edge technologies as well as syntheses of major advances within molecular biology.

Other volumes in the series include:

PCR Technology: Principles and Applications for DNA Amplification
H. Erlich, Ed.

Adhesion: Its Role in Inflammatory Disease
J. Harlan & D. Liu, Eds.

Antibody Engineering
C. Borrebaeck, Ed.

DNA Fingerprinting

An Introduction

Lorne T. Kirby

W. H. Freeman and Company
New York

Library of Congress Cataloging-in-Publication Data

Kirby, Lorne T., 1938–
DNA fingerprinting : an introduction / Lorne T. Kirby.
p. cm.
Originally published: New York : Stockton Press, c1990.
Includes bibliographical references and index.
ISBN 0-7167-7001-6
1. DNA fingerprints. I. Title.
[DNLM: 1. DNA Fingerprinting. QZ 50 K58d 1990a]
RA1057.55.K57 1992
614'.1--dc20
DNLM/DLC
for Library of Congress 91-34653
CIP

Printed in the United States of America

2 3 4 5 6 7 8 9 UB 9 9 8 7 6 5 4 3 2 1

To William and Mary Kirby, Sarah-Dean, Christopher, and Marcia Louise

CONTENTS

PREFACE

Each individual, except identical twins, possesses a unique genetic signature that can be visualized using recombinant DNA techniques. The signatures resemble the scanner bar codes on supermarket products.

These DNA "fingerprints," or profiles, have broad applications in forensic analysis, paternity testing, diagnostic medicine, plant and animal sciences, and wildlife poaching. Where applicable—for example in homicide, rape, and assault cases—the use of the new procedures should considerably reduce the weight of eyewitness testimony, which many believe is responsible for more miscarriages of justice than any other single type of evidence. As suggested by Judge Joseph Harris (1988),* County of Albany in the State of New York, this new technology could be the greatest single advance in the search for truth, conviction of the guilty, and acquittal of the innocent since the advent of cross-examination.

I had two objectives in writing this introductory text. First, there appeared to be a need at the field, laboratory, and courtroom levels for a summary of the principles and basic methodology underlying DNA analysis to identify tissues and individuals. As with any new technology, considerable care must be exercised during transfer from the experimental stage to service applications. Analysts involved in establishing and operating DNA laboratories and those individuals engaged in the legal aspects of forensic, paternity, and other identity cases should find this book a valuable source of information. A number of organizations, including government,

*_People v. Wesley_, 140 Misc. 2d 306,533 N.Y.S. 2d 643 (Co. Ct. 1988).

universities, law enforcement agencies, and private laboratories are attempting to standardize DNA fingerprinting protocols for human forensic application. In North America this process is being sponsored by the Federal Bureau of Investigation through the Technical Working Group on DNA Analysis Methods (TWGDAM). The standardization includes quality assurance, analysis materials and methods, and the assimilation of DNA profile databases. The recommendations available at the time of this printing are included in the methods sections and in Appendices I and II.

My second objective was to provide a text for students in biological and medical sciences. Students not intimately familiar with recombinant DNA technology can read this book, learn how to use the techniques, and make informed judgments on issues concerning application of this new approach to identity testing.

The chapters have been organized to first provide background principles, then practical methodology, and then case applications. An extensive glossary has been included to assist those readers not familiar with DNA fingerprinting terminology. Detailed reference lists have also been compiled; these include the articles cited in each chapter and other references relevant to the subject area. Chapter 10 was contributed by Kenneth E. Melson, Esquire, Federal prosecutor in Alexandria, VA and Lecturer at the National Law Center, George Washington University, Washington, DC. Responsibility and credit for this contribution is given to Mr. Melson. The chapter provides an up-to-date presentation of legal issues related to DNA typing in the United States.

I am most indebted to Professor Martin L. Puterman for his considerable contribution to Chapter 8, Probability and Statistical Analysis. I am grateful to the Literary Executor of the late Sir Ronald A. Fisher, F.R.S. to Dr. Frank Yates, F.R.S. and the Longman Group Ltd, London for permission to reprint Table 8-1 from their book *Statistical Tables for Biological, Agricultural and Medical Research* (6th Edition 1974).

I thank the Department of Pathology, University of British Columbia, for support, including secretarial assistance with typing of the manuscript; the extra effort by the secretaries was most appreciated. I also thank Vicky Earle of Biochemical Communications for preparing a number of excellent illustrations.

I owe a special debt of thanks to Dr. Bruce Budowle and Dr. Ron Fourney for reviewing the manuscript and for making extremely helpful, detailed suggestions. I am also grateful to Molly Creery, Dr. Adrian Drobines, Dr. Peter Gill, Mr. Mike Luchemko, and Mr. Rick Miller for a number of suggestions. Thanks are due to Chris Munro and Pat Good for preparing the camera-ready materials and to Paula Pinter for those short periods of emergency assistance.

Lastly, what would authors do without the wisdom of editors like Dr. Jim Miller, who carried out ongoing edits during the manuscript preparation, and Dr. Ingrid Krohn, my perspicacious publisher whose support and enthusiasm, like DNA fingerprints, can only be described as unique?

CHAPTER 1

Introduction

DEFINITION

DNA identification analysis, identity testing, profiling, fingerprinting, typing, or genotyping refers to the characterization of one or more relatively rare features of an individuals's genome or hereditary makeup. Every human, lower animal, and sexually reproduced plant has a characteristic phenotype or physical appearance because each possesses a unique hereditary composition. The exception to this rule is identical twins, who possess the same unique genotype but, owing to the consequences of complex developmental events, have subtly different phenotypes. The DNA of any individual is identical whether it is extracted from hair bulbs, white blood cells, or a semen specimen. These principles of individual uniqueness and identical DNA structure within all tissues of the same body provide the basis for DNA profiling.

HISTORY

Identity testing is only one aspect of recombinant DNA analysis. It is the newest, most powerful technique in forensic science, paternity testing, animal and plant sciences, and investigation of wildlife poaching. The foundation for the concept was established with the hallmark observation by Wyman and White (1980) of a polymorphic DNA locus characterized by a number of variable-length restriction fragments called *restriction fragment length polymorphisms* (RFLPs). The history of DNA fingerprinting, as such, is even more recent, dating from 1985 with the paper "Hypervariable Minisatellite Regions in Human DNA" by Alex Jeffreys et al.

Jeffreys and his coworkers were analyzing the human myoglobin gene when they discovered a region consisting of a 33-base-pair (bp) sequence repeated four times within an intervening sequence (IVS). This tandem repeat was referred to as a *minisatellite* and similar regions as being *hypervariable* because the number of tandem repeats is variable both within a locus and between loci. They also discovered that each repeat unit contains a smaller 16-bp core in common with other minisatellites. When DNA is isolated, cleaved with a specific enzyme, and hybridized under low-stringency conditions with a probe consisting of the core repeat, a complex ladder of DNA fragments is detected. This profile appears to be unique to each individual. Different core repeats were later isolated and used to produce a number of different probes useful for fingerprinting.

Within a year, two other approaches to DNA typing were published. One method, by Tyler et al. (1986), described the use of a human repeat sequence Y-chromosome-specific probe and a dot-blot technique to distinguish male and female dried blood stains. In addition, this group distinguished human from other animal blood stains using a human Alu repeat sequence probe. The system is a simple rapid broad screen, as opposed to Jeffreys' very powerful individual fingerprints and the equally powerful approach used by Kanter et al. (1986) and Guisti et al. (1986) of Lifecodes Corporation.

The Lifecodes investigators isolated and cleaved human genomic DNA and hybridized the DNA with two probes, each of which recognized a different highly polymorphic region at a specific locus. Only one or two fragments were detected as bands on the resultant autoradiograms with each probe for each individual. The main power of this procedure lies in its sensitivity and in the large number of defined alleles possible at a single specific locus. The probability of different individuals having the same alleles is low, and it becomes much lower when the DNA is hybridized with a second, third, or fourth specific-locus multiallele probe.

Nakamura et al. (1987) coined the term *variable number of tandem repeats* (VNTR) to describe individual loci where alleles are composed of tandem repeats that vary in the number of core units. These investigators isolated and characterized a number of probes suitable for single-locus VNTR profiling.

Details of the genetic principles and practice underlying the multilocus multiallele system and the single-locus multiallele system are described in later chapters.

In this book, DNA identification analysis, identity testing, profiling, fingerprinting, typing, or genotyping will refer to either of these procedures. (*DNA typing* appears to be the term of choice—at least for forensic scientists in North America.) As originally coined, DNA fingerprinting refers to the multiband print produced using multilocus multiallele minisatellite probes. Patents for this process using Jeffreys' probes, are held by the Lister Institute of Preventive Medicine, and exclusive worldwide license rights to market the system are held by Cellmark Diagnostics. DNA-Print, with patents held by Lifecodes Corporation, refers to the use of single- or double-band patterns produced using single-locus multiallele probes. It is noted that, regardless of the patents, all of the approaches are based on modifications of the original Southern blot procedure (Southern 1975).

A CASE FOR DNA

Specificity is the key to the emergence of DNA analysis. Numerous other techniques used to determine biological markers, such as HLA and blood group substances, have been successfully applied for identification purposes. All are based on exclusion, where markers are tested until a difference is found. If no difference is observed after a statistically acceptable amount of testing, the probability of a specimen match is determined.

If the DNA fragment profiles from two specimens are identical, the samples are, with very high probability, from the same source. If all bands detected in an offspring are present in the mother or putative father, paternity is usually concluded. Other factors favoring DNA analysis include the small sample requirement, the ability to rapidly replicate a sequence a millionfold or more in vitro, and the relative stability of DNA.

Genotyping is not the perfect identification test since DNA can degrade beyond recovery, and analysis at present is labor-intensive and tedious. Any one of the many simpler antigen tests available could exclude a suspect in a homicide or a putative father in a case of disputed paternity. If a blood stain is type O and a suspect AB, the stain could not be the suspect's. If a child is blood type A and the mother O, a type O male cannot be the father. The point is that DNA analysis alone can be a definitive test. Once the technique becomes routine, there is little doubt that, provided a suitable specimen can be obtained, DNA fingerprinting will be the single best test for excluding a falsely associated individual.

APPLICATIONS

DNA profiling, as already indicated, has application in a broad cross section of disciplines, including human forensic science, diagnostic medicine, family relationship analysis, animal and plant sciences, and wildlife forensic science.

According to recent statistics from the Federal Bureau of Investigation, more than 1.3 million violent crimes are reported annually in the United States (FBI 1987, Kalish 1988). The crimes include approximately 90,000 rapes (with another estimated 35,000 unreported), 20,000 homicides, 500,000 robberies, and 700,000 aggravated assaults. (The rates per capita in the United States are approximately three times greater than those in Canada and four times greater than those in Central Europe.) There is one rape each four to five minutes and one homicide each half-hour in the United States. Even more shocking is the fact that of the total number of forcible rapes, reported plus unreported, over 60 percent remain unsolved. More than one-quarter of the homicides also go unsolved; in addition, 2,000 dead children remain unidentified each year.

These statistics, together with the almost one-quarter million paternity suits filed annually, give some indication of the pressures on the justice system. Accurate identification techniques to effectively resolve legal cases are of paramount importance. DNA identity testing, with ethical considerations regarding specimen procurement and confidentiality of the results, has had a dramatic impact on the pursuit of justice in thousands of legal proceedings. The protection of human rights by quickly vindicating the innocent, the improved conviction rates for the guilty, and the potentially large economic savings to society are a few of the positive results from this powerful test.

DNA profiling is applicable in a number of areas in medicine, including:

1. Twin zygosity testing
2. Bone marrow transplantation marker analysis
3. Detection of DNA changes in tumors
4. Indication of possible contamination of fetal by maternal tissue in chorionic villus analysis
5. Pathogen identification
6. Paternity testing where family studies are being performed for antenatal diagnosis of inherited diseases.

Recombinant DNA analysis is an invaluable tool for tracing mutant genes in families for diseases such as Duchenne muscular dystrophy, cystic fibrosis, and the hemoglobinopathies. Identity profiling is only one aspect of this recently established area of "new genetics."

The use of DNA techniques to determine paternity or other family relationships for cases of disputed parentage is making rapid inroads into conventional blood marker test systems. The degree of relationship in disputed immigration cases has been resolved by DNA analysis. Cases involving missing persons have also been pursued by profiling relatives, ideally parents, to compare them with DNA profiles from unidentified bodies or amnesia victims.

DNA typing has considerable potential in the animal sciences for species and individual animal identification, for paternity testing, and for marker analysis in experimental work. These applications can be used in both domesticated and wildlife species. Perhaps the greatest constraint at present is the availability of probes specific to different animal groups.

Potential applications in the plant sciences include the determination of trait markers and variety and individual identification.

Finally, the field of investigation into wildlife poaching has been fraught with difficulties for the game warden attempting to link the poached parts of big game with animal remains at the site of illegal kills. Often, only the smoking gun is evidence convincing enough for conviction. DNA analysis has the potential to definitively link a roadside gut pile to a freezer steak, pelt, or trophy head, thus revolutionizing wildlife forensic science. Also, male and female animals can be distinguished by the use of Y-chromosome-specific probes, even when only minute pieces of tissue are available for analysis. This may be important when female big game are shot at a time when regulations stipulate that only males are to be hunted.

The aforementioned applications are discussed in considerably more detail in subsequent chapters; the unifying theme is the use of the same genetic principles and recombinant DNA technology.

REFERENCES

FBI. 1987. *Crime in the United States 1987*. FBI Law Enforcement Bulletin. August 1988, 6–9.

Giusti A, Baird M, Pasquale S, Balazs I, and Glassberg J. 1986. Application of deoxyribonucleic acid (DNA) polymorphisms to the analysis of DNA recovered from sperm. *J. Forensic Sci.* 31:409–417.

Jeffreys AJ, Wilson V, and Thein SL. 1985. Hypervariable 'minisatellite' regions in human DNA. *Nature* 314:67–72.

Kalish CB. 1988. International crime rates. Bureau of Justice Statistics Special Report. U.S. Department of Justice.

Kanter E, Baird M, Shaler R, and Balazs I. 1986. Analysis of restriction fragment length polymorphisms in deoxyribonucleic acid (DNA) recovered from dried bloodstains. *J. Forensic Sci.* 31:403–408.

Nakamura Y, Leppert M, O'Connell P, Wolff R, Holm T, Culver M, Martin C, Fujimoto E, Hoff M, Kumlin E, and White R. 1987. Variable number of tandem repeat (VNTR) markers for human gene mapping. *Science* 235:1616–1622.

Southern EM. 1975. Detection of specific sequences among DNA fragments separated by gel electrophoresis. *J. Mol. Biol.* 98:503–527.

Tyler MG, Kirby LT, Wood S, Vernon S, and Ferris JAJ. 1986. Human blood stain identificaiton and sex determination in dried blood stains using recombinant DNA techniques. *Forensic Sci. Int.* 31:267–272.

Wyman AR and White R. 1980. A highly polymorphic locus in human DNA. *Proc. Natl. Acad. Sci. USA* 77:6754–6758.

CHAPTER 2

Genetic Principles

Genetics is the study of heredity. Each individual's makeup, or phenotype, is determined by nature and modified by environmental factors. DNA identity analysis is based strictly on heredity, and only in the rare case where a human had a bone marrow transplant would the white blood cell genotype differ from that inherited. Difficulties can arise with specimens because of DNA degradation or contamination by extraneous materials, and mixed cell populations could be present in tumorous tissue. The analyst must always be cognizant of these complicating factors.

The concept of the gene was advanced by the Moravian monk Gregor Mendel in 1865 based on observations he made after crossing different varieties of garden peas; these experiments are considered the beginning of the discipline of genetics. (The term *gene* was actually coined by the Danish plant scientist W. Johannsen in the early 1900s.) Mendel formulated two laws. The law of segregation or separation states that two members of each gene pair (alleles) in a diploid organism separate to different gametes during sex cell formation. The law of independent assortment states that members of different pairs of alleles, if located on separate chromosomes or far apart on the same homologous chromosome pair, assort independently into gametes. These laws are basic to the understanding of biological family relationships and play a critical role in such contemporary issues as paternity testing and immigration disputes.

DNA STRUCTURE

The basic unit of life is the cell (Figure 2-1). Cells are microfactories in which raw materials (amino acids, simple carbohydrates, lipids, and trace elements) are received, new substances (proteins, complex lipids, carbohydrates, and nucleic acids) are produced, and wastes are removed. The thousands of different enzymes required for the myriad ongoing chemical reactions are key to the efficient functioning of cells. Each cell has the ability to self-replicate using the deoxyribonucleic acid (DNA) code as the blueprint, raw materials as building blocks, and enzymes as catalysts.

It has been estimated that the average human being is composed of approximately 100 trillion cells—a considerable amount of DNA. Within eukaryotic cells most of the DNA is nuclear; only a minor extranuclear quantity is found in the mitochondria and plant plastids. (Most mammals shed their circulating red blood cell nuclei.)

Nuclear DNA

In eukaryotes the nuclear DNA is divided into chromosomes. The human karyotype consists of 22 matched pairs of autosomes and two sex chromosomes; the female is designated XX and the male XY (Figure 2-2). One from each pair, plus a sex chromosome, is derived from each parent at the time of conception. The chromosomes in all body (somatic) cells are in the diploid state, with two sets of chromosomes per cell; gametes (egg or sperm) have only the haploid, or single, set of chromosomes. The chromosome number varies for different animals, for example: bovine, 60; equine, 64; and canine, 78. However, the same genetic principles apply as for the human.

Pairs of chromosomes are termed *homologous* because they have a common evolutionary origin (i.e., a common ancestral chromosome). A copy of each gene resides at the same position (locus) on each chromosome of the homologous pair. The genes may or may not be identical since mutations could have occurred, giving rise to altered structures and perhaps functions. If each parent were a carrier for a functionally mutant gene such as ß-globin, each would have one normal ß-globin gene at position p 11.5 on chromosome 11 and a mutant gene at the same position on the homologous chromosome. This state is termed *heterozygous*; if the alleles (alternative forms of a gene) are identical, the state is *homozygous*. Offspring of heterozygous parents could be either normal in terms of ß-thalassemia, carriers like the parents, or affected by having two copies of the mutated allele. Similarly,

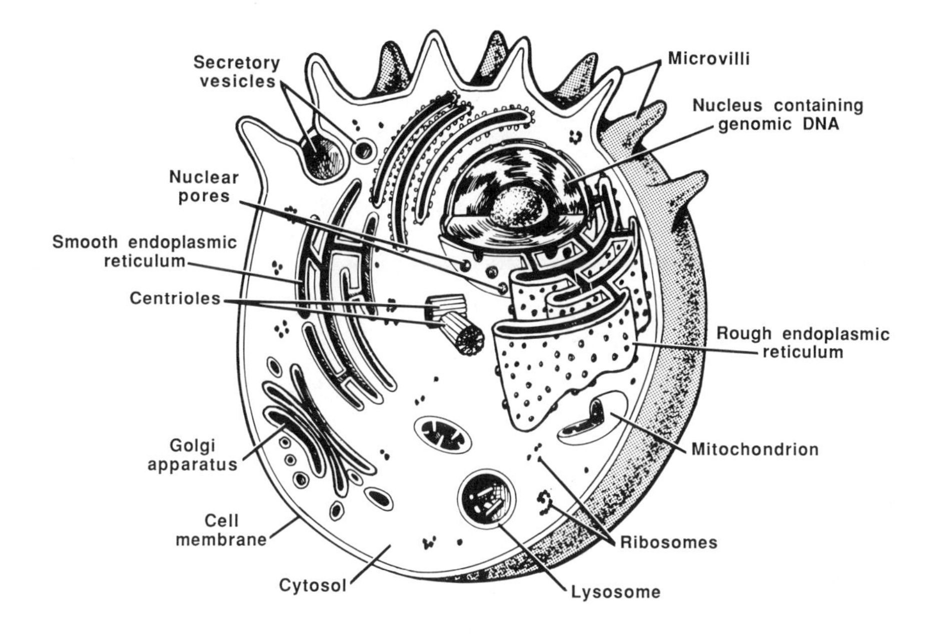

Figure 2-1. Schematic view of a cell illustrating the typical organelles. The components are not drawn to scale.

detectable structural differences in alleles at corresponding loci are key to the development of DNA identity patterns because each will appear as a different-size band on an autoradiogram profile of an individual's DNA.

According to the Watson-Crick model, the DNA macromolecule is a double helix with the complementary strands linked by hydrogen bonds (Figure 2-3). These bonds are readily broken by heating for a few minutes at 95°C to 100°C and will reform at about 65°C. The basic unit of the DNA molecule is the nucleotide, consisting of a base (adenine, guanine, thymine, or cytosine), deoxyribose sugar, and a phosphate group (Figures 2-4, 2-5). Nucleotides are covalently linked by DNA polymerase to form polynucleotide chains connected by hydrogen bonds between adenine-thymine (A=T) or cytosine-guanine (C≡G). The human genome (diploid) consists of approximately 6×10^9 bp. Strand complementarity is critical to DNA replication because the strands can be melted or separated and each can act as a template for the production of new double-stranded molecules.

A DNA molecule has control, production, and apparently nontranscribed regions, including a considerable amount of repetitive or satellite DNA. The control and production units are the regulatory and structural genes. Intergenic or spacer sequences separate the genes. Genes consist of *introns*, also referred to as intervening sequences, and *exons*, or protein-coding regions (Figure 2-6).

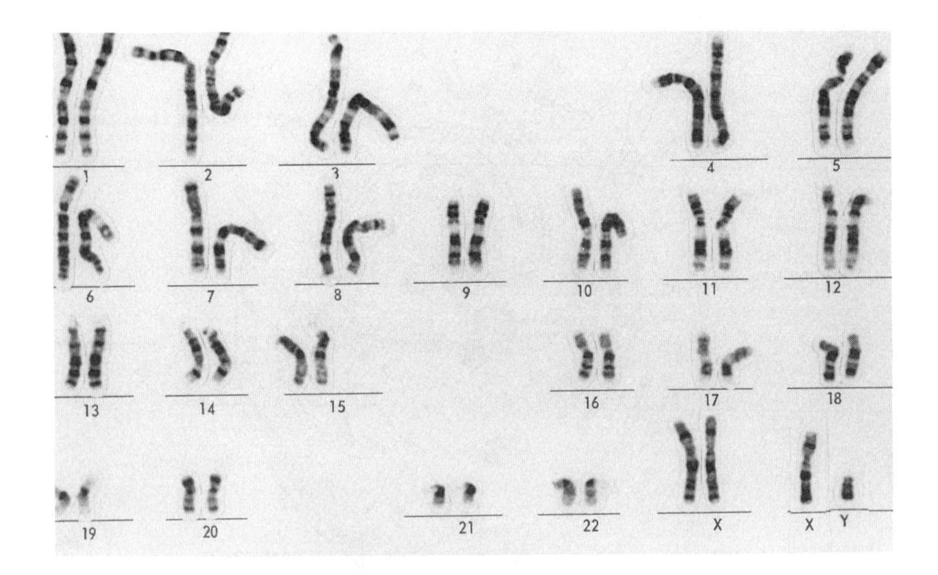

Figure 2-2. A human karyotype constructed from metaphase chromosomes. Each chromosome pair can be identified by morphology and the banding pattern produced by trypsin/Giemsa staining. Each number, 1 to 22, represents a homologous pair of autosomes. The autosomes plus XX represent a female karyotype; the autosomes plus XY represent a male karyotype. (Karyotype courtesy of Dr. T. Pantzar, Cytogenetics, Vancouver General Hospital and Children's Hospital, Vancouver, British Columbia.)

Organelle DNA

Animal mitochondria and plant mitochondria and plastids are self-replicating cytoplasmic organelles containing extranuclear genes (Figure 2-7). Each human cell contains 5,000 to 10,000 mitochondria. Each mitochondrion has up to 10 circular-duplex 16,569-bp chromosomes coding 2 ribosomal RNAs, 22 transfer RNAs, and 13 proteins, all associated with the respiratory chain and oxidative phosphorylation. The coding system is similar, but not identical, to nuclear DNA; for example, UGA is not a stop codon but codes for the amino acid tryptophan, whereas UAA, UAG, AGA, and AGG are stop codons. Most mitochondrial and plastid proteins are coded in the nuclear DNA. Operation of these organelles is a function of both extranuclear and nuclear genes.

Although both egg and sperm contain mitochondria, mitochondrial DNA (mtDNA) is apparently transmitted only via the maternal cytoplasm to the offspring. This clonal inheritance can be very useful in identity testing. The genomes (chromosomes) within mitochondria are, for the most part, identical. There are no homologous pairs, as with nuclear chromosomes; however, mutations can occur

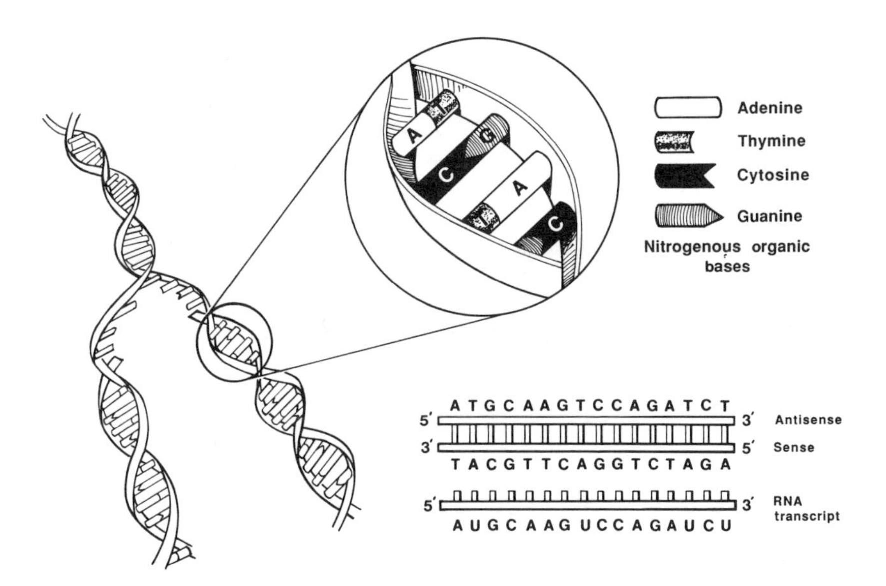

Figure 2-3. The double-stranded structure of DNA undergoing semiconservative replication. Each parental strand is a template for the synthesis of a new daughter strand formed according to the base pairing rules: A=T, C≡G. The new double-helix molecules are identical, with each consisting of an original parental and a new daughter strand. The sense (3'–5') and antisense (5'–3') designation of the molecule for mRNA transcript production is also illustrated.

that result in more than one mitochondrial line within an individual. The D-loop region, consisting of approximately 1,000 bp, is quite variable and may, therefore, prove useful for identification purposes.

Large differences exist in mtDNA among animal species. This, no doubt, has arisen because of the rapid rate of mtDNA evolution, which in vertebrates is estimated to be 10 times that of nuclear genes. These differences make this hereditary material ideal for differentiating species and for tracing organism evolution.

DNA FUNCTION

DNA provides two basic functions. It is a blueprint for assembling amino acids into proteins, and it provides a template for the reproduction (self-replication) of both somatic cells and gametes. The replication process is directly relevant to a number of aspects of DNA identity testing. This is discussed in the next section.

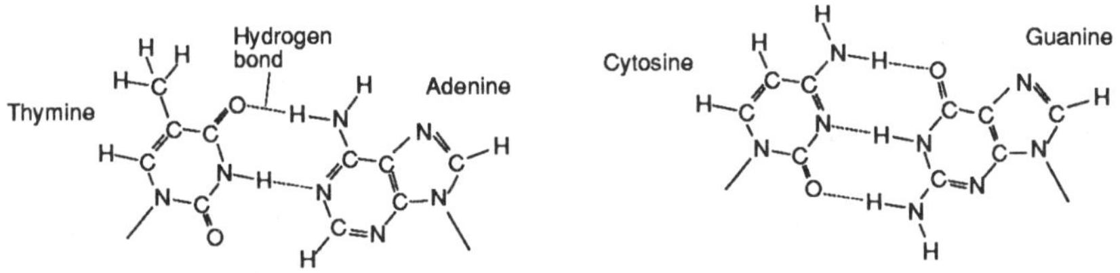

Figure 2-4. Complementary base pairing (A=T, C≡G) illustrating hydrogen bond formation. Hydrogen bonds can be broken under alkaline conditions or high temperature, resulting in denaturation to single-stranded complementary DNA molecules.

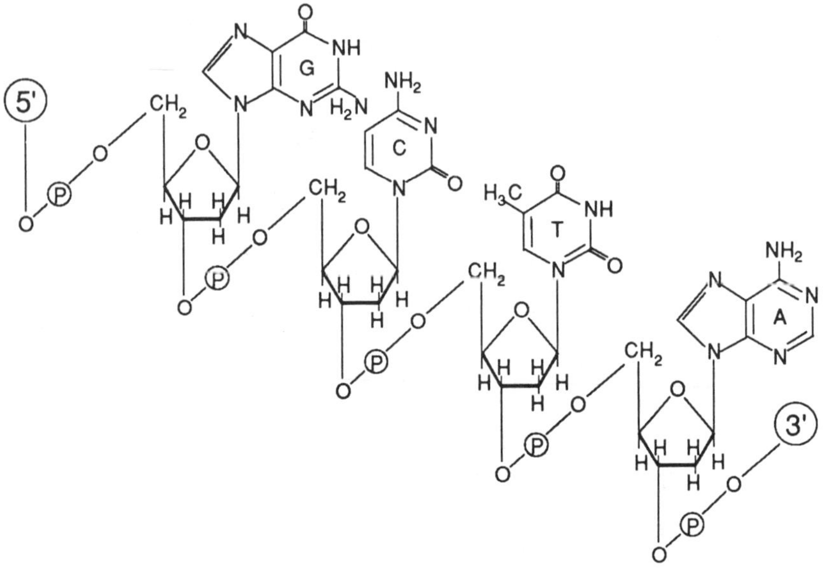

Figure 2-5. Segment of a single-stranded DNA molecule illustrating the 3'–5' phosphodiester linkages between the nucleotides. The 3' carbon of one nucleotide is linked to the 5' carbon of the next nucleotide.

The molecular flow of information from DNA through RNA to protein formation is basic to cell structure and function. The bases (A,T,C,G) of the exon nucleotides are read sequentially as triplets, with each triplet codon signifying an amino acid (Table 2-1). The amino acids in turn form enzymes and structural proteins. The processes of mRNA formation (transcription), processing, and finally translation into protein on the ribosomes are outlined in Figure 2-6.

Although each nucleated cell in an organism contains a complete DNA recipe for that body, only specific genes are active at different stages in development and

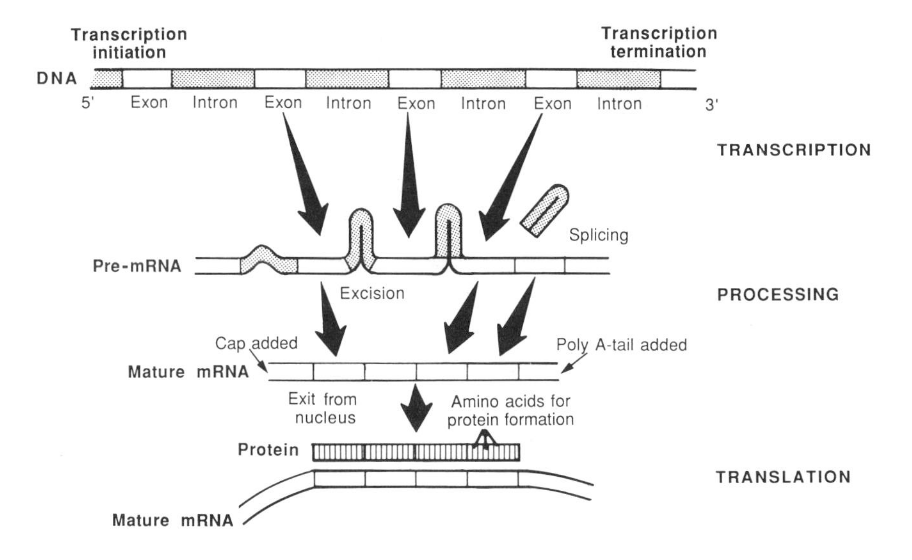

Figure 2-6. Transfer of information from DNA to protein formation in a eukaryotic cell. Non-protein-coding transcript is excised from the pre-mRNA, and a methyl cap (5') and a poly-A tail (3') are added to form the processed or mature mRNA molecule before exit from the nucleus into the cytoplasm.

in different organs, in terms of structure and metabolic functions. The codes for kidney formation and function, for example, are present in every cell; however, only the pertinent genes, influenced by factors such as microenvironment, are active in this organ at specific times during development. Similarly, certain enzymes—for example, glucose 6-phosphatase—are detected mainly in the liver, although every other organ cell contains the recipe for these proteins.

REPRODUCTION

Details of the cell division phases need not be considered here; however, the results of the overall process are basic to understanding heredity.

The DNA must replicate before cell division to provide a new set of chromosomes. One of each duplicate (chromatid) is destined for each daughter cell (Figure 2-8). Mitosis is the process whereby nuclei of the somatic forming cells divide. Two identical nuclei are first formed from each mother nucleus, followed by division of the remaining cellular contents to produce two new daughter cells. This process occurs from the time of fertilization and zygote formation through the embryonic and fetal developmental stages and eventually to the adult stage. Mitosis

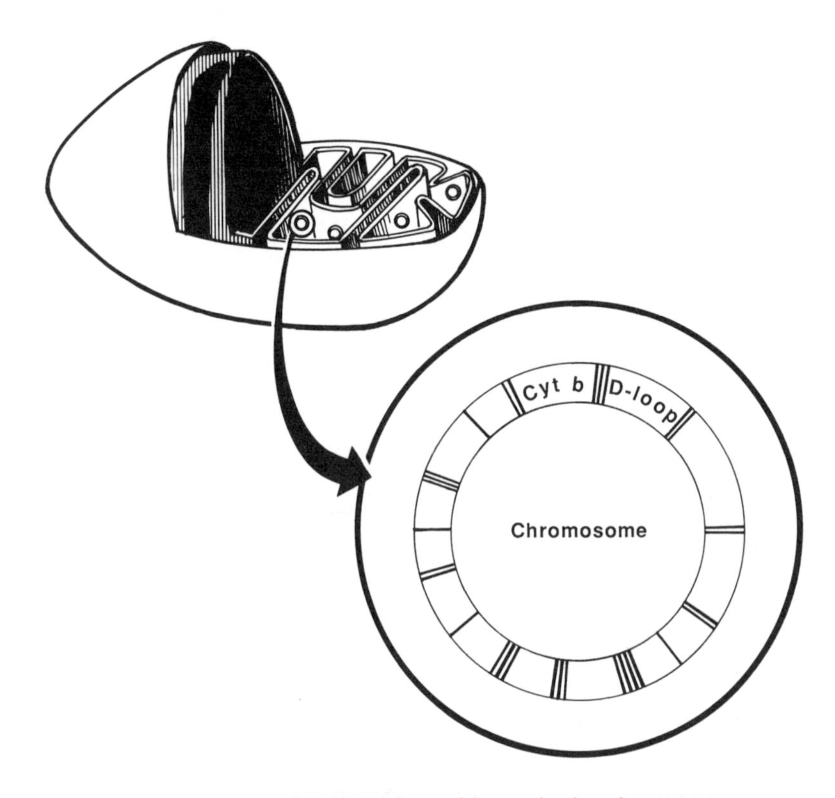

Figure 2-7. A mitochondrion with a single chromosome illustrated. Only the cytochrome b and the D-loop map regions are labeled.

is also necessary for cell replacement. Many cells are programmed for constant renewal; for example, the human red blood cell life span is approximately 120 days.

It should now be clear why DNA from any tissue in the body is suitable for identity testing—it is identical in all cells except for those with a somatic mutation.

Meiosis is the process of nuclear division resulting in the formation of reproductive cells (gametes). The process is characterized by a reduction from the diploid (2N) to the haploid (N) chromosome number. After replication, one member of each homologous chromosome pair plus one of the two sex chromosomes, all having been modified by recombination, provide the nuclear material for four new cells. The process is best illustrated diagrammatically (Figures 2-9, 2-10).

There are two successive divisions: I, a reduction division where the replicated homologues pair, exchange of segments within each pair (actually, tetrad) occurs by breaking and recombining, and the homologues separate; and II, a division similar to mitosis where the chromatids separate or segregate. Meiosis results in 2^{23}, or about 8 million, different possible combinations of chromosomes in human ova or sperm. At fertilization an ovum and sperm unite to form a diploid cell, the zygote.

Table 2-1. The Genetic Code (mRNA codons)

First base (5' end)	Second base								Third base (3' end)
	U		C		A		G		
U	UUU	-phe	UCU	-ser	UAU	-tyr	UGU	-cys	U
	UUC		UCC		UAC		UGC		C
	UUA	-leu	UCA		UAA	-stop	UGA	-stop	A
	UUG		UCG		UAG		UGG	-try	G
C	CUU	-leu	CCU	-pro	CAU	-his	CGU	-arg	U
	CUC		CCC		CAC		CGC		C
	CUA		CCA		CAA	-gln	CGA		A
	CUG		CCG		CAG		CGG		G
A	AUU	-ile	ACU	-thr	AAU	-asn	AGU	-ser	U
	AUC		ACC		AAC		AGC		C
	AUA	-met	ACA	-lys	AAA	-arg	AGA	-arg	A
	AUG		ACG		AAG		AGG		G
G	GUU	-val	GCU	-ala	GAU	-asp	GGU	-gly	U
	GUC		GCC		GAC		GGC		C
	GUA		GCA		GAA	-glu	GGA		A
	GUG		GCG		GAG		GGG		G

Amino acid abbreviations:

ala	alanine	gln	glutamine	leu	leucine	ser	serine
arg	arginine	glu	glutamic acid	lys	lysine	thr	threonine
asn	asparagine	gly	glycine	met	methionine	try	tryptophan
asp	asparic acid	his	histidine	phe	phenylalanine	tyr	tyrosine
cys	cyteine	ile	isoleucine	pro	proline	val	valine

The stop codons do not code for an amino acid and, therefore, signal gene termination (transcription ceases).

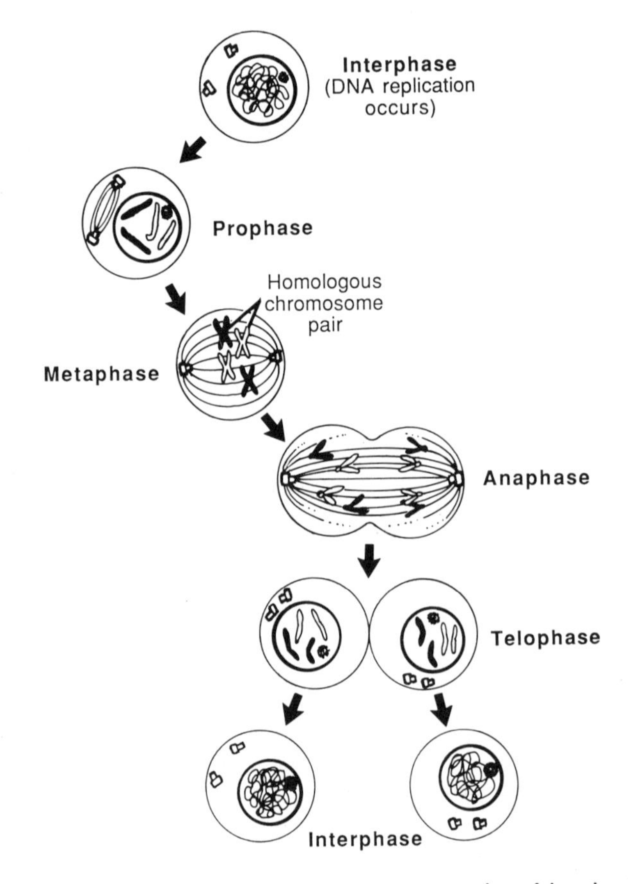

Figure 2-8. Mitosis. Diagrammatic representation of the phases of division of nuclear DNA during the formation of two identical daughter cells. The DNA replicates during interphase, and the resulting sister chromatids comprising each chromosome become visible, under magnification, during prophase. At metaphase, chromosomes line up in preparation for sister chromatid separation, which occurs during anaphase. The final phase, telophase, is characterized by the formation of two identical nuclei. The remainder of the cell contents divide to complete the production of the two new cells. Cultured human fibroblasts replicate on average once in 12 to 24 h. Mitosis (prophase to telophase) requires approximately 1 h; interphase occupies the remainder of the cycle.

The phenomenon of recombination is illustrated in Figure 2-11. This process is important in increasing offspring genetic variability in addition to that produced by the independent assortment of maternal and paternal chromosomes during meiosis. It is possible that during evolution unequal crossing over was a factor that gave rise

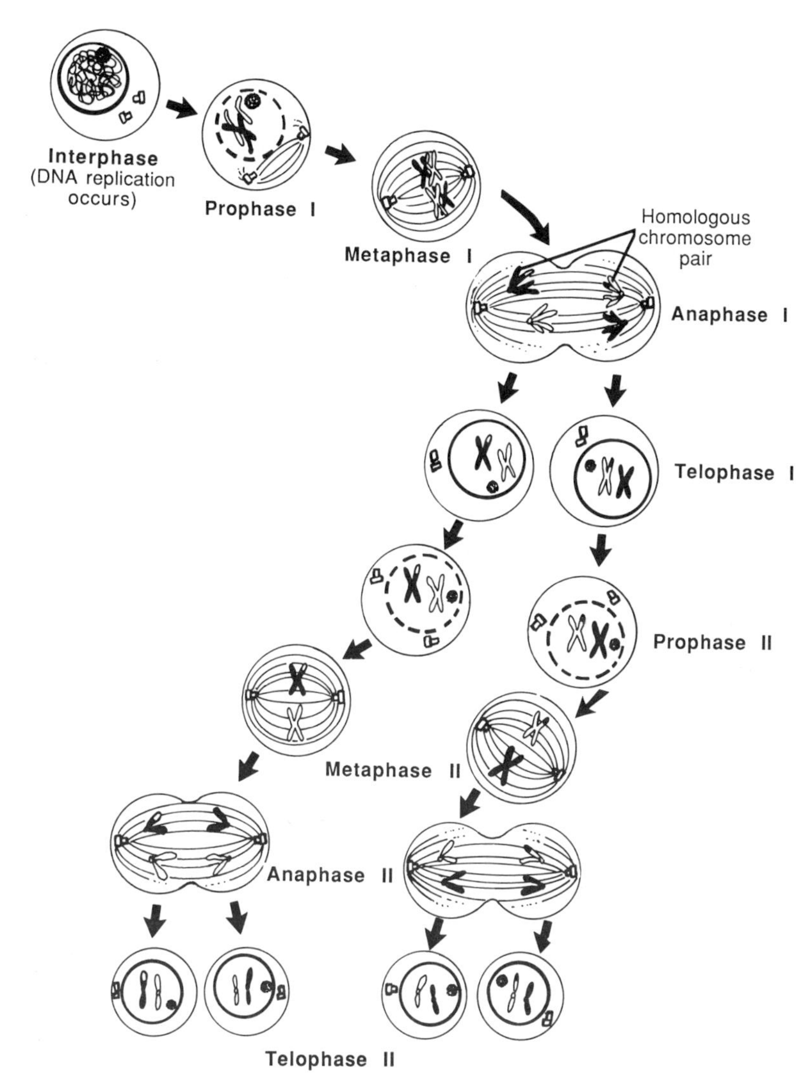

Figure 2-9. Meiosis. Diagrammatic representation of the reduction (I) and division (II) phases of nuclear DNA during the formation of four haploid cells from one diploid cell. The DNA replicates during interphase, and the resultant two sister chromatids comprising each chromosome become visible, under magnification, during prophase I. There is also an exchange of corresponding segments from homologous chromatids during prophase I. The homologous chromosome pairs line up during metaphase I, and the pairs separate at anaphase I. Telophase I ends and prophase II begins with the members of each of the homologous pairs of chromosomes in one of two nuclei. The chromosomes line up at metaphase II in preparation for sister chromatid separation, which occurs during anaphase II. Four unique nuclei are formed at telophase II.

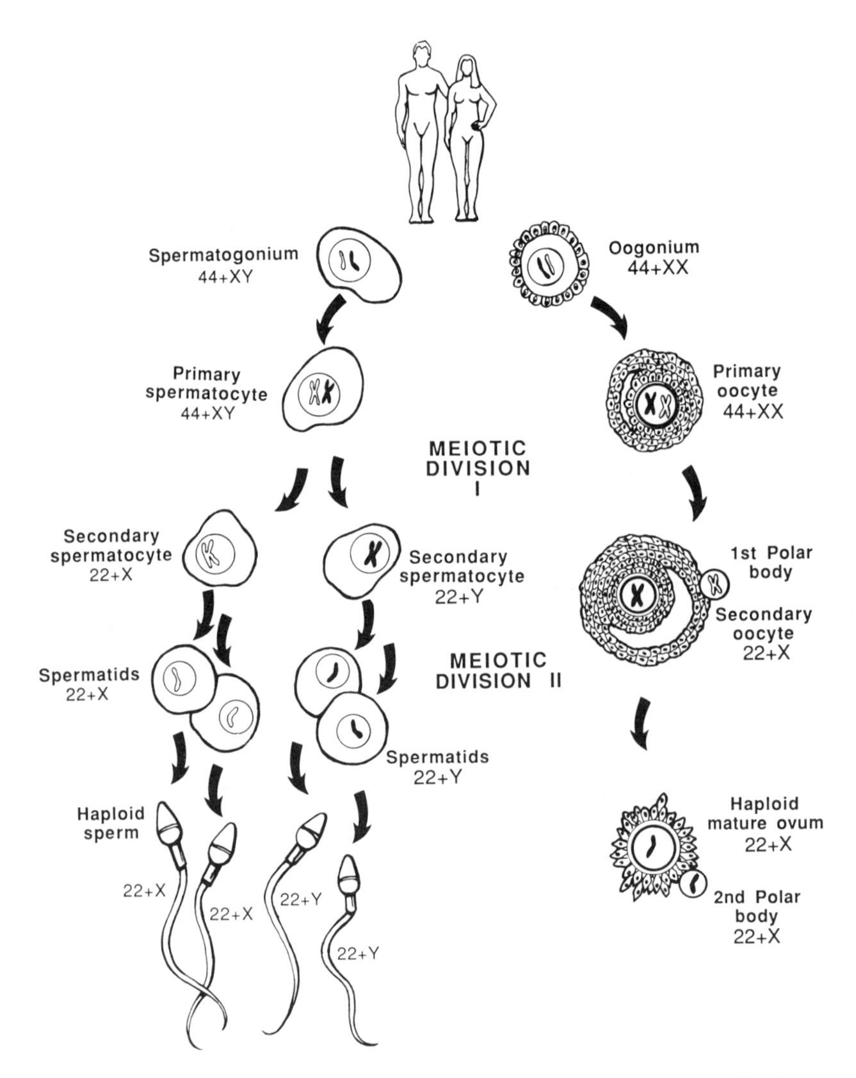

Figure 2-10. Spermatogenesis and oogenesis. Spermatogenesis in the human male is characterized by a primary spermatocyte undergoing the first stage of meiosis to form two secondary spermatocytes, and the sister chromatids of each chromosome homologue separating during the second meiotic stage to form two spermatids. Each spermatid matures to become a sperm cell. The time from primary spermatocyte to sperm formation is approximately 48 days.
Oogenesis in the human female is characterized by a primary oocyte undergoing meiosis to form one haploid ovum containing most of the cytoplasm and two nonviable polar bodies, which are discarded. The primary oocytes remain suspended in prophase I from the time of birth until sexual maturity (approximately 15 years). Meiotic divisions then resume over the reproductive period, which may last 35 years.

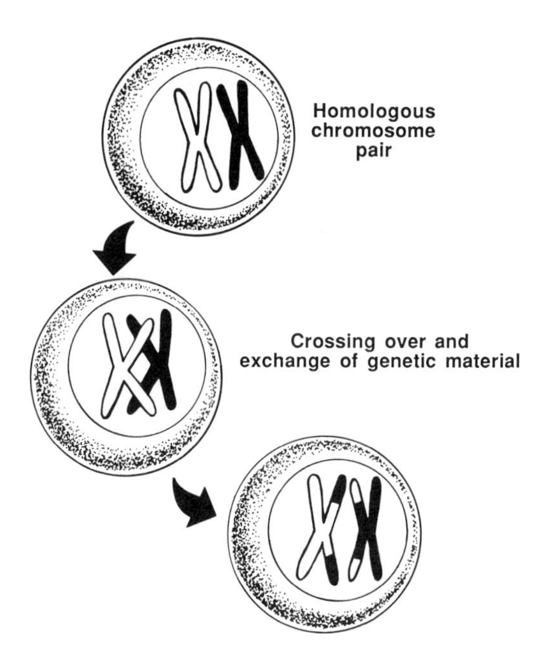

Figure 2-11. Crossing over. Sister chromatids of homologous pairs of chromosomes exchange corresponding segments during meiosis I.

to different numbers of tandem repeats in DNA sequences, thus providing the raw material for DNA profiling.

An understanding of meiosis gives insight into the common genes or DNA between different degree relatives. As discussed later, this knowledge may be important to interpreting an unknown assailant's and a victim's DNA identity patterns.

LINKAGE

Genes on the same chromosome tend to be inherited together and are therefore said to be linked. The closer the genes, the tighter the linkage and the greater the probability they will be inherited as a unit. The phenomenon of linkage is a major exception to Mendel's law of independent assortment and provides an important marker system for tracing chromosome segments from parents to offspring. When linkage equilibrium exists between two loci, the relative proportions of the possible combinations of the respective alleles should be determined only by the population allele frequencies. For example, if the respective frequencies of alleles A, B, a, and b, are 0.9, 0.6, 0.1, and 0.4, the expected combinations are AB = 0.54, Ab = 0.36, aB = 0.06, and ab = 0.04. If this is not observed, with certain allele combinations showing more association than expected, linkage disequilibrium exists. Such disequilibrium may be due to insufficient time, in terms of evolution, having elapsed to facilitate equilibrium, or to natural selection favoring certain combinations, such as the major histocompatibility complex (MHC) on chromosome 6.

It is important that the loci chosen as markers in the development of DNA identity profiles not be linked, since this is one of the main qualifications for using the multiplication rule in probability determinations (see Chapter 8).

PEDIGREE ANALYSIS

An understanding of the degree of genetic relationship, that is, the proportion of shared genes within a family (pedigree), is important for (1) genetic counseling in humans, (2) analysis of the transmission of traits in domesticated animals and wildlife, (3) determination of the true genetic relationship among humans, and among other animals when this is in dispute, and (4) wildlife population studies. Pedigree analysis may also be of considerable value in matching tissue specimens in homicide and rape cases.

Markers, in the heterozygous state, can be used to trace genes from parents to offspring. The reverse procedure may also be considered. Genotype markers in an offspring must have been inherited from the parents unless a germ cell mutation occurred.

The degree of genetic relationship between pedigree members and the proportion of shared genes are outlined in Table 2-2. The coefficient of inbreeding refers to the chance of homozygosity at any locus. The offspring of first-degree relatives, for example, have identical alleles at 25 percent of their loci; the offspring of third-degree relatives have identical alleles at 6.25 percent of their loci. If considerable inbreeding has occurred, as is the situation for many animal populations, calculation of the proportion of shared genes is rather tedious. Examples of gene transmission are outlined in Chapter 11.

MUTATION

Mutation is the process whereby DNA undergoes a stable structural change. In addition to introducing variation in evolution and providing the basis for genetic disease, mutation can cause concern in paternity testing when an offspring's DNA fragment band is not present in either biological parent. It is important, therefore, to understand the types of mutations and their influence on DNA identity profiles.

The scope of mutation ranges from the single nucleotide to the gross chromosomal structure. In terms of effect, mutations are neutral or can modify the phenotype. Single nucleotide changes involve deletion, insertion, or replacement

Table 2-2. Genetic Relationships

Relationship between individuals	Degree of relationship	Proportion of genes in common	Coefficient of inbreeding (F)*
Monozygotic twins	Identical	1	
Dizygotic twins	First	1/2	1/4
Siblings	First	1/2	1/4
Parent-child	First	1/2	1/4
Aunt/uncle–niece/nephew	Second	1/4	1/8
Half siblings	Second	1/4	1/8
Double first cousins	Second	1/4	1/8
First cousins	Third	1/8	1/16
Half-uncle niece	Third	1/8	1/16
First cousins once removed	Fourth	1/16	1/32
Second cousins	Fifth	1/32	1/64

*Chance of homozygosity

of a single base. Deletions or insertions result in code-reading frame shifts (Figure 2-12). Because the DNA code is read in triplets of nucleotide bases, every triplet in a gene distal to a mutation is altered, which in turn alters the gene code. Deletions, insertions, or inversions can also involve many nucleotides and affect a gene accordingly.

Replacement or point mutations are of three types: (1) nonsense, in which a transcription termination codon is produced, (2) synonymous, where a different code is produced but, because the code is degenerate, there is no amino acid substitution, and (3) missense, where a changed code results in an amino acid replacement.

Mutations can also involve gross duplications, deletions, or translocations of such a degree as to be obvious under the light microscope (Figure 2-13).

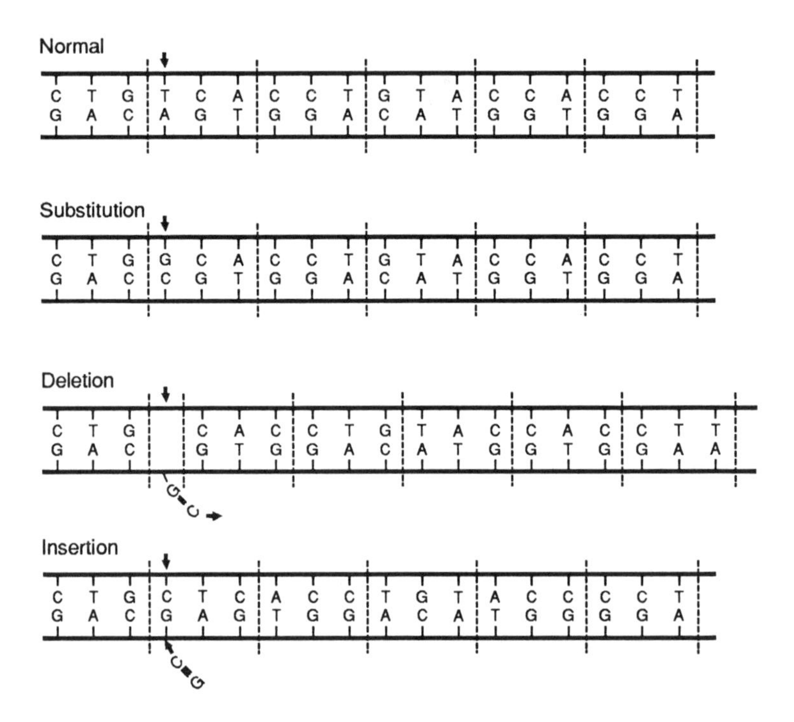

Figure 2-12. Point mutations. Examples of single nucleotide changes (mutations) in a DNA molecule and the triplet code frame shifts that result.

By definition, a mutation always results in change in a DNA molecule. This change may or may not be manifested at the phenotypic level. Fortunately, phenotypic expression is of little concern to the DNA identity analyst.

The tissue location of a mutation is critical to its influence on an individual, in terms of both medical inborn error and DNA identity analysis. Somatic mutations, as the term suggests, affect only non-gamete-forming cells. They need be considered only in extremely rare situations, such as mosaics that form because of mutation within a division or two after zygote formation. The occurrence of this type of mosaicism in the human population is so rare that it is of no practical consequence in identity profiling.

Germ cell mutations are heritable and, if transmitted, will be present in all nucleated cells of the offspring. The probability of such a mutation affecting a band in a multilocus multiallele pattern has been estimated at less than 1 in 300 for Jeffreys' minisatellite system. If there is a suggestion that a mutation has affected a single band (e.g., if all bands but one for an offspring can be traced to either parent), one or more additional probes should be hybridized to confirm paternity.

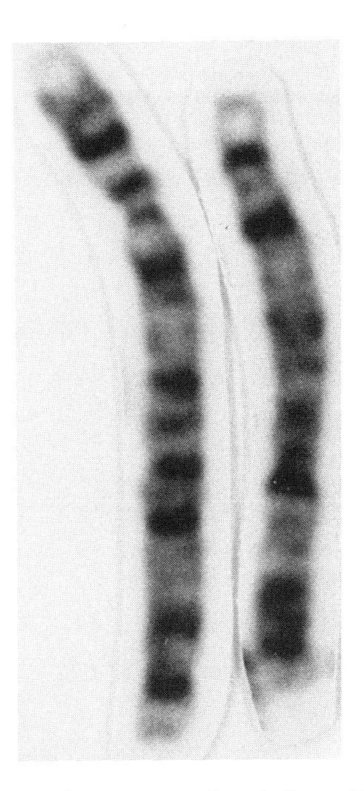

Figure 2-13. Gross chromosomal mutation. Most of the p arm of one of the chromosome 21 pair has been deleted. (Photo courtesy of Dr. T. Pantzar, Cytogenetics, Vancouver General Hospital and Children's Hospital, Vancouver, British Columbia.)

RESTRICTION ENDONUCLEASES

Restriction endonucleases cleave DNA molecules at specific recognition base sequences. In 1970 Hamilton Smith identified the first restriction enzyme, Hind II. This discovery was a key factor in the later development of recombinant DNA techniques. Hundreds of endonucleases have been isolated from more than 200 different bacterial species during the past two decades; each is symbolized by the bacterium of origin and a roman numeral indicating the series number of the enzyme from that organism (Figure 2-14). The recognition sites are palindromic: the order of the bases in a segment of one DNA strand is the reverse of that in the complementary strand. The lengths also vary, with 4- and 6-bp sequences being relatively common. The complementary strand cleavage sites may be staggered, as with Eco RI, and form sticky ends, or opposite, as in Hae III, and form blunt ends.

Enzyme	Microorganism Source	Recognition 5' → 3' Sequence 3' → 5'
Alu I	*Arthrobacter luteus*	↓ (after A C G) A C G T T A G A ↑ (after T A)
Bam HI	*Bacillus amyloliquefaciens H*	↓ (after G) G A A T T C C T T A A G ↑ (after C T T A A)
Eco RI	*Escherichia coli RY 13*	↓ (after G) G A A T T C C T T A A G ↑ (after C T T A A)
Hae III	*Haemophilus aegyptius*	↓ (after G G) G G C C C C G G ↑ (after C C)
Hind III	*Haemophilus* influenzae Rd	↓ (after A) A A G C T T T T C G A A ↑ (after T T C G A)
Hinf I	*Haemophilus influenzae Rf*	↓ (after G) G A N T C C T N A G ↑ (after C T N A)
Pvu II	*Proteus vulgaris*	↓ (after C A G) C A G C T G G T C G A C ↑ (after G T C)

Key: (A) Adenine
(T) Thymine
(C) Cytosine
(G) Guanine
(N) Any of A,T,C,G

Figure 2-14. Examples of restriction enzyme recognition sequences and cleavage sites. Note the staggered sites (↓↑) versus the blunt sites $\binom{\downarrow}{\uparrow}$.

CONCEPT OF POLYMORPHISM

Polymorphism refers to different forms of the same basic structure. If modifications of a gene exist at a specific locus in a population, the locus is polymorphic. Usually, for a locus to be considered polymorphic, the most common allele must occur at a

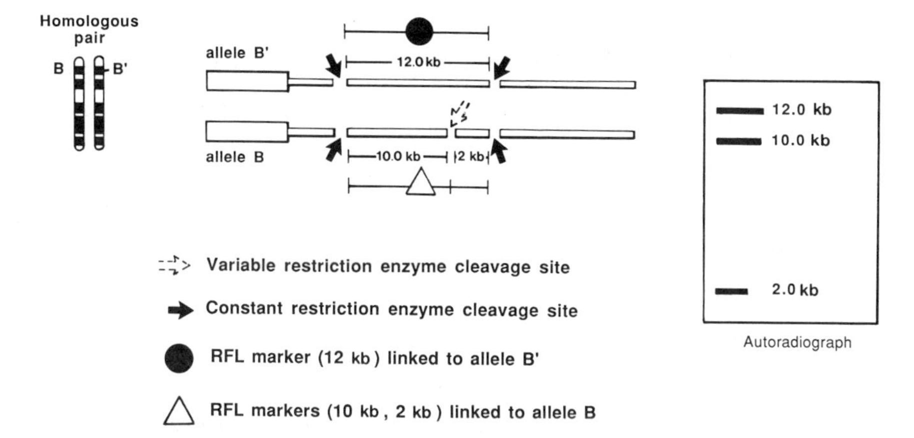

Figure 2-15. Diagrammatic representation of corresponding segments (BB') of a homologous pair of chromosomes. The variable restriction enzyme cleavage site will be recognized by the enzyme that recognizes the constant site only when the site nucleotide sequences are identical. Because the restriction fragment-length (RFL) markers (12 kb fragment and a 10 kb plus a 2 kb fragment) are closely linked to alleles B' and B, respectively, they provide indicators of the presence or absence of these alleles in a genome. The markers are detected on an autoradiograph after hybridization with a radioactive probe.

frequency of less than 99 percent at that locus. According to the Hardy-Weinberg law, at least 2 percent of the population must be heterozygous at that site.

Examples of polymorphism abound in human genetics. The ABO blood group system has been the most exhaustively studied. Four major phenotypes occur: O, A, B, and AB, with the associated genotypes OO; AA, AO; BB, BO; and AB. The symbols refer to antigens on the red blood cell surface. O is an inactive gene incapable of converting the group H antigen, whereas the A allele converts H to A antigen and the B allele converts H to B antigen. Genotype frequencies may vary considerably depending on the population. Group B is more common in Asia than in Western Europe, and Native North American Indians usually lack both A and B alleles. In determining the allele frequencies for a specific locus, careful consideration must be given to variation among populations. This is discussed in more detail in later chapters.

At the molecular level, polymorphism ranges from a single nucleotide base change to the number of tandem repeats in a repetitive DNA sequence. The changes may be neutral, with no detectable phenotypic effect, or they may result in the production of different forms of the same enzyme (isozymes) active under different environmental conditions, such as pH or temperature.

An example of neutral changes are mutations that create or abolish recognition sites for restriction enzymes in noncoding DNA. If a specific recognition base sequence is present, the restriction enzyme recognizing that site will cleave the DNA molecule and result in fragments of specific base pair lengths. If the site is absent, a different-length DNA fragment will be produced (Figure 2-15). Thus, there are fixed and polymorphic sites for endonucleases. Restriction fragment length polymorphisms (RFLPs) are different DNA fragment lengths generated within a species by the action of specific endonucleases. The fragment lengths can vary at corresponding loci for a number of reasons: (1) Some restriction sites are polymorphic; (2) the cleavage sites flank tandem repeat sequences where the number of repeats within the sequence varies; (3) insertion or deletion of nucleotides has occurred between the fixed restriction sites. Occurrences within controlling sequences or structural genes usually have serious phenotypic consequences.

The value of RFLPs as markers linked to normal genes and their disease-causing alleles within families has revolutionized the approach to antenatal diagnosis for such diseases as Duchenne muscular dystrophy, cystic fibrosis, and the hemoglobinopathies. RFLPs also provide the raw material for identity testing because of the presence of alleles with different numbers of tandem repeats of a core sequence.

RECOMBINANT DNA

Recombinant DNA technology is the process of joining two heterologous DNA molecules (Figure 2-16). Ligation of the human insulin gene to the plasmid pBR322, usually for the purpose of gene amplification, is one example of this procedure.

It is unlikely that the average genotyping laboratory will attempt to isolate new DNA probes; however, it may be necessary to amplify, purify, and characterize material received in vectors from various research centers. In terms of background knowledge, those involved with DNA analysis should have an overview of the processes involved with new probe production. The processing of vectors with inserts and DNA analysis methodology are discussed in Chapters 5 and 6.

A probe is a nucleotide sequence capable of hybridizing with its complementary chain. Probes consist of (1) genomic fragments containing complete or partial genes or intergenic sequences, (2) complementary DNA (cDNA) produced by the action of reverse transcriptase with mRNA as a template, (3) chemically synthesized sequences, and (4) mRNA. Messenger RNA probes are produced as transcripts from DNA. For the most part, probes used in identity testing consist of intergenic tandem repetitive sequences. Core units of the repeats are usually isolated from human or other animal genomes, or they are chemically synthesized. Isolation of probes from genomes usually involves screening a DNA library, followed by confirmation of the desired clone isolation. The library consists of total

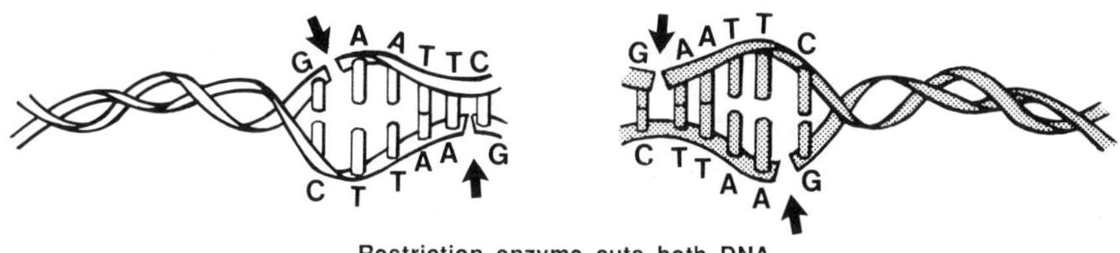

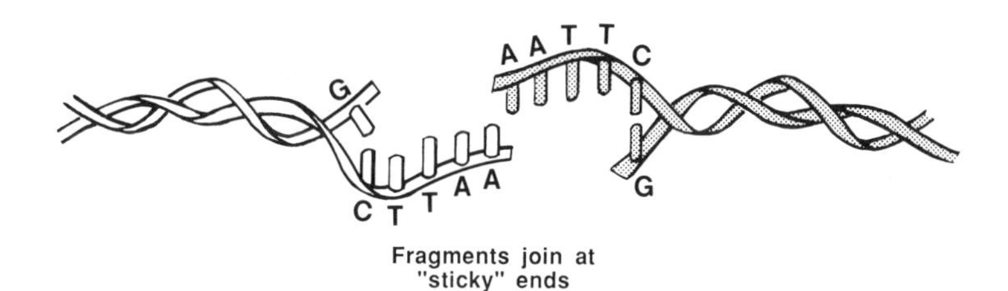

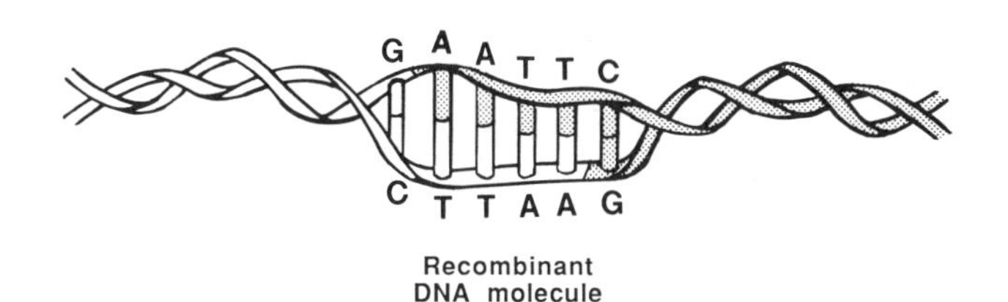

Figure 2-16. A recombinant DNA molecule. The heterologous DNA molecules (left and right in the diagram) were cleaved with the same restriction enzyme. The fragments can be joined at these sites.

or selected mRNA-directed cDNA clones, or sheared or endonuclease-digested genomic DNA clones. The biological tools required for this process include restriction endonucleases, DNA ligase, DNA polymerase, cloning vectors, and vector hosts.

DNA ligases: DNA ligases are enzymes that catalyze DNA 3'OH and 5'P termini bond formation. Any two DNA fragments produced by cleaving with the same endonuclease can reanneal by base pairing and by the action of DNA ligase.

DNA polymerases: DNA polymerases are enzymes that catalyze the formation of a complementary DNA strand.

Cloning vectors: Cloning vectors are biological carriers used to amplify an inserted DNA sequence. Plasmids, the most common cloning vectors in recombinant DNA, are double-stranded, circular DNA molecules ranging in size from 1 kb to over 100 kb. In nature, plasmids are found as extra chromosomal elements in bacteria, where they confer benefits such as antibiotic resistance to the host cell. One common cloning plasmid, pBR322, is small, (slightly more than 4 kb) and has endonuclease recognition sites within the ampicillin and tetracycline antibiotic resistance genes. These sites are useful for marker purposes to ensure that both cloning and host cell transformation have occurred.

Other cloning vehicles include bacteriophage and cosmids; both will accommodate considerably larger foreign DNA inserts than will plasmids.

Host bacterium: Different strains of *E. coli* are usually used as host cells for vector replication. Each bacterial cell will accept one vector that can replicate manifold as the culture grows.

A cDNA library is prepared by (1) isolating mRNA; (2) synthesizing single-stranded cDNA using mRNA as templates, oligo-dT (linked to the mRNA poly-A tails) as primers, and reverse transcriptase as a polymerase; (3) separating the mRNA-cDNA and using the cDNA as templates to produce the double-stranded molecules; (4) cloning the cDNA into vectors such as plasmid, bacteriophage, or cosmid; and (5) transforming host cells, usually *E. coli*, with each new recombinant vector.

A total genomic library is produced by isolating sheared or endonuclease-digested DNA, ligating the individual fragments into cloning vectors, and transforming host cells with each clone (Figure 2-17).

A library can be stored until a specific gene or other DNA fragment is required. At this time, a screening process is carried out to isolate the one or few desired clones from the perhaps million or more in the library. There are a number of approaches to this process:

1. Families of oligonucleotides, usually 17-mers, can be synthesized, with some members coding for groups of five or six amino acids within the pure protein in question. The oligonucleotides are then labeled with ^{32}P or some other tag and hybridized with the library clones. The clone or clones that hybridize are isolated, amplified, and sequenced to determine if they code for the known amino acid sequences.
2. Labeled antibodies to the pure protein can be used with expression vector libraries such as λgt11. (The clones in these libraries synthesize the protein coded by the cloned inserts.) The plaque or plaques where antigen-antibody reactions occur are isolated, amplified, and sequenced to determine, as with the oligonucleotide system, if they code for the amino acid sequences within the pure protein.

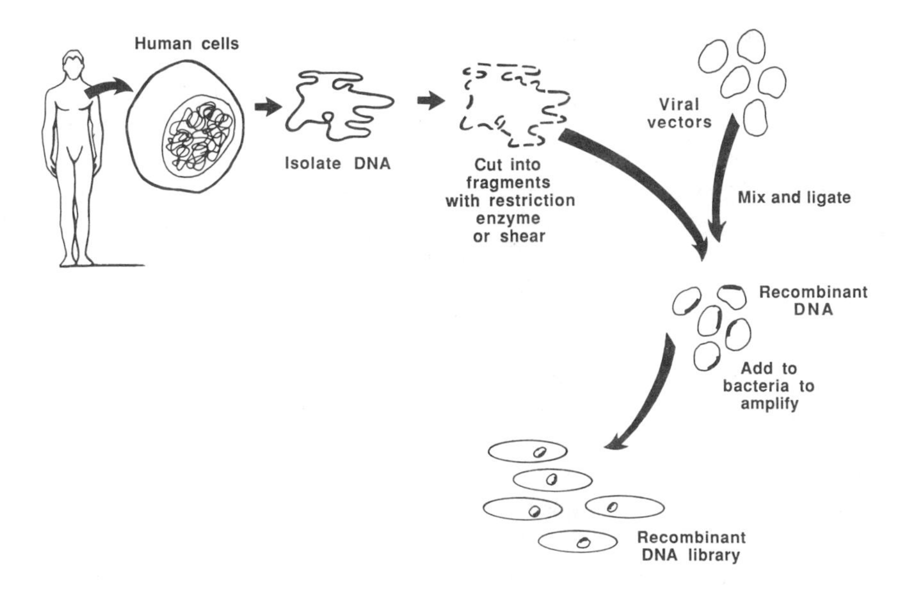

Figure 2-17. Production of a recombinant DNA library.

3. mRNA can be isolated, labeled, and hybridized to the library colonies. The hybridized mRNA is washed from the specified cDNA colony or colonies and expressed in a cell-free system. Antibodies are then used in the same manner as in the expression vector procedure.

If the requirement is to isolate clones containing noncoding DNA inserts—for example, tandem repeats—a genomic library can be screened with core probes consisting of the repeat sequences. These probes may have been previously isolated or synthesized chemically.

The desired clones, regardless of the isolation process, can later be amplified and used as probes.

BASE SEQUENCE DETERMINATION

Normal and abnormal gene base sequence information is fundamental to understanding both the transcription process and the basis of inherited diseases. In the field of DNA identity analysis, knowledge of the sequence of tandem repeats and restriction endonuclease recognition sites is also of considerable value. Minor modifications of a specific sequence in repetitive DNA may result in the code for other repetitive core units and the development of new probes. Also, choice of the best restriction enzymes for use in conjunction with each probe can be most efficiently determined when sequence data are available.

Approximately a decade ago Maxim and Gilbert (1977) and F. Sanger and associates (1981) developed processes to directly sequence DNA fragments. The principles of the Sanger technique are as follows. The DNA to be sequenced is isolated, usually cleaved by a restriction enzyme, and amplified by cloning into a vector such as phage M13. This phage is favored for sequencing since it has two forms, double- and single-stranded. The double-stranded form is used as the cloning vector, and the single strand, produced from this, provides the sequencing template. The base sequence of the cloning site region in M13 is known, and a short oligonucleotide can be chemically synthesized for use as a primer in the Sanger reaction. Alternatively, the polymerase chain reaction can be used to amplify the region of interest. This procedure requires that at least 20 base pairs of sequence, flanking the larger region to be sequenced, be known for primer production. When a ratio of primers, for example, 100:1, is used, one strand of the duplex will differentially amplify. This provides the required single-stranded raw material.

Building blocks for the sequencing reaction include four normal deoxynucleotides—dATP, dTTP, dCTP, and dGTP—as well as four dideoxynucleotides. The latter can be incorporated into a growing DNA chain; however, once they are incorporated, the chain is terminated as a consequence of their chemical structure. Tight control of each dideoxynucleotide:deoxynucleotide ratio is key to the reaction success. Four sets of reaction mixes are established. Each has the same labeled template to be sequenced, oligo-primer, and DNA polymerase, but the dideoxynucleotides differ. The end result is the production of a series of labeled DNA chains whose lengths depend on the location of the incorporated dideoxynucleotides and, therefore, a particular base in the template. The chains can be separated according to length by acrylamide gel electrophoresis and the banding pattern read to determine the original template base sequence (Figure 2-18).

Automated sequencers are now used for large-scale projects. Although the operation details differ from those just outlined, the principles are similar.

OLIGONUCLEOTIDE SYNTHESIS

The chemical synthesis of oligonucleotides is fundamental to many procedures in molecular biology. As outlined, the synthesis of the template primer for sequencing has resulted in considerably increased efficiency for this procedure. Oligo-probes are routinely used in molecular diagnosis of inherited diseases and in identity testing. Also, the synthesis of mutant genes or portions of genes is providing insight into the effects of specific nucleotide changes on the phenotypes of organisms.

Synthesis is predicated on blocking either the 3' or 5' end of the mono- or oligonucleotides. This facilitates addition of the desired individual units to the growing chain, removal of the blocking agent, and addition of another blocking

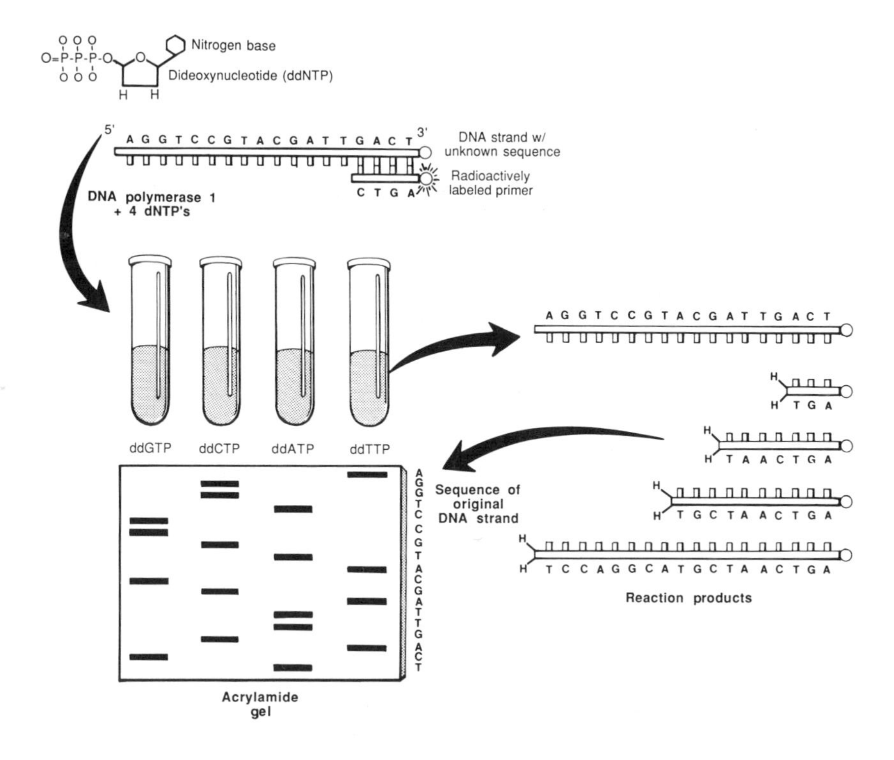

Figure 2-18. Base sequence determination in a fragment of DNA.

unit. A repeating cycle of condensation (addition), block group removal, and recondensation is continued until the desired oligonucleotide is achieved. The synthesis procedure has also been automated using solid phase chemistry. On a standard laboratory machine, one nucleotide is added per 10-min (or shorter) cycle to each of approximately ($0.2\ \mu\text{moles} \times 6.02 \times 10^{23})10^{-6} = 1.2 \times 10^{17}$) single-stranded molecules being simultaneously produced. Less than 3 h are needed to produce this amount of a 17-mer. Molecules consisting of more than 100 nucleotides are constructed without difficulty using the automated system.

REPETITIVE SEQUENCE ORGANIZATION

Eukaryotic genomes contain up to 100 times more non-protein-coding than coding DNA, much of which is repetitive. Prokaryotes, such as bacteria, consist of greater than 99 percent single-copy sequences. Human DNA is estimated to consist of over

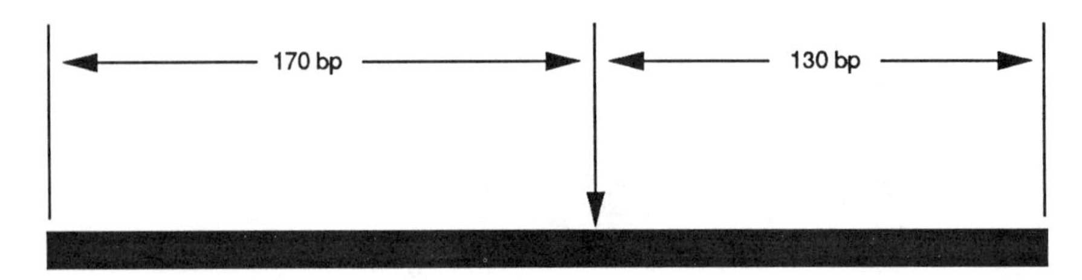

Figure 2-19. The Alu repetitive sequence. This sequence consists of a 300-bp repeat unit. The restriction enzyme Alu I cleaves (↓) each unit into 170-bp and 130-bp fragments.

30 percent repetitive sequences, and bovine DNA greater than 40 percent. As with the Alu family in humans, the number of repeated sequences can exceed one million, or about 5 percent of the total DNA.

Because of this abundance, it is worth considering for a moment the Alu repeat as an example of repetitive sequence organization. The term Alu was coined because of an Alu I endonuclease cleavage site within the 300-bp repeat (Figure 2-19). Alu sequences are usually located in spacer DNA (i.e., regions between genes) and only occasionally within introns. Other species of mammals also carry the Alu family; however, there are base sequence differences and, as with the human, the function remains unclear.

The mammalian genome is, therefore, characterized by single-copy DNA for most enzyme- and structural-protein-coding genes, as well as stretches of non-coding sequence interspersed with repetitive segments of varying lengths. Concentrations of satellite repetitive DNA are located in the heterochromatic regions adjacent to the chromosome centromeres and in the telomeres. Satellite DNA, although not transcribed, is thought to be important to the structural integrity of the chromosomes, especially during the meiotic and mitotic cycles.

Multiple copies of a few genes exist, such as histone, ribosomal RNA, and transfer RNA. All of these products are required in high concentrations by the cell.

DNA for identity profiling consists, for the most part, of tandem repeat alleles from a single locus, from loci in a hypervariable region, or from loci dispersed throughout the genome. It has been estimated that more than 1,500 of these highly polymorphic loci may be present in the human genome. Allele inheritance is in accordance with Mendelian genetics; therefore, haplotypes or allele patterns can be profiled and statistical approaches used to calculate probabilities for each profile based on known specific population allele frequencies.

NOMENCLATURE OF ARBITRARY DNA SEGMENTS

The nomenclature of arbitrary DNA fragments and loci of unknown function is summarized in "Human Gene Mapping 9: Proceedings of the Ninth International Workshop on Human Gene Mapping" (*Cytogenetics and Cell Genetics* 46–762, 1987). A four-component system is followed for most loci analysed in DNA typing: (1) D symbolizes DNA. (2) 1, . . ., 22,X,Y,XY designates the chromosomal assignment. (3) S, Z, or F designates the complexity of the DNA segment. S represents a unique DNA segment, Z represents a highly repetitive segment at a specific site, and F represents families of homologous segments found on multiple chromosomes. (4) 1, . . . is a sequential number that, in conjunction with the other three components, gives uniqueness to each segment. D4S139, for example, designates a specific locus on chromosome 4, and DYZ1 designates a specific highly repetitive sequence locus on the Y chromosome.

REFERENCES

Emery AE and Mueller RF. 1988. *Elements of Medical Genetics.* 7th ed. Churchill Livingstone, Edinburgh.

Gyllensten U. 1989. Direct sequencing of in vitro amplified DNA. In *PCR Technology Principles and Applications for DNA Amplification,* 45–60. Erlich AH, ed. Stockton Press, New York.

Gyllensten UB. 1989. PCR and DNA sequencing. *BioTechniques* 7:700–708.

Hartl DL and Clark AG. 1989. *Principles of Population Genetics.* 2nd ed. Sinaver Associates, Sunderland, MA.

Maxim AM and Gilbert W. 1977. A new method for sequencing DNA. *Proc. Natl. Acad. Sci. USA* 74:560–564.

Murray RK, Granner DK, Mayes PA, and Rodwell VW. 1988. *Harper's Biochemistry.* 21st ed. Appleton and Lange, Norwalk, CT.

Old RW and Primrose SD. 1985. *Principles of Gene Manipulation: An Introduction to Genetic Engineering.* 3rd ed. Blackwell Scientific Publications, Oxford.

Sanger F. 1981. Determination of nucleotide sequences in DNA. *Science* 214:1205–1210.

Smith HO and Wilcox KW. 1970. A restriction enzyme from *Hemophilus influenzae, I. Purification and general properties. J. Mol. Biol.* 51:393–409.

Stine GJ. 1989. *New Human Genetics.* Wm C Brown, Dubuque, IA.

Stryer L. 1988. *Biochemistry.* 3rd ed. WH Freeman Company, Publishers, San Francisco.

Suzuki DT, Griffiths AJF, Miller JH, and Lewontin RC. 1989. *An Introduction to Genetic Analysis.* 4th ed. W H Freeman Company, Publishers, San Francisco.

Thompson JS and Thompson MW. 1986. *Genetics in Medicine.* 4th ed. WB Saunders Company, Philadelphia.

Watson JD, Hopkins NH, Roberts JW, Steitz JA, and Weiner AM. 1987. *Molecular Biology of the Gene, Vols. 1 and 2.* 4th ed. The Benjamin-Cummings Publishing Company, Menlo Park, CA.

CHAPTER 3

Laboratory Organization

Laboratory organization involves both the physical establishment and its operation. It is perhaps simplest to divide the laboratory into its component sections and discuss each separately. The areas may physically overlap for a small facility, and depending on the operation specialty, some sections, such as tissue culture or probe amplification, may not be required. Unless the operation is large, dark room and cold room facilities and expensive equipment such as an ultracentrifuge and beta counter are best shared, if feasible.

The setup of a DNA analysis facility is a relatively simple process if it is incorporated into an established biochemistry program; it is considerably more involved if no such base exists. The outline presented in this chapter is only a guide; individuals contemplating the development of a new facility should visit as many established centers as possible. Discussions with sales representatives and attendance at relevant trade shows and DNA conferences are invaluable.

LABORATORY SECTIONS

Office

Office requirements for a DNA program are no different in principle from those of any other biochemistry program. At least one separate office is required, usually for the program director and, as space permits, offices for a clerk-secretary and senior technologist are useful. Everyone working in the laboratory must have at least a small partitioned desk space in a quiet location. Lockable fire-resistant cabinets are required to store sensitive records; these cabinets should be accessible, preferably located in the clerical area.

Specimen Documentation

Analysis results are worthless without proper documentation of a specimen's chain of custody (continuity). Information, including time and conditions of specimen procurement, conditions of storage and shipment, date received by the laboratory, and reason for the analysis request is also required (Figures 3-1, 3-2). These data can be manually recorded; however, entry into a computer program capable of sorting and maintaining records for long-term retrieval is almost mandatory.

Specimen Processing

Storage of unprocessed specimens may be necessary, and if at all possible, DNA should be isolated when received. The equipment required for this process includes a biohazard fume hood, water baths, vacuum apparatus, clinical centrifuge capable of 3,000 rpm (refrigeration is not necessary), microfuge, desiccator, UV spectrophotometer, electrophoresis apparatus, UV transilluminator, Polaroid or 35 mm camera station, refrigerator, and freezer (at least –20°C, –80°C preferable). A 4°C walk-in cold room is useful but not essential.

Appropriate safety precautions must be taken, especially for handling human samples. One of the extraction reagents, phenol, is very toxic, and a separate section of the laboratory with a well-vented fume hood should be set aside for the extraction process.

Lastly, arrangements must be made to incinerate waste, including tissues and organic solvents. It is not acceptable to use the public sewer or refuse collection systems for hazardous materials.

CHAIN OF CUSTODY DOCUMENT

THIS DOCUMENT MUST ACCOMPANY A SPECIMEN AT ALL TIMES

PLEASE WRITE CLEARLY

I, ______________________________ (name of person drawing blood samples) did take blood samples from the following individuals:

Mother: ______________________________

Child: ______________________________

Father: ______________________________

on this ______________ day of ______________ 199 ___ .

I identified each individual and have attached photographs and fingerprints. Each specimen is clearly labeled with the individual's name, relationship, race, date drawn, and my initials.

If you have any questions please call me at ______________________

Signature ______________________________

Witness ______________________________

Witness signature ______________________________

PLEASE COMPLETE THE FOLLOWING: (Please Print)

Mother:

Last name ______________ First name ______________ Race __________

Date drawn ______________ Date of birth ______________

Child:

Last name ______________ First name ______________ Race __________

Date drawn ______________ Date of birth ______________

Alleged father:

Last name ______________ First name ______________ Race __________

Date drawn ______________ Date of birth ______________

Is this a complete case? ______________________________

Please send results to ______________________________

Mother:

Photograph Fingerprint

Child:

Photograph Fingerprint

Father:

Photograph Fingerprint

Figure 3-1. Chain of custody (continuity) document for blood samples. (This form was redrawn with permission from Lifecodes Corporation, Valhalla, New York.)

CHAIN OF CUSTODY

THIS DOCUMENT MUST ACCOMPANY A SPECIMEN AT ALL TIMES

Section A: SAMPLE REGISTRATION ________ CASE # ______________

Accession # __________________ Sample description ____________________

Referent agency ID ___________ Additional package information _________

Section B: CASE INFORMATION

Referring agency: ____________ Agency reference _____________________

Address: ____________________ Case subject's name __________________

Section C: IN CASE OF SPLIT

Accession # ____ Volume/amount retained by referent ___________________

Split witness by _______________________________

Signature of witness ____

Section D: CHAIN OF CUSTODY RECORD

Method of delivery ____________________________

Received from ________________________________ Title _______________

Date _________________ Time _________ Signature ___________________

Given to ____________________________________ Title _______________

Date _________________ Time _________ Signature ___________________

Given to ____________________________________ Title _______________

Date _________________ Time _________ Signature ___________________

Sample materials returned to referent. Received by ______________________

Date _________________ Time _________ Signature ___________________

SECTION E: FOR EVIDENCE RECEIVED BY MAIL

I hereby certify that I received the specimens at the laboratory facility and that there is no evidence that the package has been opened.

Signature __________________________ Date ____________ Time _________

Package taped (yes/no)___ Tape intact ____ Signs of tampering ____________

Courier # _____ Courier signature ______________________________________

Samples given to ___________________________________ for accessioning.

Figure 3-2. Chain of custody (continuity) document for blood (tissue) samples after arrival at the analysis laboratory. (This form was redrawn with permission from Lifecodes Corporation, Valhalla, New York.)

General Preparation

A general laboratory preparation area with conventional benches and sinks is required for the following: to set up digestions, gel pouring, electrophoresis, Southern blotting, and reagent preparation. The required equipment includes a top-loading and an analytical balance, pH meter, incubator, heater-stirrer, vortex, leveling table, microwave oven, and a source of distilled deionized water.

Vector Handling

It is a good policy to separate the facilities for probe amplification using micro-organism vectors from those for DNA analysis. Specimens contaminated by plasmid or other microorganisms can cause considerable problems in DNA profile interpretation. This separation includes glassware, pipettes, gel apparatus, and other possible sources of DNA transfer. Although duplication of minor equipment will be necessary, the use of disposable labware is one key to success of this operation.

Bacterial culture facilities required to amplify and characterize vectors include an incubator shaker, refrigerator, –20°C freezer, spectrophotometer, microfuge, clinical centrifuge, electrophoresis apparatus, and an autoclave. An ultracentrifuge may or may not be required depending on the amplification procedure followed.

Amplification by PCR

Although minimal facilities are required for amplification of DNA using the polymerase chain reaction, it is important that this operation be isolated from the remainder of the laboratory. The thermal cycler used in the amplification process, the electrophoresis apparatus for fragment separation, and the general preparation area will occupy less than $3m^2$ of bench surface. Separation of the complete PCR system from the other laboratory areas is necessary to ensure that DNA samples are not contaminated with high concentrations of PCR products.

Hybridization and Radioisotope Handling

A number of systems are available for hybridization ranging from simple heat-sealed plastic freezer bags to Robbins type commercial holders. Hybridization is normally carried out at elevated temperatures; therefore, a precision incubator or water bath is required. The hybridization process requires labeled probes for identification purposes. Although nonradioactive tags can be used, where high concentrations of the sought after DNA sequences are present, ^{32}P remains the label of choice. Every laboratory must, therefore, be licensed to handle low- to intermediate-level radioactivity until efficient photolabeling techniques are developed.

The licensing procedure is not difficult provided proper facilities and trained personnel are available. Facilities and equipment include an appropriately lined fume hood, plexiglass shields, lockable storage, a survey meter (Geiger counter), waste disposal containers, ß-counter, and personal monitors. In Canada, The Atomic Energy Control Board regulations should be consulted regarding appropriate laboratory design. In the United States, the use of ^{32}P comes under the purview of the Nuclear Regulatory Commission (NRC). An NRC material license must be obtained by any organization using ^{32}P.

For those not trained in radioisotope use, a summary of radiation control guidelines follows.

Ordering: The current radioisotope license number, isotope, quantity of radioactivity, user's name, and specific location must be included on all purchase orders. Suppliers will not ship radioactive materials without a valid license number and they will not ship a radioisotope not listed in the license. Radioisotopes not listed can be ordered after an appropriate license amendment; this is usually obtained within a few weeks or sooner if necessary. Possession limits can also be readily increased with a license revision.

Receiving: It is imperative that upon receipt of a shipment the user be notified because

1. A contamination check must be carried out.
2. Many radioisotopes have a short half-life (14 days, for ^{32}P) and are shipped on dry ice. These materials decay rapidly, giving rise to worthless reagents and unreliable results.
3. Radioisotopes should be stored only in designated areas under controlled conditions; this usually means in the ordering laboratory.

Because facilities are not usually available for monitoring radioactive leaks, the receiving department should not open incoming packages.

This process is the responsibility of the user. Radioactive materials cannot be accepted during off-duty hours unless special arrangements are in place. The user is responsible for tracing any materials not received when expected. This usually involves checking with receiving, contacting the supplier, and if necessary, contacting the courier.

Package monitoring: A visual check of the outer box containing the radioisotope is all that can be performed by the receiving department. Detailed monitoring is undertaken by the user, who should open and inspect packages upon receipt and keep within the following guidelines:

1. Disposable gloves and a laboratory coat must be worn. Eye protection must also be provided.

2. The package contents must be scanned with a Geiger counter if the monitored isotope is capable of being detected.
3. The contents must be wipe-tested (see contamination control).
4. The package contents must be verified with the packing slip and with the original purchase order and the data recorded. The supplier must be notified immediately of any irregularities.
5. The package must be treated as radioactive waste if contamination is detected and must be processed as outlined in the waste disposal section. The supplier must be notified and the actions taken recorded.

Storage: Radioisotopes are usually stored in a refrigerator or freezer in the laboratory. The following rules apply to storage:

1. The room or storage unit should be locked during off-duty hours.
2. The room and storage unit must have proper radiation warning signs.
3. Radioactive materials must have proper shielding.
4. If there is a possibility of gas or aerosol emission, unsealed sources must be stored in an appropriately vented fume hood.
5. Liquids must be stored in double containers.
6. The radioisotope container must be labeled with the type and activity of the source as well as the arrival date and user's name.

Records: It is imperative that clear up-to-date records of (1) purchases, (2) usage, (3) waste disposal, and (4) contamination monitoring be maintained and available for inspection. All radioactive materials must be recorded in a permanent log book or computer system with the following information: date, material, quantity, storage location, and any other relevant comments such as condition if damaged, or late delivery. The disposition of the radioactive source must also be recorded, for example, quantities used in assays (by date) and waste disposal. Each entry in the log must include the recorder's name.

Laboratory working guidelines: The following guidelines apply when working with radioactive materials:

1. No food is allowed in the working area.
2. A survey meter (e.g., Geiger counter) must be readily available.
3. Weekly wipe tests should be performed. These must be undertaken if there is any possibility of contamination.
4. Work should be carried out on a tray lined with absorbent material such as paper-coated plastic.
5. A fume hood must be used if there is any possibility of volatilization of the radioisotope. For hard beta emitters such as ^{32}P, at least low density plexiglass shielding must be used.

6. A protective lab coat must be worn.
7. Rubber gloves must be worn; double-gloving may be most effective.
8. A dosimeter (TLD) must be worn when working with high-energy beta nuclides such as ^{32}P. The dosimeter is the responsibility of the user and must be exchanged at regular intervals for monitoring or returned when no longer in use.
9. Pipetting by mouth is not allowed.
10. Radioactive sources should be handled with tongs or forceps.
11. Liquids should have double containment.
12. Contamination should not be above background, although Atomic Energy Control regulations permit twice the background level.
13. Contaminated skin must be washed immediately. Abrasives are not to be used.
14. Accidental ingestion of a radionuclide must be immediately reported to medical authorities.

Protection principles: Risk factors should be understood and safety procedures carefully followed to minimize exposure to radiation:

1. There is a direct relation between the amount of radiation exposure and the length of time of the exposure. Exposure time should always be kept to a minimum.
2. The amount of exposure is proportional to the inverse square of the distance from the radiation source. If the distance is doubled, the exposure is reduced by a factor of four. The working distance from the source should be as great as possible.
3. Different shielding materials will reduce the amount of radiation transmitted. ^{32}P is a high-energy beta emitter; thus, when working with this radionuclide, at least 1 cm thickness of plexiglass shielding is recommended. Thin lead or other metal shielding is not recommended because when beta particles strike this target, a portion of the particle energy is converted into more penetrating X-radiation (Bremsstrahlung). The energy of these X-rays increases with increasing atomic number of the target material and the energy of the beta particles.

Contamination control: There are two standard laboratory monitoring procedures:

1. In wipe testing, an ethanol-wetted filter paper disk is wiped over the test area and the filter checked in a beta counter. The areas tested are marked on a diagram for reference.
2. A Geiger counter is suitable for medium- and high-energy beta emitters such as ^{32}P.

As a general rule, both wipe-testing and direct monitoring are used. The radioisotope working area must be monitored after each labeling. Monitoring of the storage and waste disposal areas should be performed weekly.

Personnel monitoring: The absorbed dose of radiation for an individual is recorded by Radiation Protection Bureau thermoluminescent dosimeters (TLDs). The maximum annual permissible doses for the general public, as outlined by the Atomic Energy Control Board (Canada), are as follows:

Organ tissue	Maximum annual dose
Whole body, gonads, bone marrow	5 mSv (0.5 rem)
Bone, skin, thyroid	30 mSv (3 rem)*
Hands, forearms, feet, ankles	75 mSv (7.5 rem)
Other single organs	15 mSv (1.5 rem)

*The dose to the thyroid of a person under the age of 16 years must not exceed 1.5 rems (15 mSv) per year.

These monitors do *not* detect radiation from low-energy beta emitters such as ^{3}H, ^{14}C, and ^{35}S but do detect high-energy nuclides such as ^{32}P. They can also be used to detect X-rays and gamma rays.

Contamination/decontamination: The presence of radioisotope in other than the assay vials—for example, on the bench top, on instrumentation, or on the user's clothing or skin—can be detrimental to personnel safety as well as assay results. The decontamination process for a minor spill involves the following steps:

1. Immediately notify all people in the area.
2. Decontaminate personnel by removing contaminated clothing, flushing the affected area with copious quantities of water, and washing with a mild soap. Repeat this procedure until any residual radioactivity is at background level.
3. Confine the spill. A liquid should be covered with absorbent material such as paper towels. A powder should be covered with dampened absorbent material.
4. Mark off the contaminated area with tape marked "RADIOACTIVE."
5. Decontaminate the area. Place the absorbed material in plastic bags and shield or dispose as radioactive waste. Wash the area and dispose of the water as radioactive waste. Repeat until any residual radioactivity is below background level.
6. Prepare a report of the incident giving an account of the spill and the action taken.

The accidental ingestion of a radioisotope can be serious and must receive the immediate attention of medical personnel. Once this priority is in hand, a spill is then treated as outlined.

Loss or theft: These situations occur very rarely and because of the low activity of the sources, are not usually of major consequence in terms of radiation hazard. Loss or theft is serious, however, because of the possibility of sloppy control in transport or storage. Immediate action must be taken to locate the lost or stolen material.

Waste disposal: A record must be maintained for all radioisotope disposal. This information is important for balancing the record books and for maintaining inventory sheets for disposal containers. The radiation level outside a waste container, as monitored with a Geiger counter, must not exceed 2.5 μSv/h (0.25 mrem/h or about 150 cpm). If this level is exceeded, increased shielding must be installed immediately.

Waste materials are processed as follows according to their physical state and activity level. For liquids:

1. Aqueous waste must be diluted to at least 0.01 of the scheduled quantity per liter of water and discharged in the sewer. It is standard procedure to flush the waste through a designated sink into the sewer for a period of time with a constant flow of water. (The scheduled quantity for ^{32}P is 10 microcuries = 370 kBq.)
2. Organic solvent waste is collected in designated containers. Radioisotopes are diluted to at least 0.01 of the scheduled quantity per liter of solvent. The disposal container must be labeled with the quantity and nature of both the radionuclides and solvents. These materials can then be processed by a chemical waste disposal facility.

For solids:

1. Hot short half-life waste, such as ^{32}P, is held for 10 half-lives in shielded labeled containers in a fume hood. After that period, less than 0.1 percent of the original activity remains, and provided the radiation level is less than 2.5 μSv/h (0.25 mrem/h or about 150 cpm), the waste can be disposed of as non-radioactive (^{32}P half-life = 14 days).
2. Low-level waste, such as gloves and test tubes used with a radioisotope, is separated into combustible and noncombustible material and collected in plastic bags held in foot-operated metal containers. When filled, the bags are sealed and labeled. If the radiation is less than 2.5 μSv/h (0.25 mrem/h or about 150 cpm) and the activity is below one scheduled quantity per kg of waste, the bags are disposed of as normal waste.

No waste materials are allowed to leave the laboratory bearing radioisotope tags or tape; all radioisotopes must have decayed to the level of normal waste designation.

Darkroom

Darkroom facilities are required for that portion of autoradiography involving the overlaying of X-ray film on the hybridized blots and development of the exposed X-rays. The only special requirement for the room is a safelight with a wavelength suitable for use with the film. Cassette holders with calcium-tungstate (not rare-earth) intensifying screens are used during exposure to the radioactive blots. The cassettes are usually stored in a –70°C freezer during exposure.

A room capable of being darkened, not necessarily a darkroom, is required to accommodate a Polaroid or 35 mm camera and the transilluminator used to observe and photograph ethidium bromide-stained gels. The skin and especially the eyes must be shielded from concentrated UV light; thus, a protective plexiglass cover is mandatory when working with this wavelength source.

Result Interpretation

A section of the laboratory should be dedicated to the interpretation of autoradiogram patterns. A digital image analysis system consisting of a video camera linked to a microcomputer is required to determine profile band positions and to compute probabilities when matches are declared.

Tissue Culture

Most identity analysis laboratories will not require tissue culture facilities; however, others such as those associated with medicine and animal sciences may find the ability to perform tissue culture a distinct advantage. The major equipment for this activity includes a laminar flow hood, CO_2 incubator, inversion microscope, clinical centrifuge, liquid nitrogen storage, refrigerator/freezer, and an autoclave. Because of the efficiency of modern laminar flow hoods, it is no longer necessary to designate isolated culture rooms, although these do provide an improved working environment.

Wash-up Area

A separate wash-up unit should include the autoclave and distilled water source for the laboratory. A dishwasher connected to distilled water for the final rinse cycle and a drying oven should also be incorporated into this area. Because high temperatures are common in wash-up rooms, an efficient air-conditioning system should be considered.

Table 3-1. DNA Laboratory Equipment

Equipment	Necessity	Equipment	Necessity
Computer and video camera (scanner)	Y	Microwave oven	Y,S
Fume hood (biohazard)	Y	Autoclave	Y,S
Water bath (2)	Y	ß-counter	Y,S
Centrifuge		Dishwasher	Y,S
(a) microfuge	Y	Drying oven	Y,S
(b) clinical	Y	Vacuum apparatus	Y
(c) ultra	O,S	Desiccator	Y
UV spectrophotometer	Y,S	Heater/stirrer (2)	Y
Electrophoresis apparatus		Vortex (2)	Y
(a) power packs (2)	Y	Leveling table	Y
(b) maxi gel (3)	Y	Precision pipettes	
(c) mini gel (2)	Y	(a) 0–20 µl (2)	Y
UV transilluminator (316 nm)	Y,S	(b) 0–200 µl (2)	Y
Polaroid or 35 mm camera		(c) 0–1000 µl (2)	Y
with appropriate filters	Y,S		
Refrigerators (2)	Y	Plexiglass shields (4)	Y
Freezer (non-cycling)		Geiger Mueller counter	Y
(a) –20°C	Y	Safelight	Y,S
(b) –70°C	Y,S	Cassette holders with intensifying screens (8)	Y
Cold Room (4°C)	O,S	Speed Vac concentrator	
Balances		centrifuge (Savant)	Y
(a) top-loading	Y	Heat sealer (plastic bag)	Y
(b) analytical	Y,S	Tube rotator	Y
pH meter	Y,S	Ice machine	Y,S
Incubator	Y,S	DNA thermal cycler	Y,S
Distilled/deionized		Kodak M35A X-Omat	
water apparatus	Y,S	processor	O,S
Bacterial incubator/shaker	Y,S		

Code
Y = necessary O = optional
S = share (n) = number required

EQUIPMENT SUMMARY

A summary of the equipment, excluding tissue culture, required for a DNA laboratory is presented in Table 3-1. Those items designated S are not needed on an ongoing basis and if reasonably accessible, should be shared with adjacent laboratories. Numerous scientific supply companies stock specialty equipment for DNA analysis; thus, availability should not be a problem.

OPERATION

Personnel

As with many types of technology, unless the director of a new service laboratory is intimately familiar with the details of the methodology and has sufficient time to thoroughly train and assist technologists unfamiliar with the techniques, it is folly to begin operations before a fully trained analyst is available. At present, DNA analysis is simply too time-consuming to be assimilated by trial and error in a service setting. Once the laboratory is established, it is the duty of the organization to train new personnel.

Because of the many steps involved in recombinant DNA technology, including documentation, DNA isolation, digestion, electrophoresis, transfer, probe labeling, hybridization, autoradiography, interpretation, and confirmation of interpretation, at least two technologists should be available at all times. To accommodate holidays, sick days, and training time, at least two full-time individuals and one part-time person are required. The director of a newly established service laboratory must be very much in evidence during the initial months of operation to assess personnel. Through careful monitoring of the various technical processes, it is possible to determine whether an analysis is progressing properly. A failure, especially at a later stage such as hybridization, could result in the loss of at least two working days. A DNA laboratory is no place for a technologist who lacks the patience or incentive to problem-solve and try again.

Once the technologists are in place, lest some readers believe that little else is required, I will recount an anecdote—somewhat embellished, but true. I was fortunate one summer to have a number of assistants working with me. As the summer progressed, it became obvious that one after another of my staff began arriving earlier each morning. Others would stumble in later, groaning about the early and still dark hour. At first I was baffled by the apparent burst of employee enthusiasm and was perhaps somewhat immodest in my interpretation of the phenomenon as evidence of the unswerving loyalty I had engendered. I even had the fleeting thought of reducing salaries to enable the hiring of more technologists who would, no doubt, soon be queuing at my door after hearing of the perfect laboratory and its director. To my chagrin, a delegation arrived at my desk one morning with an ultimatum. Either more equipment and space would immediately be provided or I would be directing a one-man operation. Those who arrived first in the day had done so to secure the scarce equipment and bench space. Humbled, I rectified the situation before rumors about the rather ill-equipped laboratory and its parsimonious director spread. Top-quality production cannot be expected from first-rate technologists in third-rate facilities.

	(1)	(2)	(3)
Tissue specimen ↓			
Documentation ↓	1/4	1/4	Day 1
DNA isolation→Characterization	7.5 → 2	6 → 2	Day 1/2
↓ (Amplification)	(4)	(4)	
Restriction endonuclease digestion ↓	8	2	Day 2/3
Electrophoresis ↓	16	2	Day 3/4
Southern transfer ↓	6	2	Day 4
← Probe labeling	8	2	Day 4/5
Hybridization ← ↓	18	2	Day 5/6
Wash ↓	2	2	Day 6
Autoradiography→Strip blot (→ Probe labeling) ↓	48 → 1	2 → .5	Day 6/7/8
Interpretation	1	1	Day 7 or 8

Code:
(1) = Average time (hours) for each component of the analysis.
(2) = Net technologist time (hours).
(3) = Overall days involved.

Figure 3-3. Flow chart for specimen processing.

Work Flow

Details of recombinant DNA methodology are discussed in Chapter 6. In terms of laboratory organization, it is also important to have some concept of the time required to perform the various steps in DNA analysis. A flow chart for specimen processing from documentation to result interpretation is outlined in Figure 3-3. On average, DNA profiling requires one to two weeks to complete. (Because DNA techniques are rapidly changing, the time period could soon be reduced to less than three days.) The average time required for each component of the analysis of a single whole blood specimen, the net technologist time, and the overall days involved, are listed in columns (1), (2), and (3).

A single specimen is analyzed only under special circumstances; automated extractors handle up to eight samples and at least this number can simultaneously be extracted manually. Fifteen extracts can be electrophoresed on a single gel and at least two blots hybridized with the same labeled probe. It is, therefore, possible to analyze large numbers of specimens simultaneously on an assembly line basis, resulting in considerably reduced cost per sample. The degree to which this is possible for a specific laboratory will depend primarily on sample flow. A large paternity testing center could implement the high-production approach; alternatively, a small forensic laboratory may simultaneously analyze only two or three samples for a homicide or rape case. One technologist should process the equivalent of three to five paternity cases at three specimens each per week. The actual time frame from blood collection to result interpretation is one to three weeks.

Once a DNA sample has been fixed on a nylon membrane, the blot can be hybridized with a number of different probes. This again results in considerably increased efficiency. The only caveat concerns the endonuclease used for DNA digestion because the enzyme must be compatible with the specific probe used to properly interpret the results.

Safety

Safety is one of the most important aspects of laboratory operation. Two areas of direct ongoing concern in the DNA laboratory include radiation hazards (previously discussed) and biological hazards. The handling and disposal of biological specimens, especially those of human origin, are of concern to the technologist, other laboratory workers, and the public at large because of the possible presence of hepatitis, HIV, or other contaminants. It is prudent to treat all samples as hazardous and to adhere to the following guidelines:

1. Disposable gloves and cover gown must be worn. Face protection is also required.
2. Specimen containers must be opened with caution and in an area of the laboratory suitable for disinfecting with 1% hypochloride solution (5% Lysol is normally used for bacterial contamination) and 70% ethanol. A biological safety hood should be available for use with specimens considered infectious.
3. Disposal areas, such as sinks for diluting biological fluids, should be disinfected daily with 10% hypochlorite.
4. Used disposable labware should be soaked in 2.5% hypochloride before disposal by autoclaving or incinerating.
5. Food or beverage must not be stored or consumed in the laboratory.
6. Personal cleanliness is imperative. Open cuts or other skin abrasions must be thoroughly covered; if this is not possible, the affected technologist should not work with biological specimens.
7. Any accidental cut or prick caused by an instrument used to handle biologicals should receive immediate medical attention.
8. Solid waste and concentrated fluids must be autoclaved or incinerated. Organic solvents must be labeled and disposed of at an appropriately licensed facility.
9. Some reagents such as ethidium bromide are mutagenic and must be handled with considerable caution.

REFERENCES

Bushong SC. 1984. *Radiologic Science for Technologists: Physics, Biology, and Protection.* 3rd ed. C.V. Mosby Co., St. Louis.

Hellriegel D, Slocum Jr. JW, and Woodman RW. 1989. *Management.* 5th ed. West Publishing Company, St. Paul, MN.

Johnson G, Scholes K, and Sexty RW. 1989. *Exploring Strategic Management.* Prentice-Hall, Englewood Cliffs, NJ.

Lillie DW. 1986. *Our Radiant World.* Iowa State University Press, Ames.

Von Sonntag C. 1987. *The Chemical Basis of Radiation Biology.* Taylor and Francis, Philadelphia.

Whetten DA and Cameron KS. 1984. Developing Management Skills. Scott, Foresman and Company, London.

CHAPTER 4

Specimens

The proper handling of specimens for direct storage or DNA extraction and characterization is one of the most important aspects of the profiling procedure. Because DNA typing is not yet a routine test, some laboratories may perform only the isolation portion of the overall analysis and leave the other methodologies to specialized centers. Profiling may never be required for many forensic specimens and only intermediate storage needed. It is essential that smaller centers have at least the facilities to isolate, characterize, and store DNA.

A broad range of DNA sources exists. Fresh tissue usually includes whole blood, buccal epithelial cells, and hair follicles. Under special circumstances, in the medical setting, the genotyping of amniotic fluid cells, chorionic villus samples, and tissue culture fibroblasts may be required. Dried specimens usually include blood and semen stains, tooth pulp, and bone marrow. Animal trophy heads and pelts are also sources of dried DNA. Preserved or unpreserved human autopsy specimens, and tissues from animal gut piles and frozen meat, are other possible sources of DNA.

As with any biological test, the quality of the results can be no better than the quality of the input sample. If the DNA is highly degraded or contaminated, it may be unusable; thus, every effort should be taken to collect, record, transport, store, and isolate materials using meticulous techniques.

COLLECTION

The specimen of choice is 1 ml or more of fresh whole EDTA blood. Anticoagulants other than EDTA may be acceptable; however, there are reports that heparin interferes with the activity of certain restriction enzymes. The quantity of DNA isolated from 1 ml of blood is usually sufficient for the necessary testing and a considerably smaller sample will often suffice (Table 4-1). There are occasions, however, when the DNA yield is low and a repeat specimen is required; for this reason it is prudent to collect an additional sample if possible. Buccal epithelial cells obtained from mouth swabs, and hair follicles are two other general sources of fresh DNA; however, the DNA may require amplification by the polymerase chain reaction to provide sufficient material for analysis. Cells from amniotic fluid surrounding the fetus or chorionic villi from the placenta are obtained by medical procedures to test for genetic anomalies. Profiling can be carried out on these specimens as a check for maternal-fetal cell mixtures or for paternity testing. Fibroblasts derived by culture of these tissues and skin biopsies are also a source of DNA. At least 10 mg of CVS or 10 ml of amniotic fluid are normally obtained in a medical procedure. The CVS should provide sufficient DNA without the necessity of tissue culture; however, amniotic fluid cells must be cultured to provide at least two T_{25} flasks of fibroblasts unless a PCR protocol is followed.

DNA extraction should begin as soon as possible after the fresh tissue has been collected. Tissue can be stored under cool conditions for a few days without excessive DNA degradation; however, if the sample cannot be extracted within at least 48 h it should be frozen at –20°C to –80°C. The quality and quantity of extractable DNA decreases with time and depends on storage conditions after collection (Figure 4-1), and consequently, modified isolation procedures may be required.

Tissues can be placed directly in lysis buffer, then refrigerated, and later extracted. This procedure has the advantage of preventing DNA degradation, but once fixed, the specimen is useful only for DNA analysis.

Dried stains should be collected in clean envelopes and maintained in a dry, cool or frozen state if possible. As with fresh specimens, the DNA should be extracted soon after collection. As a general rule, a stain greater than the size of a dime is desirable. A few microliters of blood may be sufficient if the DNA is to be amplified; otherwise, approximately 50 µl should be obtained if possible. A 10 µl semen stain is usually sufficient. A 15 mg piece of dried tissue, such as skin (hide), should be collected in wildlife poaching cases even though a few milligrams are

Table 4-1. DNA Content of Tissues

Source	**DNA content (approximate)**
1. Amniotic fluid	65 ng/ml (1 x 10^4 cells/ml at 16 weeks gestation)
2. Blood (mammalian)	40 µg/ml (1 µl = 4 x 10^3 to 11 x 10^3 WBC, 1 WBC = 6.6 pg DNA)
3. CVS	8 µg/mg
4. Fibroblast culture	6.5 µg/T_{25} flask (approximately 1x 10^6 cells)
5. Hair roots	250 ng/plucked hair root
6. Liver	15 µg/mg
7. Muscle	3 µg/mg
8. Skin (whale)	3 µg/mg
9. Sperm	3.3 pg/cell

sufficient. It may be noted at this point that DNA suitable for VNTR analysis can be isolated from dried stains of blood and semen more than a decade old, provided the materials have been maintained under satisfactory storage conditions. These materials may be critical to the reopening of criminal cases in which innocence is claimed by those convicted. The analysis of old stored specimens may also be of considerable value in situations where a suspect's guilt was never proven because of insufficient evidence.

Autopsy tissue, animal organ pieces from gut piles or carcasses of wildlife, and frozen specimens should be wrapped in aluminum foil, placed in clean plastic bags, and immediately frozen. If lysis buffer is available, tissue pieces can also be placed in this solution. A 15 mg portion is ample for organs such as liver and kidney. Muscle contains considerably less DNA; therefore, collection of up to 25 mg of this tissue is not unreasonable. It is prudent to sample more than one organ because DNA quality may vary depending on the action of DNAases (Figure 4-1).

All samples must be clearly labeled as to source, tissue type, and date collected. Labels should be placed inside the storage containers if possible. A separate record should be kept outlining possible contamination with other DNA sources, storage temperature, and any other relevant information. An incorrectly or insufficiently labeled specimen may be as worthless as an unlabeled specimen; thus, considerable care is mandatory at this stage.

Because specimen collection is an integral part of DNA profiling, those individuals involved in the collection process should have some understanding of analysis principles and techniques to ensure the DNA is not degraded because of poor handling. The quantity of material required will vary, depending primarily on the analysis approach and the specimen type. If DNA is to be amplified, only nanogram amounts are needed; otherwise, as much as reasonably feasible should be obtained because it may be necessary for an independent laboratory to validate the analysis results. Inactivation, or at least reduced activity of DNAases, is

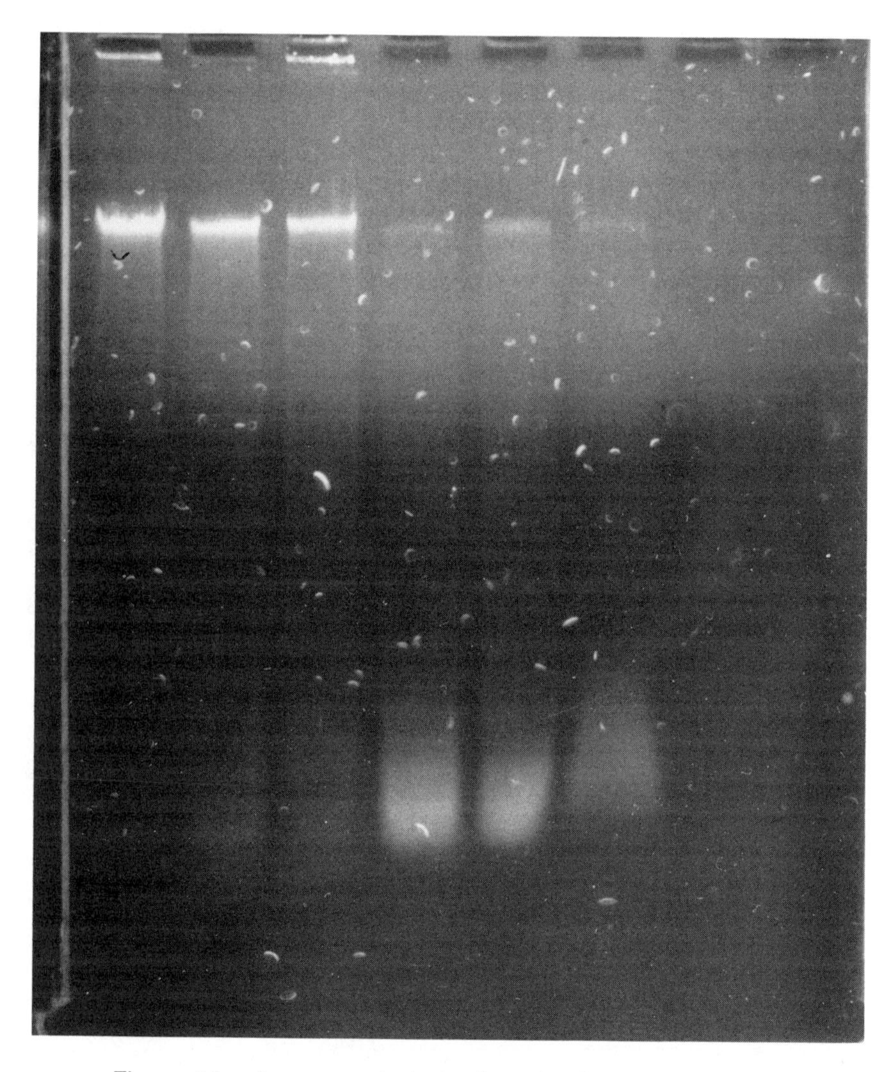

Figure 4-1. Agarose gel electrophoresis of moose genomic DNA stained with ethidium bromide and viewed under UV irradiation. The lanes from left to right are (1), (2), and (3), DNA isolated from a frozen muscle specimen; (4), (5), and (6), DNA isolated from muscle stored in 0.85% saline. Note the high MW DNA (bands at the top of the gel greater than 20 kb in length) derived from the frozen tissue and the low MW DNA (diffuse material a few hundred bp in length) from the saline-preserved specimen.

important to reduce DNA degradation. Storing at low temperature and humidity or placing the tissue in lysis buffer are methods to achieve this end.

STORAGE AND TRANSPORT

Carelessness or ignorance of proper handling procedures during transport and storage can result in a specimen unfit for analysis. This could be especially critical in forensic work where the quantity of sample is often limited to small stains or other minute pieces of tissue that cannot be duplicated.

DNA is usually stored in one of three states: (1) as nonextracted tissue; (2) as a lysate with the tissue placed in a 100 mM Tris-HCl, 40 mM EDTA, 0.2% SDS solution at pH 8.0; or (3) fully extracted pure DNA. Depending on the tissue source, nonextracted material may be kept at room temperature (RT), refrigerated at 4°C, frozen at –20°C to –195°C (liquid nitrogen), or fixed in a solution such as saline or alcohol. A lysis mixture with up to 25 mg of tissue per ml of solution is usually refrigerated at 4°C for long periods (months) or held at RT for shipping purposes. Fully extracted DNA can be shipped at RT, stored at 4°C for months, mixed with chloroform (5 μl/ml of DNA solution) and kept indefinitely, or frozen. Repeat freeze-thawing is not recommended because of the possibility of shearing the DNA.

Shipping via courier with one- or two- day delivery service is advisable because packages can normally be traced if problems arise. The mail system, although less expensive, is to be used at the sender's peril.

DNA EXTRACTION

Principles

The primary objective of the isolation process is to recover the maximum yield of high molecular weight DNA devoid of protein and other restriction enzyme inhibitors (Sambrook 1989). When DNA is isolated from whole blood, the nonnucleated cells are first lysed and discarded together with the plasma. The remaining white cells are pelleted, washed, and lysed. Protein is removed by phenol extraction (or a nonorganic NaCl or LiCl process) and chloroform is added to eliminate traces of phenol. (Care must be exercised to ensure the phenol solution pH is maintained above 7.8 because DNA will partition into the organic phase in the acid range.) In the final step, DNA is precipitated with alcohol from a salt solution of a moderate concentration of monovalent cations.

Cells from blood, semen stains, and other tissue cells (muscle, liver, and CVS) are lysed and the protein further digested by the action of proteinase K. Deproteinization and DNA precipitation processes are performed as with whole blood.

A number of DNA extraction procedures have been developed. These procedures can be performed using either automated or manual systems.

Automated System

Automation of a repetitive process is a key factor in developing an efficient operation when large numbers of specimens are handled. The Applied Biosystems Model 340A Nucleic Acid Extractor (Figure 4-2) is used by a number of laboratories. This instrument consists of a series of eight extraction vessels on a rocker arm located in a temperature-controlled chamber. Each vessel contains a separate sample. The vessels are constructed with an inlet-outlet port at the bottom through which organic solvents and alcohol from prepackaged reagent containers are introduced. Waste products are removed through these ports, and filters are inserted to collect the nucleic acids. The liquid flow system is pressurized with helium and the overall instrument operation controlled by a computer program.

The automated process cycles as follow. Lysis buffer plus proteinase K are pumped into the vessels and are heated to 50°C. Sample specimens consisting of pelleted white cells (from up to 10 ml of whole blood) or other tissues are manually added. (Small volumes of whole blood can also be directly added.) The cells are lysed, nucleases are inactivated, and proteins are digested. Phenol and chloroform are pumped into the vessels for protein and other nonnucleotide extraction; this addition and the organic phase removal are repeated two or more times as required. An ethanol-sodium acetate solution is drawn into the vessels, and a disposable filter is manually inserted into each to collect the DNA and RNA precipitated by the ethanol-salt mixtures. Cycles of ethanol wash the precipitates on the filters. The filters are later removed and are stored frozen or placed directly in buffer to redissolve the nucleic acids. The vessels are purged with hot 6 N nitric acid at the end of the cycle to remove residual traces of nucleic acids before a new cycle begins. If unforeseen delays do not develop, three extraction cycles with eight specimens processed per cycle can be accommodated per 24 h period. A modified instrument capable of accommodating a larger number of smaller volume specimens is eagerly awaited.

The quantity and quality of DNA extracted are at least comparable to manual extractions. RNA mixed with the DNA is removed by treating the sample with RNAase. If interest on investment, depreciation, service contract, reagents, and technologist time are considered, it is doubtful that the automated system will prove less costly than a manual approach; however, DNA extraction is a tedious process and automation has the advantage of freeing a technologist for other less repetitive work.

Manual System Organic Solvent Extraction

Whole blood (Method 1). This procedure follows a protocol outlined by L. Madisen et al. (1987) and D.I. Hoar (personal communication). Advantages include a rapid sample pretreatment to the lysis stage and storage at this point for extended time periods. By incorporating a slight modification, frozen samples and fresh specimens a week or more old can usually be extracted with acceptable DNA recovery. The

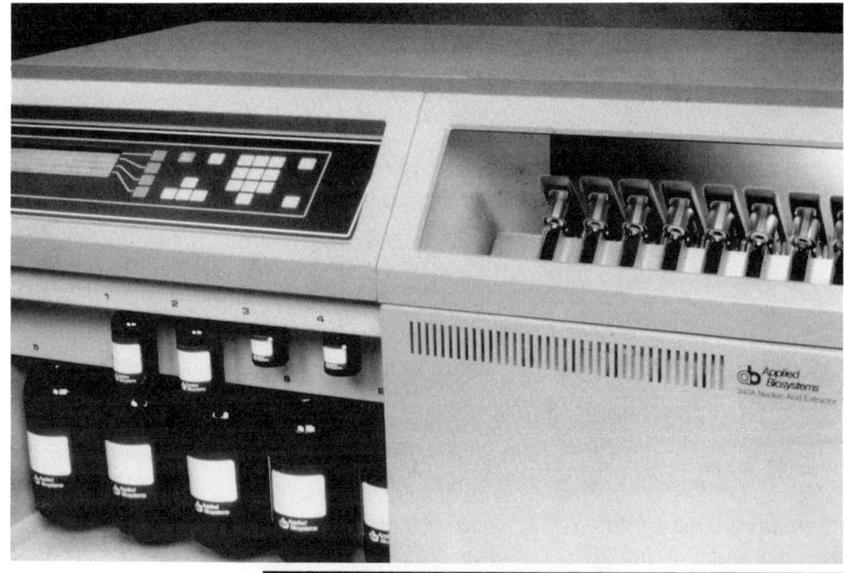

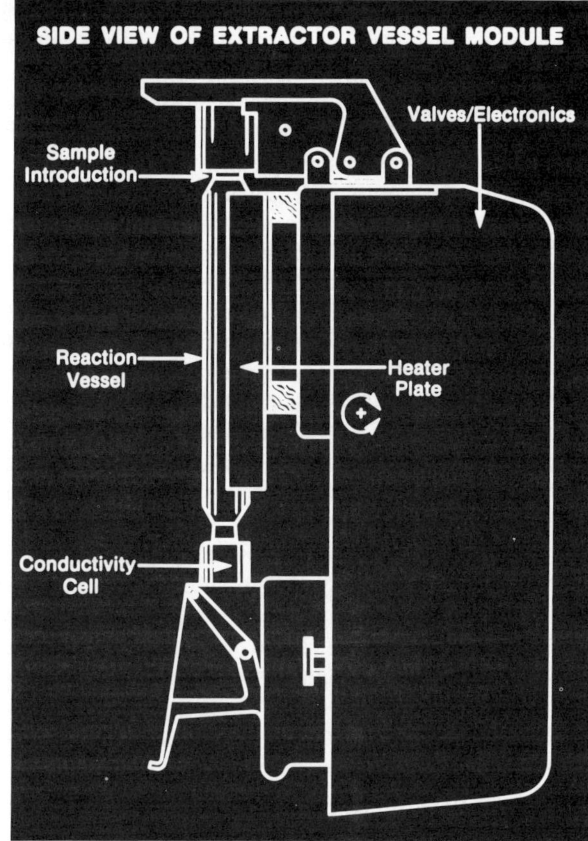

Figure 4-2. Model 340A Nucleic Acid Extractor. A diagram of one of the eight extractor vessel modules is also illustrated. (Courtesy of Applied Biosystems, Foster City, California.)

specimen volumes can be scaled as appropriate; however, the description below is based on one 7 ml tube of anticoagulated whole blood.

Lysis of nonnucleated cells is achieved by adding five volumes (35 ml) of ammonium chloride:Tris at 37°C to one volume (7 ml) of blood and incubating for 5 min. (Overincubating, especially if the white cell membranes are fragile as with non-fresh or frozen specimens, may result in considerably reduced DNA recovery.) The mixture is centrifuged at approximately 900 × g in a swinging bucket centrifuge (refrigeration is not required) for 10 min to form a loose pellet of nucleated cells. The supernatant is aspirated, leaving about 5 ml above the nucleated cell pellet. The pellet is resuspended by adding 10 ml of normal saline and recentrifuging. The supernatant is aspirated, leaving only the pellet, and the pellet is rewashed with saline. The cells are resuspended in 2 ml high TE, then lysed by injecting, with an 18 gauge needle, 2 ml of nucleated cell lysis mixture. The resultant lysate is stable for years.

DNA is extracted from the aqueous lysate by adding an equal volume (approximately 4 ml) of TE-saturated phenol, mixing vigorously to an emulsion, and rotating for 10 min. The organic and aqueous layers are separated by centrifuging at 900 × g for 5 min and removing the upper aqueous phase with a large-bore pipette. The curdy interface is reextracted by mixing with high TE and repeating the phenol procedure. The pooled aqueous layers are again phenol-extracted, then extracted once with chloroform-isoamyl alcohol, centrifuged, and the DNA precipitated from the upper layer. A volume of 4 M ammonium acetate, equal to one-tenth of the aqueous volume, is first mixed with the aqueous layer, and a volume of isopropanol, equal to the aqueous plus ammonium acetate, is added. (Precipitation of DNA using ammonium acetate and isopropanol can result in the formation of residues that adversely affect some enzymes. This precipitation procedure is not recommended if the DNA is to be used in processes such as cloning. Sodium chloride or sodium acetate and ethanol, as described later, should be used.) The white fluffy DNA strands are swirled together with a curve-tipped Pasteur pipette. (After the isopropanol addition, the solution can be cooled to approximately –20°C and microfuged at about 10,000 rpm for 15 min.) The DNA is washed with a gentle stream of 70% ethanol and allowed to air-dry. The pipette tip, with adhering DNA, is broken, placed in a microfuge tube with 500 µl of low-TE buffer and rotated overnight at 4°C. The resultant DNA solution can now be quantitated and the quality determined.

If the blood sample is frozen or not fresh, the white blood cell membranes may have weakened and be susceptible to lysis, resulting in DNA loss at the nonnucleated cell lysis stage. This problem can be circumvented by centrifuging the blood, discarding the plasma, and resuspending the packed cells in two volumes of high-TE buffer. The resultant suspension is lysed as described above, except 1 M NaCl is added to the lysis buffer. The remainder of the procedure follows that described for fresh samples; however, additional phenol and chloroform:isoamyl extractions may be required.

The yield for human DNA approximates 40 µg/ml whole blood.

Whole blood (Method 2). The following method, modified from a Federal Bureau of Investigation protocol, is useful for rapidly extracting large numbers (over 30 per day per technologist) of small volumes (under 1.0 ml) of blood. Signer et al. (1988) also describe a procedure for small samples.

Blood is collected in EDTA anticoagulant vacutainers and stored for up to 48 h at 4°C. (EDTA also chelates cations required for DNAase activity, thus, reducing the possibility of DNA degradation.) 700 µl of blood are aliquoted into a 1.5 ml microfuge tube and frozen at –20°C or –80°C if possible. Any remaining sample is frozen for future use. (The freeze and thaw processes should lyse the red blood cells but not the nucleated white cells.) The 700 µl sample is thawed, and 800 µl of 1 × SSC added. The tube is gently vortexed, and microfuged (10,000+ rpm) for 1 min. The supernatant is decanted and, for precious specimens, saved until the quantity of DNA isolated from the pellet has been determined. (Low yields may be due to lysed white cells and release of DNA into the supernatant. If required, this solution can later be extracted.) One ml of 1 × SSC is added to the pellet and the resultant mixed by gentle vortexing making certain the pellet is resuspended. The tube is microfuged for 1 min, and as much of the supernatant as feasible pipetted off without disturbing the pellet. This supernatant is saved as above, and 375 µl of 0.2 M sodium acetate (pH 5.2) plus 25 µl 10% SDS are then added to the pellet. (For large pellets, up to twice the amount of 10% SDS is added.) The tube is vortexed, 15 µl of proteinase K solution (at a concentration of at least 0.4 U/µl) are added, and gently mixed by inversion. (SDS causes the white cells to rupture and initiates protein denaturation; proteinase K reduces the proteins to their component amino acids.) The mixture is incubated by rotating at 56°C for 1.5 h to 2 h. The pellet should fully dissolve. If the pellet remains intact after 1 h, it should be manually agitated and incubation continued until the pellet is fully dissolved. When the digestion process is completed, 120 µl of phenol are added to the tube. The tube is vortexed for 30 sec, and microfuged for at least 2 min. The upper aqueous phase containing DNA (do not include the denatured protein interface layer) is then transferred to a new microfuge tube. Because of the potential loss of DNA, transfer of a small portion of the organic phase at this stage is preferable to leaving any of the aqueous material. 120 µl of phenol/chloroform/isoamyl alcohol solution are added to the transferred phase, and the tube is vortexed for 30 sec, and microfuged for 2 min. The aqueous supernatant is removed to a new tube, ensuring that none of the organic layer is transferred. (Chloroform-isoamyl alcohol in the aqueous solution may prevent efficient DNA precipitation with ethanol.) One ml (2.5 volumes) of cold 95% ethanol is added and the resultant mixed for 10 min at RT on a rotator. The tube is microfuged for at least 1 min, and the supernatant is carefully decanted. 180 µl of LTE are added to the white DNA film or pellet and the resultant is gently vortexed, and incubated for 10 min at 56°C on a rotator. 20 µl of 2.0 M sodium acetate are added, the tube is vortexed for 10 sec, then 500 µl of cold 95% ethanol are added. The tube is inverted by hand a few times, then allowed to sit for 15 min. The tube is microfuged for 30 sec, and the supernatant is decanted. 500

μl of 70% ethanol are added to wash the pellet. The tube is microfuged for at least 1 min, the supernatant is decanted, and the tube rim blotted dry. The tube is placed under vacuum for approximately 5 min or until any residual ethanol is removed. (Alternatively, residual alcohol can be removed by placing the tube in a Savant Speed-Vac centrifuge for approximately 5 min.) Overdrying must not occur because the DNA may then be very difficult to resolubilize. The DNA is resolubilized by adding 100 μl of LTE, gently vortexing, and incubating at 56°C overnight. The DNA is gently vortexed for 30 sec at the end of the incubation period. If the DNA remains undissolved, it should be reprecipitated with EtOH in an attempt to remove residual salts or other contaminants.

The presence of a brownish pellet after the vacuum drying stage suggests that DNA recovery will be low because of bound protein. Protein is retained as part of the interface layer and will require redigestion beginning at the 0.2 M sodium acetate, 10% SDS step. Redigestion is required only when the initial pellet does not completely dissolve during the SDS/proteinase K incubation.

Some analysts may prefer to follow a simpler organic extraction procedure rather than the modified FBI protocol just outlined. In the FBI protocol (Budowle 1990), after the proteinase K digestion step is completed, the tubes are cooled to RT and 120 μl of phenol/chloroform/isoamyl alcohol (25:24:1) are added to each. The tubes are vortexed for 20 sec and centrifuged (10,000 rpm) for 2 min. The supernatants are transferred to new tubes and the DNA is precipitated as previously described.

At least 50 μg of DNA per ml of blood can be recovered using this technique. Lower recoveries may be due to degradation before processing or to loss during extraction; this latter possibility is likely and should be resolvable by paying careful attention to the suggestions outlined in the procedure.

Tissue culture, amniotic fluid, and buccal cells. The cells are first pelleted from an aqueous suspension by centrifuging at 900 × g for 10 min. The pellets can be washed by resuspending in normal saline followed by recentrifuging. A volume of high-TE buffer is added to each pellet at the rate of approximately 0.5 ml per 10^6 cells, for example, the cells from one T_{75} flask of fibroblasts. The cells are thoroughly resuspended and an equal volume of nucleated cell lysis buffer injected. The resultant lysate is stored at this stage or the phenol and chloroform:isoamyl alcohol procedure is carried out as described for whole blood. Prior to phenol extraction, the lysate can be digested with proteinase K using the procedure described for hair roots.

Because of the small quantities, DNA is precipitated from aqueous solution in microfuge tubes. The process is accomplished by adding one-tenth volume of 4 M ammonium acetate (many analysts prefer to use 2 M sodium chloride or sodium acetate), two volumes of cold (–20°C) 95% ethanol, and centrifuging under 4°C refrigeration at 10,000+ rpm for 15 min. The ethanol is carefully decanted and the tubes allowed to drain before gently adding 500 μl of 70% ethanol. The tubes are recentrifuged for 5 min, the alcohol decanted, and the tubes placed in a desiccator under vacuum for 20 min or until dry (do not overdry). Alternatively, any residual alcohol can be removed by placing the tubes in a Speed-Vac for 5 min. The DNA is resuspended in low TE.

At least 5 μg of DNA should be recovered from 10^6 cells. RNA may also be present as indicated by spectrophotometric and electrophoresis analysis. If this presents a concern, RNAase treatment is carried out.

Hair roots, biopsy, and autopsy tissues. Biopsy and autopsy tissues may include skin, CVS (Figure 4-3), tooth pulp, muscle, bone marrow, and organ pieces such as liver and kidney. Minced tissue is placed in a nucleated cell lysis mixture at the rate of approximately 10 mg per ml. Just before starting the extraction process, self-digested proteinase K is added to the sample at the rate of 50 μg/ml of reaction solution. The mixture is placed on a rotator in a 37°C incubator for 30 min or until the tissue has dissolved. An equal volume of high-TE buffer is added and the resultant is extracted once or twice with phenol as previously described. The aqueous layer is extracted once with an equal volume of phenol-chloroform (one-half volume phenol, one-half volume chloroform-isoamyl alcohol), centrifuged, and the aqueous layer extracted with an equal volume of chloroform-isoamyl alcohol. The DNA is precipitated as described for whole blood or tissue culture cells.

Because of the close proximity of maternal and fetal tissues, isolation of DNA from CVS can be tedious. Dry (1988) describes a simple technique to accomplish this isolation task.

DNA recovery should range from approximately 1 μg per four plucked hair roots to 3 μg per mg of muscle, 8 μg per mg CVS, and 15 μg per mg of organs such as liver.

Fixed tissues. Pathology laboratories use formalin as the long-term fixing agent of choice for autopsy specimens. Tissues can be fixed in other agents such as alcohol; however, this is usually for only short storage periods.

Attempts have been made to isolate DNA from formalin-fixed materials; however, usually only small molecular weight (under 5 kb) fragments are recoverable (Impraim 1987). This fixed tissue is not generally suitable for DNA typing unless short DNA stretches are amplified using the PCR system.

Specimens fixed in 50% ethanol or normal saline with 0.2% sodium azide (to prevent bacterial contamination) yield good quality DNA provided the DNA is not degraded before fixing. There is one report (Amos 1989) of good DNA recovery from skin specimens stored for up to one month in a saturated saline solution. Preservation for at least one year is possible in a 20% dimethyl sulfoxide (DMSO) solution saturated with salt. DMSO increases cell membrane permeability.

Stains. The procedure for isolating DNA from dried blood or semen stains is similar to that followed by Gill et al. (1985, 1987, 1987*a*) and the FBI *(Promega GenePrint*

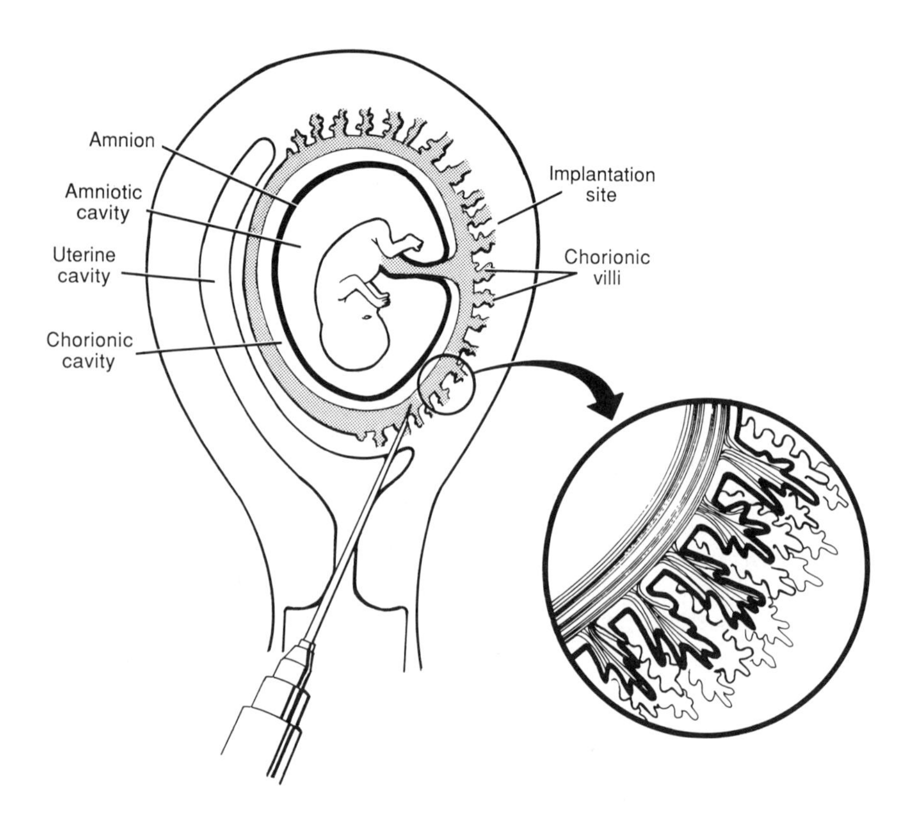

Figure 4-3. Chorionic villus sample (CVS). Chorionic villi can be aspirated as illustrated to provide a source of fetal DNA.

DNA Typing Technical Manual, Promega Corporation). A 100 μl blood sample will on average produce a 2 cm (3/4 inch)-diameter stain. (The area of a circle, 3.14 $\times r^2$, varies as the square of the radius *r*; therefore, a 2 cm-diameter circle contains four times as much specimen as one of 1 cm diameter.) The stain size will vary depending on the substrate. A 100 μl stain on thin cotton sheeting would probably be twice this size.

300 μl of stain extraction buffer containing 2% SDS and 0.1 volume (30 μl) of DTT (0.06 g/ml) are added to cut-up stain material equivalent to 100 μl in a 1.5 ml microfuge tube. 30 μl of proteinase K (1 mg/100 μl) are then added. The tube is incubated overnight at 56°C. Centrifugation is carried out to pellet the stain matrix, for example, cloth or paper, and the supernatant is carefully pipetted into a separate tube. Alternatively, (1) the bottom of the reaction tube can be punctured, placed in a fresh container and the combined unit centrifuged to collect the liquid, or (2) a hole can be punched in the lid of the microfuge tube, the cut-up stain matrix pieces placed

in the lid and the tube centrifuged. If desired, the matrix material can be washed with stain extraction buffer, recentrifuged, and the resultant supernatant combined with the original for DNA precipitation.

The aqueous DNA solution can be extracted with organic solvents and the DNA precipitated with EtOH as described in the previous procedures. Because of the small volumes, considerable care must be exercised when separating the aqueous from the organic layers. If difficulties arise, a small portion of the organic material should also be removed with the aqueous layer and recentrifuged. The organic fraction should remain compact in the tip of the microfuge tube while the aqueous layer is removed by pipette.

Approximately 2.5 µg of DNA per 100 µl of good quality bloodstain should be recoverable.

Sperm and female-cell mixtures. Semen specimens, if obtained from vaginal swabs, are highly contaminated with female cells. Sperm nuclei can be preferentially separated from this mixture by preliminary treatment with extraction buffer, SDS, and proteinase K, but without DTT. (DTT must be present to rupture the sperm nuclear membranes by breaking the protein disulfide bridges.) The nuclei are pelleted and the female DNA is removed in the supernatant. The FBI extraction procedure as outlined in the *Promega Gene Print DNA Typing Technical Manual* is as follows (Figure 4-4).

The swab containing the cell mixture is removed from the applicator stick and is placed in a 1.5 ml microfuge tube that has a depression in the cap. 450 µl of HEPES-buffered saline or PBS together with 50 µl of 20% sarkosyl are added to the tube and the tube is rocked overnight at 4°C. A hole is punched in the tube cap, the swab is placed in the cap, and the unit is centrifuged for 3 min to recover the fluid and cells remaining in the swab. The supernatant is discarded and the swab is returned to the tube. The following components are added to the tube: 200 µl Tris-NaCl-EDTA, 50 µl 10% SDS, 240 µl sterile distilled H_2O, 10 µl proteinase K solution at 1 mg/100 µl. The reaction mixture is incubated at 37°C for 2 h. At the end of the incubation, a hole is punched in the tube cap, the swab is placed in the cap, and the unit is centrifuged for 5 min. The supernatant (containing the female DNA) is removed and saved for further processing. The following components are added to the sperm cell pellet in the tube: 200 µl Tris-NaCl-EDTA, 125 µl 10% sarkosyl, 5 µl DTT at 0.06 g/ml, 150 µl sterile distilled H_2O, 20 µl proteinase K solution at 1 mg/100 µl. The reaction mixture is incubated at 37°C for 2 h. The aqueous DNA solutions can be extracted with organic solvents and the DNA precipitated with EtOH as outlined in the previous procedures. Note that prior to the addition of ethanol, the salt concentration of the aqueous phase should be adjusted to greater than 100 mM. This can be accomplished by adding 10 µl of 5 M NaCl.

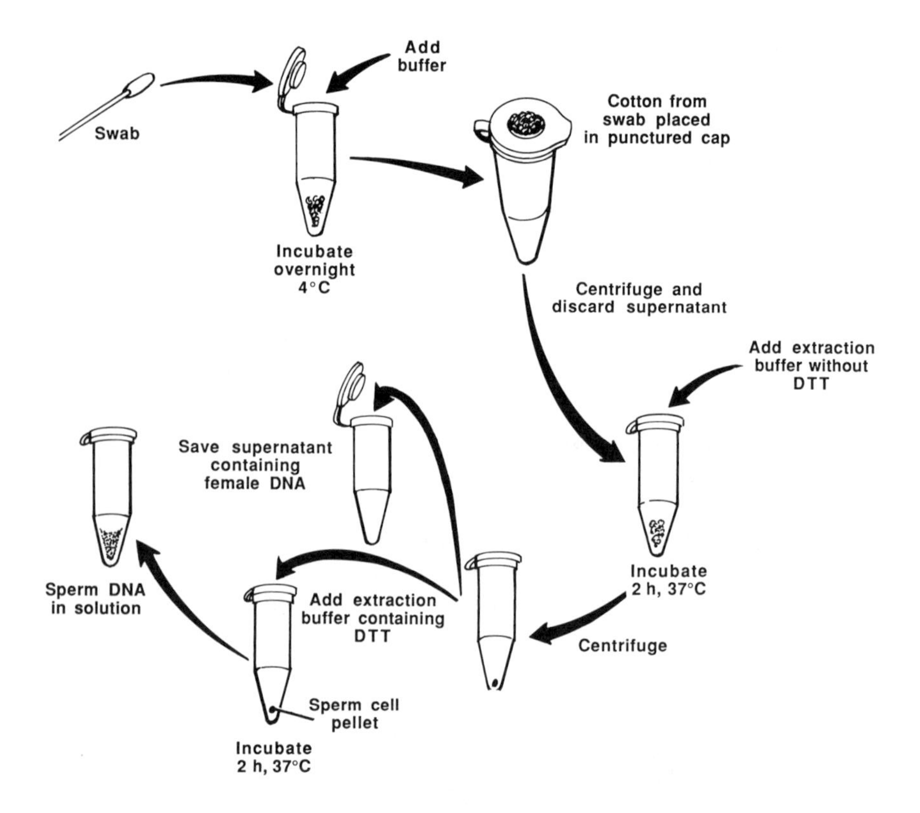

Figure 4-4. Isolation of sperm DNA and female-cell DNA from stain mixtures.

Plant tissues. Total plant cellular DNA can be efficiently extracted using a protocol followed by Saghai-Maroof et al. (1984). Approximately 750 mg of freeze-dried tissue are powdered and dispersed in 15 ml of extraction buffer consisting of 50 mM Tris (pH 8.0), 0.7 M NaCl, 10 mM EDTA, 0.1% hexadecyltrimethylammonium bromide, and 0.1% 2-mercaptoethanol. The mixture is gently swirled and is incubated at 60°C for 30 to 60 min. Ten ml of chloroform-octanol (24:1) are added and the contents mixed by inverting the vessel. The resultant is centrifuged at 5,125 × g in a clinical centrifuge for 10 min at RT and the aqueous phase is saved. Two-thirds by volume of isopropanol is then added to the aqueous material, and the DNA is hooked-out with a curved-tip Pasteur pipette. The pipette with the DNA is transferred to a tube containing 20 ml of 76% EtOH-10 mM NH_4OAc for 20 min. The DNA is finally dissolved in 1.5 ml of 10 mM NH_4OAc-0.25 mM EDTA.

Recovery with this procedure is approximately 275 μg DNA per gram of freeze-dried tissue.

Manual System Nonorganic Solvent Extraction

Nontoxic DNA extraction procedures using NaCl or LiCl have been developed as a substitute for phenol and chloroform to salt out protein from cell lysates. Although not yet widely used, these methods have the advantage of eliminating biohazardous organic solvents and one of the methods (Grimberg 1989) is very efficient for small forensic specimens.

Whole blood. The 6 M NaCl technique developed by Miller et al. (1988) involves the following steps:

1. The white blood cell buffy coat is isolated from 10 ml of blood by slow-speed centrifugation.
2. The white cells are resuspended in 3 ml of lysis buffer (10 mM Tris-HCl, 400 mM NaCl, 2 mM Na_2 EDTA, pH 8.2).
3. The protein is digested by overnight incubation at 37°C (or for 3 h at 50°C) with 0.2 ml 10% SDS and 0.5 ml proteinase K at 1 mg/0.5 ml 1% SDS, 2 mM Na_2 EDTA.
4. 1 ml of 6 M NaCl is added, the tube is agitated for 15 sec and the resultant centrifuged at 2,500 rpm for 15 min.
5. The supernatant containing the DNA is transferred to a new tube and two volumes of RT ethanol are added.
6. After the tube has been inverted several times, the DNA is spooled out and dissolved in LTE buffer by incubating at 55°C for 1 h.

A simple and efficient procedure for the isolation of genomic DNA from blood and other tissues has been developed by Lifecodes Corporation (Grimberg 1989). Nuclei are isolated by mixing one volume of whole blood with four volumes of an ice-cold cell lysis solution consisting of 0.32 M sucrose, 10 mM Tris-HCl pH 7.6, 5 mM $MgCl_2$, and 1% triton X-100. The lysate is centrifuged at $900 \times g$ for 5 min at 4°C, and the nuclear pellet is washed with the lysis solution followed by recentrifugation. The pellet is then rinsed with an ice-cold pellet lysis solution consisting of 10 mM Tris-HCl pH 8.0, 10 mM NaCl, and 10 mM EDTA. The nuclei are resuspended at the rate of 1 ml of whole blood to 250 µl of the Tris-HCl, NaCl, EDTA rinse solution containing 1 mg/ml of proteinase K. The mixture is incubated for 2 h at 65°C. Periodic agitation of the incubation mixture is necessary to ensure that uniform lysis of the nuclear membranes occurs and that optimum recovery of the DNA is achieved.

This isolation procedure is especially useful for small-volume forensic specimens. After incubation of the nuclei with proteinase K, the DNA can be directly digested with restriction enzymes without additional purification. In one protocol (Grimberg 1989), 0.5 µg of DNA were digested with 10 units of enzyme, in buffer supplemented with 10 mM $MgCl_2$, over a 2 h period. A 7.5 M LiCl solution was added to the digestion mixture (at the rate of 0.5 volume to 1.0 volume of digestion mixture), and the resultant was placed on ice for 10 min. After centrifugation at 13,000 × g for 10 min, the supernatant was transferred to a new tube, 2 volumes of EtOH were added, and the resultant was kept at RT for 30 min. The DNA was pelleted by centrifugation. The pellet was then resuspended in buffer, and the solution was stored or further analyzed.

One disadvantage of this isolation technique is that conventional UV spectrophotometric procedures at 260 and 280 nm cannot be used to quantitate the DNA because of interference from the proteolytic digestion products. Spectrofluorometric quantitation using Hoechst dye (Labarca 1980), or quantitation on agarose gels stained with ethidium bromide can, however, be readily carried out.

Stains. A modification of the nonorganic whole blood procedure has been developed by Lifecodes to isolate DNA from bloodstains and vaginal swabs (Turck 1990). Sufficient DNA was recovered from a 1 μl (approximately 25 ng DNA) bloodstain on cotton cloth to give a detectable Hae III digest D2S44 locus band pattern. The sperm DNA lysates prepared from vaginal swabs were free of contaminating female DNA. In contrast, male DNA was detected in the female lysates.

Bloodstain fabric is cut into pieces and placed in a microfuge tube. Cells are lysed with 1 ml of cell lysis solution (see whole blood procedure) during a 5 min incubation on ice. The lysate is centrifuged at 13,000 × g for 30 sec and the supernatant is discarded. The cell lysis cycle is repeated with the fabric nuclei pellet and is repeated a second time with only pellet lysis solution (see whole blood) in place of cell lysis solution. Depending on the quantity of fabric, 40 to 300 μl of pellet lysis solution containing 1 mg/ml proteinase K are added to the tube. The mixture is transferred to a new tube for storage or direct restriction enzyme digestion as described by Grimberg et al. (1988).

Sperm and Female-cell mixtures. A differential lysis is required for vaginal swabs to separate female from male DNA. Female DNA lysate is prepared according to the same protocol outlined for bloodstains. After the lysate is removed, traces of female DNA are eliminated by washing the fabric-sperm pellet three times with 1.5 ml of pellet lysis solution at 65°C. The sperm cells are lysed by adding a solution consisting of pellet lysis solution, 5 mM DTT, and 1 mg/ml proteinase K and incubating for 2 h at 65°C. The male DNA lysate is transferred to a new tube for storage or further analysis as described for whole blood.

DIALYSIS

Contaminants in DNA extracts may result in incomplete restriction enzyme digests and reduced fragment mobility during electrophoresis in agarose gels. If this presents a problem, a simple dialysis step as described by Gill (1987) should be incorporated into the DNA purification procedure. Aliquots of DNA extracts, for example, pellets from 100 µl bloodstains suspended in 10 to 25 µl of distilled water, are applied to the surface of a Millipore membrane filter (type VMWP-0.05 µm pore size). A number of samples can be applied to a single filter. The filter is then floated in approximately 100 ml of dialysis buffer (10 mM Tris, 10 mM NaCl, pH 7.6) in a sealed container at RT for 2 h. The extracts are removed and the filter spot washed by adding 5 µl to 10 µl of distilled water. The wash solution is added to the original filtered extract. Samples may increase in volume during dialysis; therefore, the original extract volume should be minimal.

Alternatively, purification and concentration of n-butanol-extracted DNA can be accomplished using a Centricon ultrafiltration device (Amicon) or other similar systems (von Beroldingen 1989). Caution must be exercised when using these devices to ensure that complete recovery of DNA from the filter unit is achieved.

RNAase TREATMENT

A considerable quantity of RNA is often precipitated during the DNA isolation process. RNA can cause falsely elevated spectrophotometric readings during DNA quantitation. Also, when endonuclease-digested DNA, contaminated with RNA, is electrophoresed on an agarose gel, the DNA may be carried forward by the RNA. The increased distance moved results in an underestimation of DNA fragment sizes. RNA moves rapidly relative to DNA fragments when electrophoresed and is observed ahead of the dye front on ethidium bromide stained agarose gels (see Figure 7-5).

To overcome the problems caused by contaminating RNA, DNA samples are treated by first dissolving in low-TE buffer. Heat-treated RNAase is added at 40 µg per ml and the mixture allowed to digest for 1 h. RNAase must be heat-treated prior to use to eliminate any DNAase contaminants. RNAase is removed by extracting with chloroform:isoamyl alcohol (24:1), and the DNA is precipitated using the sodium acetate or sodium chloride and a cold 95% ethanol process.

QUANTITATION

Absorbance at 260 nm wavelength and fluorescence after staining with ethidium bromide are common techniques for measuring DNA concentration (Sambrook 1989).

For accurate determinations by spectrophotometric analysis, the sample must not be contaminated with proteins, phenol, or other nucleic acids. The DNA must also be completely solubilized. Heating at 37°C to 56°C for 10 min or longer and additional rotation in a diluted solution may be useful, especially if particles of undissolved DNA remain. To quantitate, a 25 µl aliquot of DNA sample is added to 475 μl of low-TE buffer and heated to 37°C for 15 min. The spectrophotometer is blanked at 260 nm with low TE and the absorbance of the DNA solution recorded. An absorbance optical density (OD) of 1.0 corresponds to 50 μg double-stranded DNA per ml and 40 μg for single-stranded DNA or RNA. DNA sample concentration in μg/ml = OD_{260} × dilution factor × 50. For a 1:20 dilution, the original DNA sample concentration = $OD_{260} \times 10^3$ μg/ml. An estimation of the purity of a sample can be obtained by calculating the ratio of the OD at 260 and 280 nm. The spectrophotometer is blanked at 280 nm with low-TE buffer and the diluted sample absorbance read. For a pure preparation of DNA, OD_{260}/OD_{280} = 1.8 and for RNA 2.0. Ratios significantly less than these indicate contamination of the sample with protein or phenol. Contaminants will invalidate the concentration results and may cause incomplete DNA digestion with certain restriction enzymes. Specimens with values greater than 1.5 have proven sufficiently pure for RFLP analysis. Contamination of DNA by RNA will also result in incorrect concentration determinations.

Ethidium bromide staining acts by intercalating between stacked DNA bases. UV irradiation absorbed by DNA and transmitted to the dye is emitted as a visible orange-red fluorescence. The amount of fluorescence is proportional to the quantity of DNA; thus, quantitation estimation can be made by comparison with DNA standards. Electrophoresis of DNA on agarose gels containing 0.5 μg/ml of ethidium bromide is a convenient method for this analysis; however, DNA band distortion has been observed, especially with higher DNA concentrations. DNA mobility in the gel can be reduced by as much as 15% due to the ethidium bromide. To overcome these potential problems, staining is delayed until the electrophoresis is completed. The gel is then immersed in electrophoresis buffer or water containing 0.5 μg/ml ethidium bromide for approximately 45 min at RT. Destaining with distilled water or a 1 mM $MgSO_4$ solution for about 1 h at RT can be performed to reduce the background caused by unbound dye. Contaminating RNA can also be detected as a fluorescent smear ahead of the gel dye front.

Forensic specimens often contain insufficient DNA for both quantitation and profile analysis. A novel technique requiring only subnanogram amounts of human genomic material has recently been developed to circumvent this problem (Waye 1989). The procedure involves immobilizing the DNA on a nylon membrane and hybridizing with a highly repetitive primate-specific alpha satellite probe p17H8 (locus D17Z1). The assay is simple and requires less than 4 h for completion. Also, because this probe does not hybridize with nonprimate DNA, contaminants of nonhuman origin will not interfere with the quantitation estimates.

DNA CONCENTRATION

The concentration of DNA solutions often requires adjusting to facilitate efficient handling. Concentrated solutions are viscous and difficult to pipette; this is remedied by adding low-TE buffer or distilled water. Dilute solutions require reprecipitation. This can be accomplished by adding one-tenth volume of 4 M ammonium acetate or 2 M sodium acetate or sodium chloride and two volumes of –20°C 95% ethanol and placing the DNA solution in a –20°C or lower temperature freezer for approximately 1 h. The resultant is microfuged under refrigeration at 10,000+ rpm for 15 min, decanted, and the supernatant discarded, ensuring that all remaining drops are removed. 500 μl of 70% ethanol are gently added. The tube is recentrifuged for 5 min and the supernatant is carefully decanted. Any remaining drops of solution are removed and the remaining contents evaporated to dryness in a vacuum dessicator. Overdrying must not occur because resuspension of the DNA in low-TE buffer or distilled water may then be very difficult. Alternatively, a DNA solution can be concentrated using a Centricon ultrafiltration device (von Beroldingen 1989).

QUALITY DETERMINATION

The degree of DNA degradation, as evidenced by long strands sheared or digested into smaller pieces, can be estimated by electrophoresis of the sample in an agarose gel as described previously. Large molecular weight DNA appears as a band, perhaps with sharp spikes, whereas partially degraded material forms a long smear of large to small fragments. DNA degraded to fragments a few hundred base pairs or less in length appears as a diffuse spot near the dye fragment (Figure 4-1). Slight degradation may be difficult to detect and appears only as slightly increased orange-red shading ahead of the main band of high molecular weight DNA.

The concentration of fragments of known size after DNA is digested with specific restriction enzymes and hybridized with specific probes also provides an indication of degradation.

EFFECTS OF ENVIRONMENTAL FACTORS INCLUDING CONTAMINATING DNA ON DNA PROFILES

Over 2,000 specimens were analyzed by the Laboratory Division of the Federal Bureau of Investigation (Budowle 1990) to determine the influence of sunlight, temperature, specimen substrate, chemical contaminants, and contaminating DNA on DNA profiles. Stringent hybridization and wash conditions were used as in the standard FBI protocol. Probe YNH24 (D2SHH) was used in most of the analyses.

The DNA from specimens exposed to sunlight over an eight week period was too degraded for analysis. Control material maintained in the dark provided good quality DNA. DNA from samples subjected to daily temperatures of up to 41°C in sunlight was too degraded for analysis. DNA from control material maintained in the dark gave good quality profiles.

Stain substrates including cotton, nylon, glass, wood, and metal have no obvious affect on DNA quality. Specimens were subjected for five days to unleaded gasoline, motor oil, detergent, acid, base, salt, bleach, and soil. Only the soil exposure resulted in DNA degraded to the degree that production of a profile was not possible.

Cross-reactivity of the probes used in genomic analysis and other possible sources capable of contaminating DNA are important considerations in quality control. Semen and bloodstains were contaminated with the bacterium *Staphylococcus epidermidis* and the yeast *Candida valida* (these are commonly found in vaginal swabs). DNA profiles contained only human genetic patterns. Specimens were also contaminated with the ubiquitous *Escherichia coli* and *Bacillus subtilis*. Again, only human profiles were detected when the DNA was analyzed. Analysis of mixed specimens including combinations of semen, blood, urine, saliva, and vaginal secretions resulted in the expected mixed profiles with the exceptions that insufficient DNA was present in urine stains. The quantity of DNA in saliva was also insufficient for the preparation of a profile. A number of nonhuman primates—gorilla, Japanese macaque, spider monkey, celebes ape, and the debrazza monkey—as well as cockatoo, scarlet macaw, ferret, dog, cat, rabbit, cow, horse, goat, burro, pig, chicken, and sheep were tested. Only the primate DNA cross-reacted.

Over 400 of the 2,000 specimens analyzed consisted of evidentiary stains from case materials analyzed for non-DNA markers during the previous one- to two-year period. The best quality portions of the stain material had already been removed and the remainder stored at –20°C. Satisfactory DNA profiles were prepared form approximately 60% of the specimens.

ANALYSIS REAGENTS

Whole blood reagents (Method 1 or 2)

(i) Ammonium chloride-Tris (NH_4Cl:Tris) [RT] [Storage temperature]
900 ml 0.155 M NH_4Cl
100 ml 0.170 M Tris pH 7.65

(ii) Normal saline (NaCl) [RT]
0.85% NaCl

(iii) High-TE buffer [RT]
Tris 100 mM, pH 8.0
EDTA 40 mM

(iv) Low-TE buffer [RT]
Tris 10 mM, pH 8.0
EDTA 1 mM

(v) Nucleated cell lysis mixture [RT]
Tris 100 mM, pH 8.0
EDTA 40 mM
SDS 0.2% (sodium dodecylsulfate)

(vi) Phenol-Tris-Cl (pH 8.0) saturated [–20°C] [4°C for in-use material]
0.1 M Tris-Cl pH 8.0, containing 0.2% B-mercaptoethanol
Good quality phenol
0.1% 8-hydroxyquinoline
(Note the separate section on phenol preparation.)

(vii) Chloroform:isoamyl alcohol [RT]
Chloroform 24 parts
Isoamyl alcohol 1 part

(viii) Phenol/chloroform/isoamyl alcohol (25:24:1)
(Note the separate section for preparation.)

(ix) Proteinase K (see hair roots, biopsy and autopsy tissues section)

(x) SSC (see tissue culture, amniotic fluid, and buccal cells section)

Tissue culture, amniotic fluid, buccal cells reagents

(i) RNAase (heat treated) [–20°C]
Sigma type XIA RNAase at 2 mg/ml of 1 × SSC, pH to 5.0 with citric acid
Heat at 80°C for 10 min

(ii) SSC (20×) [RT]
NaCl 3.0 M
Na citrate 0.3 M

(iii) Nucleated cell lysis mixture (see whole blood reagents)

Hair roots, biopsy and autopsy tissue reagents

(i) Proteinase K (self-digested) [-20°C]
Sigma P0390 or P4914 at 10–20 U/mg protein
Add 10 mg of proteinase K (at 20 U/mg) to 1.0 ml of high-TE buffer plus 0.05% SDS (incubate at 37°C for 30 min; aliquot and store frozen)

(ii) Phenol-chloroform [4°C]
0.5 volumes TE-saturated phenol, 0.5 volumes chloroform:isoamyl alcohol (24:1)

(iii) Nucleated cell lysis mixture (see whole blood reagents)

Stain reagents (organic solvent extraction)

(i) Stain extraction buffer [RT]
Tris 0.01 M, pH 8.0

EDTA 0.01 M (disodium salt)
NaCl 0.1 M

(ii) DTT (dithiothreitol) [–20°C]
0.06 g per ml of stain extraction buffer

Sperm and female-cell mixture reagents (organic solvent extraction)

(i) Phosphate buffered saline (PBS) [RT]

NaCl	137 mM
KCl	2.7 mM
KH_2PO_4	1.5 mM
Na_2HPO_4	8.1 mM
pH 7.3 to 7.5	This solution should be autoclaved.

(ii) HEPES-buffered saline [RT]

HEPES	10 mM
NaCl	150 mM
pH 7.5	This solution should be autoclaved.

(iii) Sarkosyl—20% [RT]
Sarkosyl (N-lauroylsarcosine) 20 g
Bring to a volume of 100 ml with sterile distilled H_2O.
This solution should be filter sterilized through a 0.2 µm filter.

(iv) Tris-NaCl-EDTA (TNE) [RT]

Tris base	10 mM
NaCl	100 mM
$Na_2EDTA\ 2H_2O$	1 mM
pH 8.0	

Dialysis Reagents

(i) Dialysis buffer
10 mM Tris-HCl pH 7.6
10 mM NaCl

Phenol preparation. Only molecular-biology-grade liquid phenol should be used. Material that is a pink or yellow color should be rejected, and crystalline phenol is not recommended because it must be redistilled to remove oxidation products that cause RNA-DNA cross-linking or the breakdown of phosphodiester bands. Phenol must be equilibrated to a pH greater than 7.5 because DNA will partition into the organic phase if it is isolated at acid pH (Sambrook 1989).

The phenol (stored at –20°C) is melted at 65°C and hydroxyquinoline is added to give a final concentration of 0.1%. Hydroxyquinoline will chelate metal ions and inhibit RNAase as well as act as an antioxidant. An equal volume of 0.5 M Tris Cl buffer pH 8.0 is added, and the phenol buffer mixture is transferred to a separatory funnel. After thorough mixing, the phases are allowed to separate and the lower phenol phase is retained. An equal volume of 0.1 M Tris-Cl pH 8.0 is added to the phenol, the resultant is thoroughly mixed, the phases are allowed to separate, and the lower phenol layer is retained. This cycle with 0.1 M Tris-Cl is repeated until the pH of the phenol phase is greater than 7.5. 0.1 volume of 0.1 M Tris-Cl pH 8.0 containing 0.2% B-mercaptoethanol is added to (layered on top of) the phenol. The resultant can be stored for up to four weeks at 4°C in a light-proof bottle.

Phenol/chloroform/isoamyl alcohol preparation. A mixture consisting of phenol/chloroform/isoamyl alcohol in the ratio of 25:24:1 is often used to separate proteins from DNA (Sambrook 1989). The chloroform/isoamyl alcohol solution is added to the equilibrated phenol (prepared as just described) and 0.1 volume of 0.1 M Tris-Cl pH 8.0 containing 0.2% B-mercaptoethanol is added (layered on top). This mixture can be stored in a light-proof bottle at 4°C for up to four weeks.

REFERENCES

Amos B. 1989. Preserving tissues without refrigeration. *Fingerprint News* 3:20.

Bahnak BR, Wu QY, Coulombel L, Drouet L, Kerbiriou-Nabias D, and Meyer D. 1989. A simple and efficient method for isolating high molecular weight DNA from mammalian sperm. *Nucleic Acids Res.* 16:1208.

Budowle B and Baechtel FS. 1990. Modifications to improve the effectiveness of restriction fragment length polymorphism typing. *Appl. Theoret. Electro.* (in press).

Budowle B, Baechtel FS, and Adams DE. 1990. Validation with regard to environmental insults on the RFLP procedure for forensic purposes. *American Chemical Society Series* (in press).

Cariello NF, Keohavong P, Sanderson BJS, and Thilly WG. 1988. DNA damage produced by ethidium bromide staining and exposure to ultraviolet light. *Nucleic Acid Res.* 16:4157.

Dry PJ. 1988. A quick and easy method for the purification of DNA chorionic villus samples. *Nucleic Acids Res.* 16:7730.

Gill P, Jeffreys AJ, and Werrett DJ. 1985. Forensic application of DNA "fingerprints". *Nature* 318:577–579.

Gill P. 1987. A new method for sex determination of donor of forensic samples using a recombinant DNA probe. *Electrophoresis* 8:35–38.

Gill P, Lygo JE, Fowler SJ, and Werret DJ. 1987*a*. An evaluation of DNA fingerprinting for forensic purposes. *Electrophoresis* 8:38–44.

Grimberg J, Nawoschik S, Belluscio L, McKee R, Turck A, and Eisenberg A. 1989. A simple and efficient nonorganic procedure for the isolation of genomic DNA from blood. *Nucleic Acids Res.* 17:8390.

Impraim CC, Saiki RK, Erlich HA, and Teplitz RL. 1987. Analysis of DNA extracted from formalin-fixed, paraffin-embedded tissues by enzymatic amplification and hybridization with sequence-specific oligonucleotides. *Biochem. Biopys. Res. Commun.* 142:710–716.

Labarca C and Paigen K. 1980. A simple, rapid, and sensitive DNA assay procedure. *Anal. Biochem.* 102:344–352.

Madisen L, Hoar DI, Holroyd CD, Crisp M, and Hodes ME. 1987. DNA banking: The effects of storage of blood and isolated DNA on the integrity of DNA. *Am. J. Med. Genet.* 27:379–390.

Miller SA, Dykes DD, and Polesky HF. 1988. A simple salting out procedure for extracting DNA from human nucleated cells. *Nucleic Acids Res.* 16:1215.

Saghai-Maroof MA, Soliman KM, Jorgensen RA, and Allard RW. 1984. Ribosomal DNA spacer-length polymorphisms in barley: Mendelian inheritance, chromosomal location, and population dynamics. *Proc. Natl. Acad. Sci. USA* 81:8014–8018.

Sambrook J, Fritsch EF, and Maniatis T. 1989. *Molecular Cloning: A Laboratory Manual.* Vols. 1, 2 and 3. 2nd ed. Cold-Spring Harbor, Cold-Spring Harbor, New York.

Signer E, Kuenzle CC, Thomann PE, and Hubscher U. 1988. DNA fingerprinting: Improved DNA extraction from small blood samples. *Nucleic Acids Res.* 16:7738.

Turck A, Schall B, Maguire S, Belluscio L, Nawoschik S, McKee R, Grimberg J, and Eisenberg A. 1989. A simple and efficient nonorganic procedure for the extraction of DNA from evidentiary samples. In Proceedings of the 13th International Congress of the ISFH, New Orleans (in press).

von Beroldingen CH, Blake ET, Higuchi R, Sensabaugh GF, and Erlich H. 1989. Applications of PCR to the analysis of biological evidence. In *PCR Technology Principles and Applications for DNA Amplification*, 209–223, Erlich HA ed., Stockton Press, New York.

Waye JS, Presley LA, Budowle B, Shutler GG, and Fourney RM. 1989. A simple sensitive method for quantifying human genomic DNA for forensic specimen extracts. *BioTechniques* 7:852–855.

CHAPTER 5

DNA Amplification

Amplification of DNA may be necessary to increase the quantity of sample available for profiling, to reduce the analysis time, or to produce probes for the hybridization process (Higuchi 1989, Li 1988, Marx 1988, Mullis 1990, Paabo 1989, Saiki 1986).

Stretches of nucleotides up to at least 3,000 bp from any DNA-containing samples may be efficiently amplified by the polymerase chain reaction (PCR). Alternatively, living tissue can be placed in culture, and fibroblasts, epithelial type cells, or lymphoblasts grown. The culture process differs considerably from the PCR approach in that the total genome is reproduced. Also, tissue culture is usually at least a two-week procedure, whereas the polymerase chain reaction requires only a few hours. Cultured cells can be used for enzyme and other biochemical tests, and storage in liquid nitrogen is a standard practice for regrowth at a later time.

Probe material, that is, DNA capable of hybridizing with its complementary region in the genome, must be amplified, aliquoted, and stored to provide an ongoing source for use with each profile analysis. Probe amplification has been mainly carried out in bacterial culture; however, probes can be chemically synthesized as discussed in Chapter 2 or amplified by the PCR system.

POLYMERASE CHAIN REACTION

At least 10 to 50 ng of high molecular weight genomic DNA are required for VNTR analysis using single-locus probes, and at least 0.5 to 1.0 μg required if multilocus probes are used. If only a small quantity of DNA is available, amplification using the PCR may be the only feasible option for obtaining sufficient material for analysis. PCR has revolutionized the approach to the recovery of DNA from a variety of sources. Microgram quantities of DNA can be produced in vitro by the amplification of picogram starting amounts. Single-copy genomic sequences greater than 2 kb in length have been amplified more than 10 millionfold in a few hours. Amplified material can also be directly sequenced without the necessity of incorporating DNA fragments into vectors such as M13 (Gyllensten 1989, 1989*a*).

Availability of oligonucleotide primers is the key to the amplification process. One primer is annealed to the flanking end of each DNA target sequence complementary strand, and thermal stable Taq polymerase is added to mediate the extension. The system requires a reaction buffer, the nucleotides dATP, dCTP, dGTP, and dTTP, and a means of thermal cycling the reagent mix. Twenty-five or more amplification cycles can be performed, as illustrated in Figure 5-1. Each cycle consists of template denaturation, primer annealing, and extension of the region between the primers. Synthesis proceeds across the target sequences flanked by the primers with the extension products of one primer acting as a template for the other primer. The amount of DNA synthesized in each successive cycle is doubled, resulting in an exponential accumulation (2^n where n = number of cycles).

Kit reagent systems are available from Perkin-Elmer Cetus. The company also markets a thermal cycler to automatically perform the rapid temperature changes and incubations. Temperatures up to 95°C are used in the reaction. This is made practical by the use of the highly thermal stable Taq polymerase from the thermophilic bacterium *Thermus aquaticus*.

Since PCR is an extension system, short stretches of template-flanking base sequence must be known. With this information, oligonucleotide primers, for example, 25-mers, can be synthesized. This requirement limits the universal application of the system at present; however, with the rapid progress in base sequence determination for many animal and plant genomes it is anticipated the limitation will be short-lived.

Degraded DNA from formalin-fixed paraffin-embedded tissues, dried blood and semen stains, and hair follicles and shafts can be successfully amplified. A misincorporation error rate as low as 0.25% is usual. Unlike molecular cloning systems, where bacterial repair mechanisms will salvage damaged molecules, thus introducing artifacts, the polymerase chain reaction will be blocked at molecular cross-links and only intact molecules will be amplified.

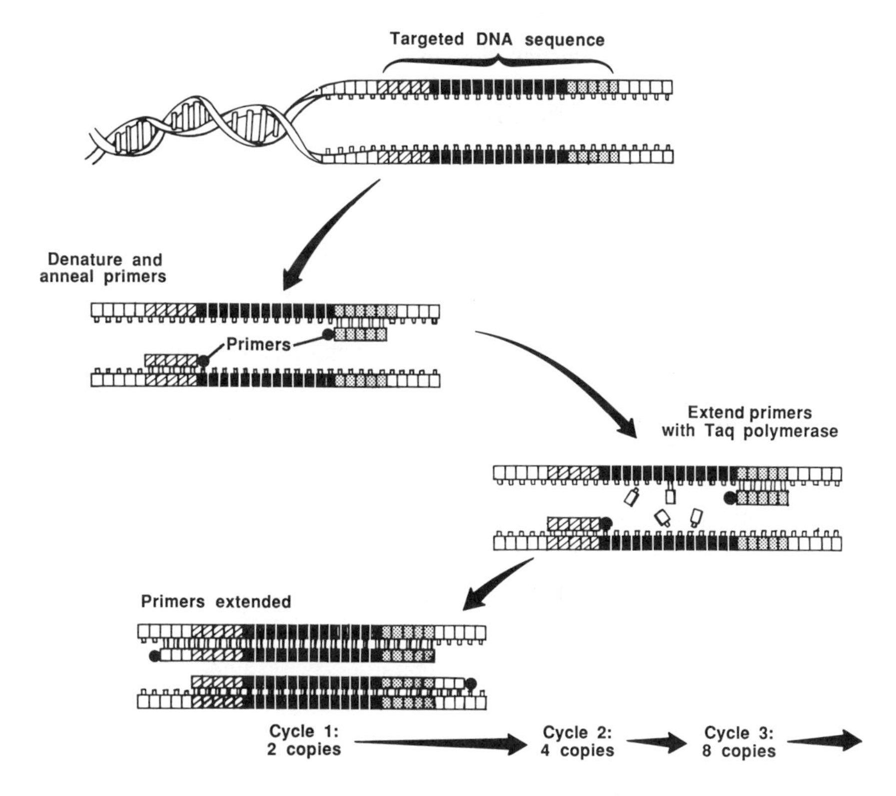

Figure 5-1. Amplification of DNA using the polymerase chain reaction (PCR). Each cycle consists of heat denaturation of the target molecules, annealing of an oligonucleotide primer to each target complementary DNA strand, and extension of the primers using Taq polymerase. The newly synthesized molecules are denatured to provide double the number of templates for the next cycle.

Amplified Fragment Length Polymorphisms (AMP-FLPs)

A high-resolution rapid technique employing rehydratable polyacrylamide silver stained gels has recently been developed to separate and detect, using autoradiography, PCR amplified RFLPs (Allen 1989, Budowle 1990). (The acronym AMP-FLP has been coined for the term amplified fragment length polymorphism.) VNTR loci amenable to amplification using the PCR system include D17S30, D1S80, and the 3' HVR region of the apolipoprotein B gene. The AMP-FLP approach should be especially useful when insufficient genomic material is available for identity analysis. Less than 10 pg of DNA/mm of fragment band width are required for detection. The requirement for radioactive labeling has also been eliminated, thus making the technique available to a broader cross section of laboratories. The resolution ranges from 4 bp for RFLPs under 500 bp to approximately 50 bp for fragments in the 2- to 3-kb range. These results can be achieved within a single working day.

Advantages of PCR

The advantages of in vitro DNA amplification include the following:

1. Minute amounts (from as little as a single cell) of DNA template are required. It should be noted, however, that reducing the number of template molecules increases the possibility of obtaining false results because of cross-contamination with exogenous DNA. This is especially of concern in evidentiary specimens where contamination prior to laboratory submission may occur.
2. Sufficient DNA is usually available from the original specimen (for example in homicide and rape cases) for retesting.
3. Crude DNA preparations such as whole blood can be processed.
4. DNA degraded to fragments only a few hundred base pairs in length often provides templates suitable for amplification. This includes material from formalin-fixed and paraffin-embedded tissues.
5. Because of the large number of copies of specific amplified fragments, nonradioactive labeled probes will usually provide sufficient signal for detection.
6. If sequence-specific oligonucleotide (SSO) probes are used, a simple dot-blot or reverse dot-blot hybridization protocol can be followed. A +/– (present/absent) detection system is then possible, thus negating the need for measuring the lengths of DNA fragments. This approach may, however, limit the informativeness of the results, especially for specimens consisting of mixed body fluids.
7. Contaminant DNA, such as fungal or bacterial sources, will not amplify if primers are designed specifically for human sequences.

Pitfalls in PCR

A number of potential pitfalls must be considered when using PCR-amplified DNA:

1. The target template may be refractory to efficient amplification because of inhibitors in the DNA extract. This problem may be resolved by reextracting the DNA or by preparing a more dilute DNA solution.
2. Amplification by the PCR may fail on occasion because of genetic variation in the primer binding region in genomic DNA (Fujimura 1990). This could

lead to mistaken conclusions about identity. The use of more than one set of primers could help to protect against this source of error.

3. False positive results can occur because the genomic sample is contaminated with DNA from other sources or because the buffers or other supplies are contaminated with amplified DNA from previous reactions. Meticulous care must be exercised to ensure that this carry-over does not occur. Controls containing no DNA must always be included in the amplification protocol. To overcome the problems associated with reagents contaminated with previously amplified DNA, the PCR mixture can be treated with UV irradiation before adding the template DNA and Taq polymerase (Sarkar 1990). The irradiation does not affect the single-stranded primers but renders double-stranded DNA contaminants incapable of acting as templates.
4. To date, only a few marker systems, such as HLA-DQα (with six alleles), have been sufficiently well-characterized for use in identification analysis. The power of discrimination in terms of individual identity is low relative to that of the conventional VNTR systems.

A PCR Protocol

Since the specific oligonucleotide primers used in a PCR reaction are the key to the success or failure of the amplification, no universal set of reaction conditions exists. R. Saiki (1989), however, suggests the following general protocol should at least provide a good base which can be modified as necessary.

The reaction is usually carried out in a 50 μl or 100 μl volume containing 10^2 to 10^5 copies of template DNA (approximately 0.1 μg of human genomic material); 50 mM KCl; 10 mM Tris-HCl (pH 8.4); 1.5 mM $MgCl_2$; 100 μg/ml gelatin; 0.25 μM of each primer; 200 μM of each dATP, dCTP, dGTP, and dTTP; and 2.5 units of Taq polymerase. A few drops of mineral oil can be added to the surface of the reaction mixture to act as a seal. The reaction is performed in a set of water baths, or more conveniently in a DNA Thermal Cycler programmed to denature at 94°C for 20 sec, anneal at 55°C for 20 sec, and extend at 72°C for 30 sec. Up to 30 of these cycles, at 3.75 min per cycle, can be performed; the additional time is required for the heating and cooling processes. There is approximately a one millionfold (2^{20}) amplification of the DNA templates after 20 PCR cycles.

PROBE AMPLIFICATION

Both DNA and RNA are suitable as probe material for hybridization to and identification of complementary sequences. Oligonucleotides such as 20-mers are synthesized. Large DNA probes traditionally have been ligated into vectors such as plasmids and amplified in bacterial culture; however, with the advent of PCR it is possible to undertake direct amplification as described in the previous section.

Service laboratories can purchase some commonly used probes from scientific supply companies. Alternatively, a bacterial stab containing the vector may be requested from another laboratory. If the vector is a plasmid, the stab is first plated, a colony selected and transferred to liquid culture, and the plasmid isolated from the amplified material. It is possible to use the vector plus insert as a probe; however, the insert is usually excised, isolated on agarose gel, quantitated, and stored for use as required.

Details of one procedure for plasmid processing from an *E. coli* stab to characterized probe is outlined below.

Media preparation. LB medium (L-broth) consists of 5 g/L Bacto-yeast extract, 10 g/L Bacto-tryptone and 10 g/L NaCl. The pH is adjusted to 7.5 with 5 M NaOH. If plates are to be poured, 1.5% Bacto-agar is also added. The mixture is autoclaved, and filter-sterilized ampicillin at 50 µg/ml or tetracycline at 20 µg/ml added as required when the medium has cooled.

Colony isolation. Plates are streaked using standard format and incubated at 37°C overnight. Individual colonies are located and an isolate transferred to a new plate, stab, or liquid culture as appropriate.

Mini-prep/plasmid amplification. The following amplification and isolation technique should result in yields of over 1 µg of plasmid per ml of medium. This alkali lysis method is a modified procedure from Sambrook (1989).

1. Inoculate 10 ml of medium containing the appropriate antibiotic with a single bacterial colony. Incubate at 37°C overnight with vigorous shaking.
2. Fill Eppendorf tubes with the culture and centrifuge for 1 min at 10,000+ rpm. The remaining culture medium can be stored overnight at 4°C.
3. Remove the medium from each tube by aspiration, leaving the bacterial pellet as dry as possible.
4. Resuspend the pellet by vortexing in 50 µl of an ice-cold solution of 50 mM glucose, 10 mM EDTA, and 25 mM Tris Cl (pH 8.0). Add lysozyme at the rate of 8 mg/ml to a second 50 µl aliquot of the suspension buffer. Do not vortex after lysozyme addition because this action could cause shearing and inactivation of the large lysozyme molecules.
5. Store for 5 min at RT. The top of the tube need not be closed during this period.
6. Add 200 µl of a freshly prepared ice-cold solution of 0.2 M NaOH and 1% SDS. Close the tube top and mix the contents by inverting the tube rapidly two or three times. Do not vortex. Store the tube on ice for 5 min.
7. Add 150 µl of an ice-cold solution of potassium acetate (pH about 4.8) prepared as follows: to 60 ml of 5 M potassium acetate, add 11.5 ml of glacial acetic acid and 28.5 ml of H_2O. The resulting solution is 3 M with respect to potassium and 5 M with respect to acetate. Close the cap of each tube and vortex gently in an inverted position for approximately 10 sec. Store on ice for 5 min.

8. Centrifuge for 5 min at 10,000+ rpm at 4°C.
9. Transfer the supernatant to a fresh tube.
10. Add an equal volume (approximately 450 µl) of phenol/chloroform (SEVAG solution). Mix by vortexing. After centrifuging at 10,000+ rpm for 2 min, transfer the supernatant to a fresh tube. This step may be repeated if necessary.
11. Add two volumes (approximately 900 µl) of ethanol at RT. Mix by vortexing. Stand at RT for 2 min.
12. Centrifuge for 5 min at 10,000+ rpm at RT.
13. Remove the supernatant. Stand each tube in an inverted position on a paper towel to allow all the fluid to drain away.
14. Dissolve the pellet in 0.5 ml of 0.25 M Na acetate pH 5.2.
15. Add 1 ml of 95% EtOH at RT.
16. Store for 5 min at RT.
17. Microfuge for 5 min at 10,000+ rpm and discard the supernatant.
18. Add 1 ml of 70% ethanol. Vortex briefly, then recentrifuge.
19. Remove all of the supernatant, and dry the pellet briefly in a vacuum desiccator to remove any residual ethanol.
20. Add 50 µl of LTE (pH 8.0) containing DNAase-free pancreatic RNAase (20 µg/ml). Vortex briefly and allow to remain overnight at 4°C.
21. Remove an aliquot of the solution to a new Eppendorf tube. Add the appropriate buffer and restriction enzyme and incubate. Store the remainder of the preparation at –20°C.
22. Analyze the DNA fragments in the restriction digest by gel electrophoresis.

Characterization. It is essential to ensure that the correct plasmid insert has been isolated. This can be accomplished by measuring the size of the plasmid on ethidium bromide-stained agarose gel and by excising the insert with the appropriate restriction endonuclease and measuring the size of the linearized plasmid and the insert. An estimation of the quantity of insert produced can be determined by comparing with standards of known concentration on the stained gel.

Storage

1. Bacterial containing vector: Agar plates can usually be stored for a few weeks at 4°C provided drying is prevented by sealing the plates with parafilm. Storage for approximately 1 yr can be achieved in stab culture slants. Screw-cap vials are prepared with 2 ml of agar medium. A loop from a rapidly-growing liquid culture is stabbed into the agar vial and the stab is incubated (with cap loosened) overnight at 37°C. The cap is tightened, sealed with paraffin or parafilm, and stored in the dark at RT. Long-term storage (for years) is possible in glycerol-media. An aliquot of rapidly-growing culture is added to a screw-cap vial containing glycerol producing a 15% glycerol suspension. The cultures are stored at –20°C to –70°C.

2. Probe: Aliquots of insert excised from the vector and isolated on low melting point agarose gel can be stored frozen for extended periods.

TISSUE CULTURE

Tissue culture for fibroblast or lymphoblast production is not a difficult procedure (Barnes 1984, Freshney 1987, Nuzzolo 1983, Paul 1975); however, the process can be extremely frustrating if precious biopsy material either fails to grow or grows very slowly. Bacterial, fungal, or mycoplasma contamination can be devastating if allowed to spread unchecked.

The following tissue culture outline is a guide for growing fibroblasts, the cells most commonly cultured in service laboratories (Figure 5-2). To obtain optimal growth from chorionic villi (CVS) or amniotic fluid cells, procedure modifications are necessary. It may be necessary to add growth factor Ultroser-G and to establish the cultures in small vessels. Lymphoblast cultures should be established if large quantities of cells are needed to provide an ongoing source of DNA for controls. Because lymphoblast transformation may require the use of agents such as Epstein-Barr virus (EBV), this type of system is subject to more stringent safety regulations and may require more elaborate facilities (Moss 1978).

Cell culture lines are available from two large collections in the United States. The American Type Culture Collection established in 1925 in Rockville, Maryland carries approximately 3,000 lines from over 50 species. Recombinant DNA materials are also handled by this organization. The cost for a T_{25} flask or its equivalent is approximately 50 dollars. The combined NIGMS Human Genetic Mutant Cell Repository at the Coriell Institute for Medical Research in Camden, New Jersey was established in 1972 and carries approximately 6,000 lines, most of which are human, but other species are also handled. The cost for a T_{25} flask is 75 dollars.

Preventive measures are the key to reducing to a minimum the likelihood of culture contamination. If contamination occurs, it is important that the source be determined; thus, meticulous records are mandatory. Contamination results in a period of extra stress for the laboratory. If the proper procedures are followed, decontamination should proceed in a manageable fashion.

Maintenance

1. Culture room: The technologist in charge of tissue culture is responsible for the general tidiness of the facility. The culture room should not be cluttered with supplies, and any areas prone to collecting dust should be cleaned periodically.
2. Laminar flow hood: Up-to-date records must be kept for the gauge readings. Filters should be changed at least every six months or sooner if required.

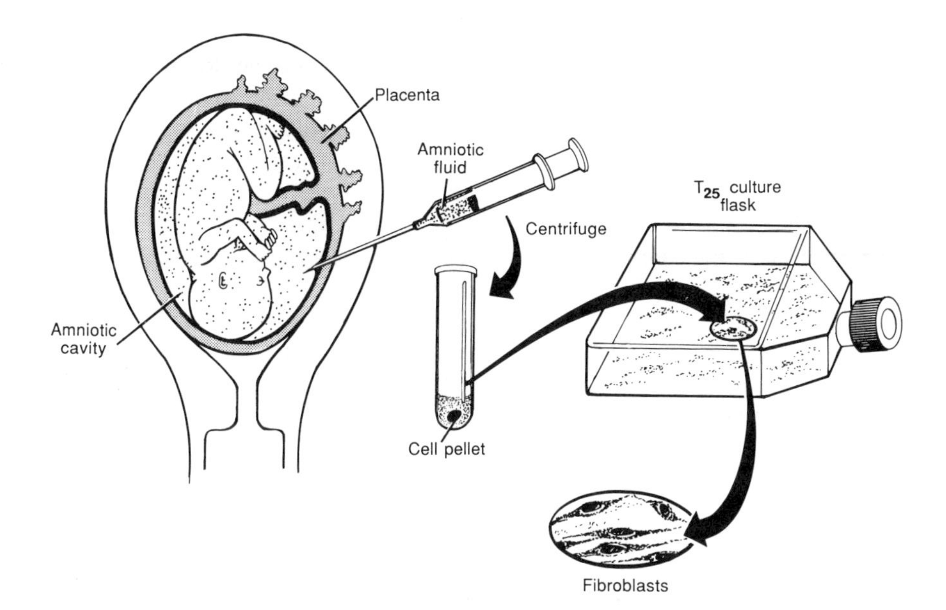

Figure 5-2. Amniocentesis for the purpose of isolating and culturing amniotic fluid cells. Approximately 10 ml of amniotic fluid are removed from the amniotic cavity and centrifuged to pellet and fetal cells. The cells are resuspended in tissue culture medium and dispensed into tissue culture flasks. Fibroblasts gradually form a monolayer over the lower surface of the flask.

3. Incubator: The incubator should be operated with humidity on; consequently, extra precautions must be taken concerning fungal contamination. Disinfectant such as Rocal should be added to the water reservoir in the bottom of the incubator. The reservoir should be changed monthly. All gauges and the thermometer inside the incubator should be checked daily. The CO_2 level should be tested weekly.
4. Liquid nitrogen level: This must be checked twice per week.
5. Supplies: Ample supplies should be stocked to provide for the possibility of back orders.

Records. A complete record for each flask of cells should be maintained. It should be possible to trace the history of each flask to facilitate contamination or other checks. These permanent records can be kept in a log book or in a computer program.

Contamination. The prevention of contamination is one of the keys to successful tissue culture. New biopsies and cultures received from other laboratories should be monitored carefully. All newly prepared media must be tested for contamination. The small quantity of medium remaining in "empty" storage bottles should

be kept until no longer needed for tracing contamination. Bench and apparatus swabs should be performed weekly. Aseptic techniques must be strictly followed and periodic (usually monthly) contamination checks performed on all glassware and other autoclaved materials. During seasons of high fungal spore counts (usually the spring and fall wet period), extra precautions must be taken. A separate lab coat should be worn in the tissue culture room, and tissue culture materials must be protected during transfer to the culture room during these periods.
The detection of contamination involves the following materials and procedures.

Reagents:

1. Tryptic soy broth (GIBCO T2096) or other bacterial growth medium
2. Blood agar plates
3. McKonkey plates
4. Gram stain

Procedure:

1. Add approximately 1 ml of medium with or without cells to tryptic soy broth.
2. Incubate with the cap tube slightly loose at 37°C for 48 h (preferably in a 5% CO_2 incubator).
3. Plate on a blood agar plate.
4. Check the plate at 24 h, 48 h, 72 h, and 7 days. (Penicillin-streptomycin in the culture medium could inhibit some bacteria and detectable growth may not be observed for up to 7 days.)
5. Recheck the broth before discarding the agar plates. If there is any appearance of cloudiness (indicating the presence of bacteria), re-plate on a blood agar plate and a McKonkey plate. Make a smear and do a gram stain.

Mycoplasma should be suspected in slow-growing cultures. Mycoplasma test kits are available; some laboratories analyze all of their cell lines for this organism.

If contamination is observed in a tissue culture flask or a broth culture,

1. Immediately tighten all culture flask caps.
2. Visually check the medium in all flasks.
3. If a number of flasks contain cloudy media (indicating contamination), transfer the non-cloudy flasks to a separate incubator. Autoclave the contaminated flasks and any stored contaminated media after removing a small portion of medium for bacterial identification. Autoclave all pipettes and apparatus. Clean the entire incubator with Rocal and rinse with 70% alcohol. Shelves should be cleaned by placing them under UV light overnight or if possible by autoclaving.
4. If only one or two flasks appear contaminated, and the original T_{25} also appears contaminated, autoclave all contaminated material. The remaining

flasks should remain tightly capped and untouched until contamination is proven to be negative; otherwise, the instructions in (3) should be followed.

5. If contamination appears only in flasks with antibiotic-free medium, autoclave the contaminated flasks and re-culture the cells from the original stock.
6. If a slow-growing microorganism is suspected in any of the remaining flasks not showing obvious signs of contamination, 2 ml aliquots of media should be tested by long-term bacterial culture (greater than seven days). If a slow-growing organism is present, decontaminate the tissue culture system as outlined in (3).

Sterilization. Autoclaved materials should be placed in an oven at approximately 300°C for 2 h to 3 h before use. Unused autoclaved materials should be re-autoclaved every two weeks.

Culture Medium

Reagents:

1. Minimum essential medium (MEM) with Earl's salts 10× concentration (GIBCO 330-1430 without $GluNH_2$)
2. Nonessential amino acids (GIBCO 320-1140)
3. 7.5% sodium bicarbonate
4. Antibiotics and anti-fungal agents (These should be used only in primary cultures.) Penicillin–streptomycin–fungizone (20 ml lyopholized vial, GIBCO 60-5245)
5. Fetal calf serum (GIBCO 200-6140)
6. Sterile distilled water

For the MEM working medium, mix the following:

1. 50 ml — MEM 10×
2. 5 ml — nonessential amino acids
3. 15 ml — 7.5% $NaHCO_3$

   The freshly prepared medium, without antibiotics and fungizone, should be tested for bacterial and fungal contamination. When the medium is found to be free of contaminants, the antibiotic/fungizone is added. The medium is now cleared for use.
4. 75 ml — penicillin–streptomycin–fungizone, 0.1 ml gentamycin
5. 75 ml (12.5%) — Fetal calf serum

   120 ml (20%) — FCS for amniotic fluid cells
6. 450 ml — Sterile water

To reduce pH changes, medium should be stored in full bottles.

Biopsy setup

1. Place the biopsy in a sterile petri dish with approximately 4 ml of medium.
2. Mince into small pieces.
3. Pour 2 ml of medium into each of two T_{25} tissue culture flasks. (This will wet the bottom interior surface.) The two flask lines must always be treated separately and with different sources of medium. They should also be handled on different days.
4. Transfer several pieces of biopsy sample, with a bent-tip Pasteur pipette, into each of the flasks. Add about 2 ml of medium into each flask (sufficient to cover the tissue), and leave the caps loose. Problems arise, on occasion, because small tissue pieces fail to anchor to the flask surface. This problem is resolved by (a) anchoring the pieces by covering them with a sterile narrow glass slide, or (b) reducing the quantity of medium in the vessel to ensure that only the vessel surface is wet.
5. Incubate the original petri dish for bacterial culturing.
6. Leave the flasks in the incubator for at least one week without disturbing.
7. Add new medium when the pieces of tissue become anchored to the flask surface. Do not remove the original medium. After the culture is established, the medium will require approximately weekly replacement. If the medium turns yellow before one week, it should be changed immediately. (Proceed with caution in this situation because microorganism contamination may be present.)
8. After one to two weeks (depending on the growing conditions of the cells), trypsinize and transfer the cells to a new T_{25} flask. Transfer only one flask at a time. Add about 5 ml medium to the original T_{25} flask and continue to grow. After the cells become confluent in the new T_{25} flask, trypsinize and transfer to a T_{75} flask. When this flask is confluent, trypsinize and transfer to three T_{75} flasks. If sufficient flasks are available, it is advisable to leave one of the flasks with the original medium. This is a safety precaution to reduce the risk of contaminating all cultures. Cultures destined for pellet formation or liquid nitrogen storage should be grown in antibiotic-free medium. The flasks are checked and the medium changed as necessary until culture confluency is reached.
9. When the cells become confluent, harvest and freeze. Add antibiotic-free medium to the original flask and reincubate. This step serves the dual purpose of continuing the cell line and checking for contamination. A few drops of the frozen DMSO medium-containing cells should be added to a T_{25} flask and grown to test for viability.
10. Some cultures grow slowly. Provided microorganism contamination is not the cause, this situation is remedied by (a) increasing the FCS content to 20%, and (b) when changing the medium, leaving approximately one-third of the original medium and adding two-thirds new material. Contamination

by mycoplasma or other slow-growing organisms could be the cause of slow culture growth. If this is suspected, contamination tests should be performed.

Trypsinizing procedure. It is important that different cell lines be grown under the same conditions and harvested at the same stage of development. This is to ensure that the rate of cell metabolism is not stimulated or reduced in one line relative to another due to different environmental conditions.

Reagents:

1. Hanks' balanced salt solution (HBSS) without Mg or Ca (100 ml; GIBCO 31-4170)
2. 2.5% lyophilized trypsin (20 ml vial; GIBCO 610-5095)
3. Antibiotics: penicillin–streptomycin–fungizone (GIBCO 600-5245)
   Mix the following:
   100 ml HBSS
   10 ml trypsin (vial)
   1 ml (optional) antibiotics

To trypsinize,

1. Decant the medium from the flask.
2. Wash the flask contents twice with 5 ml HBSS for a T_{25} flask and 15 ml for a T_{75} flask.
3. Decant the HBSS.
4. Pour 3 ml of working trypsin solution into a T_{25} flask, and 6 ml into a T_{75} flask.
5. Incubate at 37°C for 5 to 10 min, gently rotating the flask at intervals. When the cells begin to lift from the flask surface, proceed with the transfer.
6. Gently flush the cells from the flask base with a sterile pipette.
7. Divide the cells into the appropriate number of flasks (usually three), add approximately 15 ml of medium to each, and incubate. Ideally, the medium should be changed within one day to eliminate traces of trypsin.

Cell Storage. Five vials of cells should be frozen and stored in liquid nitrogen for each culture. All cultures from the same line number must not be frozen and stored at the same time. A new cell line should never be completely removed from culture and stored until a vial of the frozen material is thawed and tested for viability.

Reagents:

1. MEM working medium
2. Dimethyl sulphoxide (DMSO) solution (Fisher #D-128)
3. Working trypsin solution
4. HBBS

Procedure:

1. Trypsinize one T_{75} flask (see the protocol for trypsinizing), transfer the contents to a sterile centrifuge tube, and centrifuge for 5 min at 1,200 rpm in a clinical centrifuge.
2. Decant the supernatant.
3. Wash the cells by resuspending in HBSS, centrifuging, and discarding the supernatant.
4. Add 1.5 ml of working medium containing 10% DMSO at 4°C. Resuspend the cells in this medium and transfer into a liquid nitrogen vial.
5. Add a few drops remaining from step (3) into a T_{25} flask. Add working MEM, and incubate to test the cells for viability.
6. Immediately place the vial in a liquid nitrogen biological freezer (set at position D) for at least three hours. If two flasks are to be frozen, treat as two separate samples. Do not freeze more than two vials of an identical cell line on the same day.
7. Place the vials on a cane and lower into the liquid nitrogen for storage. Ensure that the vial caps are tight since a leak of liquid nitrogen into a vial could result in an explosion when the tube is removed from the freezer tank. Testing for leaks can be accomplished by dipping the vials into alcohol-methylene blue before freezing. Observe whether the dye leaks into the vial.
8. Record the date of freezing, number of vials, and location of vials.

Thawing cells:

1. Remove one vial from a cane in the liquid nitrogen container. (Observe the safety precautions for liquid nitrogen.)
2. Thaw quickly by warming to 37°C. This should require only a few min.
3. Resuspend the cells immediately in a sterile centrifuge tube containing HBSS.
4. Immediately centrifuge and decant the supernatant.
5. Resuspend the cells in working MEM for culture (approximately 15 ml for a T_{75} flask and 5 ml for a T_{25} flask) and transfer to a T_{75} flask. If the cells have a history of slow growth, as found in some lines with metabolic disorders, transfer to a T_{25} flask.
6. Change the medium after 24 h.

Pelleting Procedures

1. Decant the medium from a T_{75} flask and gently wash the cells with 20 ml of normal saline to remove traces of fetal calf serum. Repeat this procedure twice for amniocytes.

2. Trypsinize the cells (see the protocol for trypsinizing).
3. When the cells have lifted from the flask surface, transfer to a 12 × 75 mm glass tube and centrifuge at 1,500 rpm for 5 min.
4. Decant the supernatant.
5. Wash the cells with 7 ml normal saline, recentrifuge, and decant supernatant.
6. Repeat step (5).
7. Store the pellets in a –70°C freezer.
8. Record the number of pellets prepared by date.

REFERENCES

Allen RC, Graves G, and Budowle B. 1989. Polymerase chain reaction amplification products separated on rehydratable polyacrylamide gels and stained with silver. *BioTechniques* 7:736–744.

Barnes DW, Sirbasku DA, and Sato GH, eds. 1984. *Cell Culture Methods forMolecular and Cell Biology.* Alan R Liss, New York.

Budowle B and Allen RC. 1990. Discontinuous polyacrylamide gel electrophoresis of DNA fragments. In *Molecular Biology in Medicine*, Volume 7. Mathew C, ed. Humana Press, London (in press).

Freshney IR. 1987. *Culture of Animal Cells: A Manual of Basic Technique.*2nd ed. AR Liss, New York.

Fujimura FK. 1990. Genotyping errors with the polymerase chain reaction. *N. Engl. J. Med.* 322:61.

Gyllensten U. 1989. Direct sequencing of in vitro amplified DNA. In *PCR TechnologyPrinciples andApplicationsforDNA Amplification*, 45–60. Erlich HA, ed. Stockton Press, New York.

Gyllensten U. 1989*a*. PCR and DNA sequencing. *BioTechniques* 7:700–708.

Higuchi R. 1989. Simple and rapid preparation of samples for PCR. In *PCR TechnologyPrinciples andApplicationsforDNA Amplification*, 31–43. Erlich HA, ed. Stockton Press, New York.

Li H, Gyllensten UB, Cui X, Saiki RK, Erlich HA, and Arnheim N. 1988. Amplification and analysis of DNA sequences in single human sperm and diploid cells. *Nature* 335:414–417.

Marx J. 1988. Multiplying genes by leaps and bounds. *Science* 240:1408–1410.

Moss DJ, Rickinson AB, and Pope JH. 1978. Long-term T-cell-mediated immunity to Epstein-Barr virus in man. I. Complete regression of virus-induced transformation in cultures of seropositive donor leukocytes. *Int. J. Cancer* 22:662–668.

Mullis KB. 1990. The unusual origin of the polymerase chain reaction. *Sci. Am.* 262:56–65.

Nuzzolo L and Vellucci A. 1983. *Tissue Culture Techniques.* Warren H Green, St Louis.

Paabo S, Higuchi RG, and Wilson AC. 1989. Ancient DNA and the polymerase chain reaction. *J. Biol. Chem.* 264:9709–9712.

Paul J. 1975. *Cell and Tissue Culture*. 5th ed. Churchill Livingstone,Edinburgh.

Saiki R, Bugawan TL, Horn GT, Mullis KB, and Erlich HA. 1986. Analysis of enzymatically amplified ß-globin and HLA-DQα DNA with allele-specific oligonucleotide probes. *Nature* 324:163–166.

Saiki RK. 1989. The design and optimization of the PCR. In *PCR Technology Principles and Applications for DNA Amplification*, 7–18. Erlich HA, ed. Stockton Press, New York.

Sambrook J, Fritsch EF, and Maniatis T. 1989. *Molecular Cloning: A Laboratory Manual*. Vols 1,2, and 3. 2nd ed. Cold-Spring Harbor, Cold-Spring Harbor, New York.

Sarkar G and Sommer SS. 1990. Shedding light on PCR contamination. *Nature* 343:27.

CHAPTER 6

Analysis Techniques

Conventional DNA analysis techniques include cleavage of DNA by restriction enzymes, fragment electrophoresis, Southern transfer, probe labeling, probe-genomic fragment hybridization, and print detection (Figures 6-1, 6-2) (Cawood 1989, Sambrook 1989, Berger 1987). Details of the assay conditions may vary considerably depending on the specific probes hybridized. Endonuclease digestion, electrophoresis, and Southern transfer are not required with simple dot-blot procedures.

The quality of the final result can be no greater than the quality of the input DNA specimen and the attention of the analyst to assay details. The format of the analysis blot must be carefully considered to include control specimens and a broad range of size markers. The analyst must also be certain about the sizes of the profile fragments to accurately determine if matches exist between crime evidence and suspect specimen or offspring and putative parent specimens and to calculate the match probabilities.

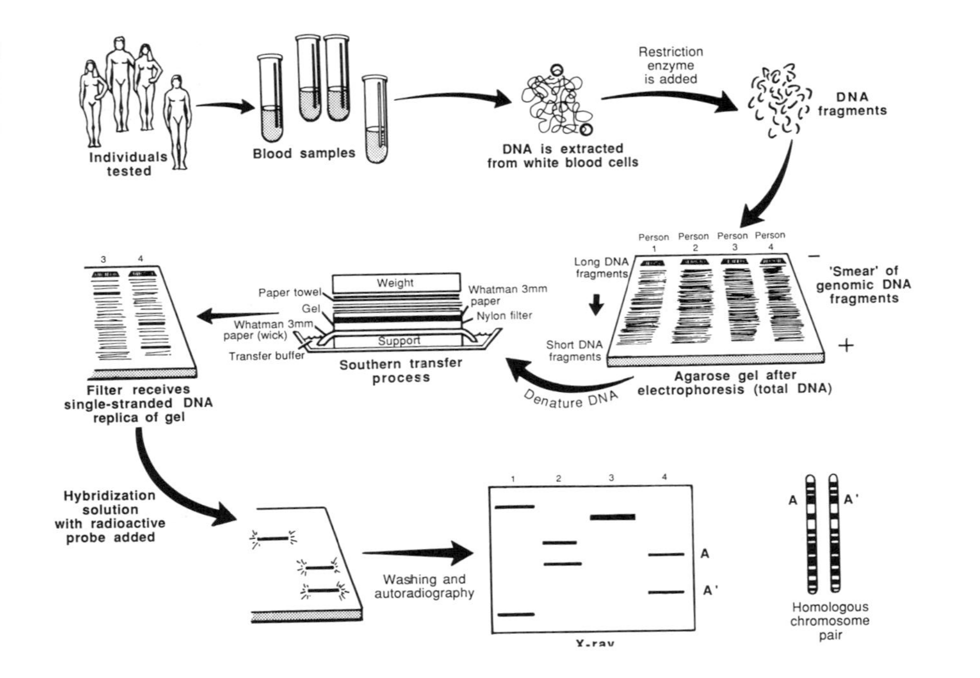

Figure 6-1. Schematic illustration of the single-locus multiallele DNA fingerprint analysis procedure.

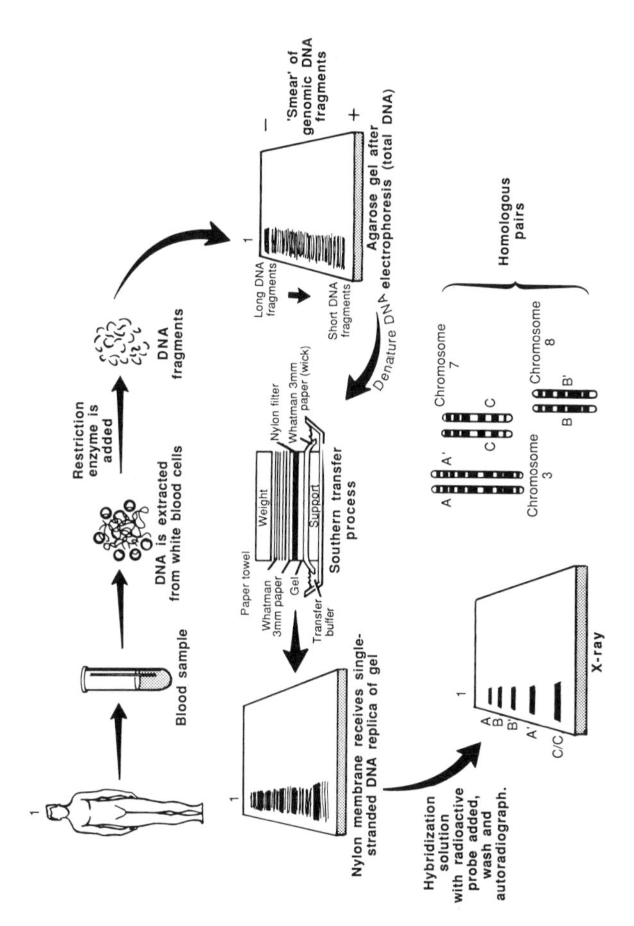

Figure 6-2. Schematic illustration of the multilocus multiallele DNA fingerprint analysis procedure.

RESTRICTION ENZYME CLEAVAGE

Restriction enzymes cleave DNA at specific recognition base sequences. It is important to choose an enzyme with sites flanking the repeats when fragments consisting of different numbers of tandem repeats are to be characterized for DNA profiling. Cleavage within a repeat sequence will result in the production of small fragments that may be unresolvable. The choice of enzyme, in this respect, is accomplished either by trial and error or by knowledge of the base sequence of the fragment flanking regions.

The optimum reaction conditions vary for each enzyme; consequently, suppliers usually provide information sheets for the user. Digestion temperature and buffer salt concentration are the critical features. The reaction mixture can be overlaid with a few drops of paraffin oil to prevent vapor formation and changes in the buffer concentration. This applies mainly to enzymes such as Taq I that require high reaction temperatures (65°C in this example).

Unless specifically indicated otherwise, three different strength ionic buffers will accommodate most enzymes (Table 6-1). The enzyme Sma I is one example of an exception to this rule because it requires a buffer consisting of 20 mM KCl, 10 mM Tris Cl (pH 8.0), 10 mM $MgCl_2$, and 1 mM dithiothreitol. Buffers are usually prepared at 10× concentration and stored at –20°C.

One unit of enzyme is the amount required to digest 1 μg of λ DNA in 1 h. To ensure complete digestion, both enzyme concentration and reaction time are usually increased. Reactions are normally performed with a twofold enzyme excess and 4 h to overnight incubation. All enzymes should be stored in a manual-defrost freezer to eliminate the temperature cycling of frost-free units. They should always remain on ice until the reaction mix is incubated. A DNA specimen can be cleaved with more than one enzyme provided the required buffer salt concentration and reaction temperatures are similar. If different buffers are required, the lower ionic strength reaction is undertaken first and the concentration later adjusted before adding the second enzyme. The addition of spermidine to the reaction mixture is optional; however, if enzyme inhibitors are of concern, inclusion of this reagent may be beneficial.

A typical reaction procedure, based on a 50 μl volume, is as follows (this volume can be adjusted downward as appropriate):

20 μl DNA specimen in LTE
5 μl 10× buffer
5 μl 10× BSA at 1 μg/μl (final concentration = 100 μg/ml)
2 μl 0.1 M spermidine (final concentration 4 mM)
1 μl enzyme (15 units)
17 μl distilled water

Table 6-1. Buffers for Restriction Endonuclease Digestion

Buffer	NaCl	Tris Cl (pH 7.5)	$MgCl_2$	Dithiothreitol
Low	0	10 mM	10 mM	1 mM
Medium	50 mM	10 mM	10 mM	1 mM
High	100 mM	50 mM	10 mM	1 mM

It is important to note that concentrated enzyme solutions must be diluted in the final volume. A 10× enzyme solution, for example, must be diluted at least 1:10.

The reactions are incubated in microfuge tubes for 4 h to overnight. One-fifth or greater by volume of stop/loading dye (such as sucrose 40%, EDTA 25 mM, bromphenol blue 0.02%, in TE buffer; or, glycerol 50%, EDTA 0.1 M, bromphenol blue 0.1%, in TE buffer) is added to stop the reactions and prepare the digest for electrophoresis. Before stopping the reaction, it is prudent to remove an aliquot (less than 5 µl) of the reaction mix for electrophoresis on a mini-gel to ensure that complete DNA digestion has occurred.

GEL BLOT ORGANIZATION

Forensic

The arrangement of blot lanes is an important consideration. The following format should provide the information required in a forensic case.

Size marker	*Male control*	*Female control*	*Size marker*	*Victim*	*Crime*	*Suspect*	*Size marker*

Parentage

The following information is required in parentage analysis.

Size marker	*Control*	*Mother*	*Offspring*	*Putative father*	*(Putative father plus offspring)*	*Size marker*

If a match is declared, a mixture of putative father plus offspring may be of value to ensure that the appropriate bands superimpose. Considerable caution must be exercised in this respect to ensure that the specimen concentrations are identical, contaminants are not present, and no DNA degradation has occurred; otherwise, band shifting could result and the bands will not necessarily superimpose. As discussed later, the use of monomorphic "control" markers is one excellent way to circumvent such potential problems. An extra control may also be included. If more than one putative parent requires testing, additional lanes for these individuals plus mixtures may be necessary.

Size Markers

A size-marker lane should be included every four to five lanes to facilitate correction of electrophoretic variations. Many different size markers are available from commercial suppliers. The choice of marker depends on the specific application; however, there is an advantage to using standards having molecular structures similar to those of the samples being analyzed.

Two examples of markers are illustrated in Figure 6-3. Both are suitable for sizing linear double-stranded DNA. λ DNA Hind III spans a wider range than the 1-kb ladder but with considerably fewer bands.

ELECTROPHORESIS

The separation of the DNA fragments produced by restriction enzyme cleavage is carried out by conventional sub-marine gel electrophoresis in agarose gels or by modified electrophoresis techniques (Fowler 1988, Signer 1988, Uitterlinden 1989). The rate of movement of linear double-stranded molecules in the gel is inversely proportional to the log of the molecular weight and varies directly with the applied voltage. A gel concentration of 1.0% is usually used for mammalian genomic DNA. Reducing the concentration increases the size range separation efficiency. A 0.7% gel is efficient for the 0.8- to 10-kb size range, whereas, 0.9% is best for the 0.5- to 7-kb range. Considerable care must be exercised as the agarose gel concentration is decreased because gels become progressively more fragile, with the distinct and frustrating possibility of breaking during handling.

Both mini-gels and maxi-gels are in common use. Mini-gels measure approximately 7 × 10 cm and contain eight wells for sample application of up to 25 μl each. The gels are usually run at high voltage (100 V) for periods of usually under 1 h. These gels are useful for determining whether a DNA digest is complete and in general for determining DNA quality. Approximately 10 ng of undigested DNA can be detected on mini-gels after staining with ethidium bromide. If digestion is complete, a maxi-gel is then set up.

Maxi-gels measure 14 × 16 cm (FBI standard) or 20 × 16 cm (RCMP standard) and accommodate 15 or more specimens. To gain increased separation efficiency, these gels are normally run at considerably lower voltage (30–33V) overnight (16 h). If the electrophoresis is continued for longer periods, buffer ion depletion can

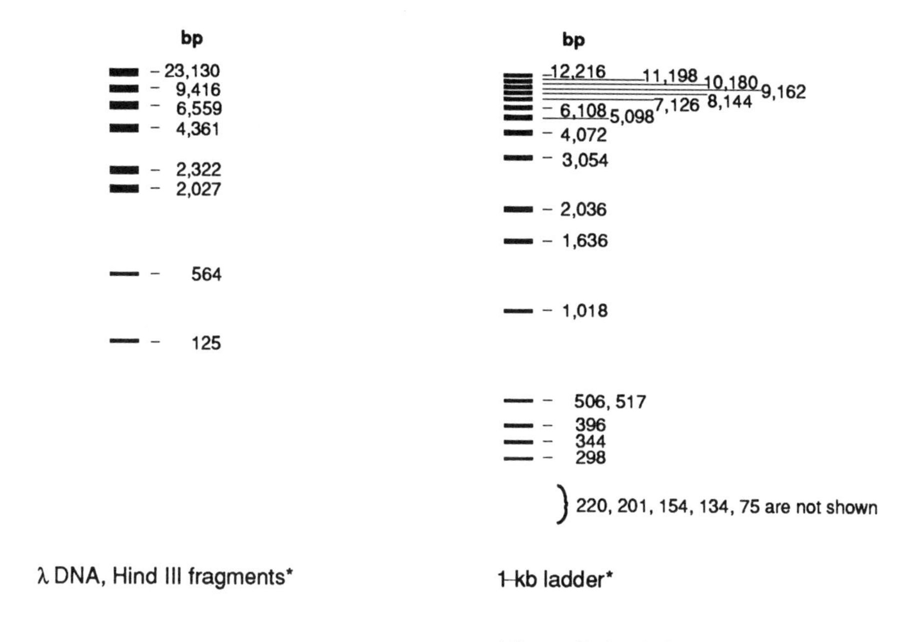

Figure 6-3. DNA molecular size markers. The markers illustrated are suitable for sizing linear double-stranded DNA. Approximately 2 μg are applied per lane to visualize the markers with ethidium bromide staining and approximately 1.5 ng/lane if radiolabeled tags are hybridized. The 12 fragments (1,018 to 12,216 bp) of the ladder are composed of 1,018 bp repeats. The additional bands represent vector DNA fragments.

occur resulting in problems such as DNA band shifting. A human DNA digest may consist of millions of fragments (over 25 million ($7 \times 10^9/(4)^4$) for Hae III); thus, efficient separation is necessary to gain maximum resolution.

Equipment

Gel apparatus components, consisting of a powerpack, buffer tank with cover, casting tray, holder, combs, and a leveling table are available from numerous supply companies. For most purposes, horizontal systems with submerged gels are satisfactory and are much easier to handle than a vertical gel apparatus.

Reagents

Agarose, Tris borate buffer (1 × TBE) or Tris acetate buffer (1 × TAE), ethidium bromide stain, and a tracking dye/loading buffer are the key components of the reagent system. Electrophoresis-grade standard melting point agarose is satisfactory for most analyses. Low melting point agarose is required if fragments are to be excised from the gel. The agarose is mixed with TBE or TAE buffer at approximately 50 ml for mini-gels and 250 ml for maxi-gels. The bed gel should be 3- to 5-mm thick (Chambers 1990). The mixture is heated to boiling in a microwave oven or on a hot plate. When the agarose has dissolved, the mix is cooled to about 50°C and ethidium bromide (10 mg/ml concentration) added to a final concentration of 0.5 µg/ml of mix. (Ethidium bromide is mutagenic; thus, it must be handled with caution. The stock solution should be stored at 4°C in a lightproof container.) Some analysts prefer to overlay gels with ethidium bromide stain when the electrophoresis is completed because this reagent can cause slowing and distortion of linear duplex DNA fragments as they move through the gel. Such mobility shifts could lead to problems when comparing profile patterns, as well as causing difficulties with the establishment of allele frequencies in population data bases (Waye 1990). Also, ethidium bromide migrates toward the cathode (opposite to DNA) during electrophoresis. During extended electrophoresis, a considerable quantity of the ethidium bromide may be removed from the gel, making the detection of small DNA fragments difficult (Sambrook 1989).

Electrophoresis is carried out in the same buffer used to prepare the gel. Although recirculation is not necessary with TBE, curvature (smiling) of DNA bands is considerably reduced with recirculation. TAE must be recirculated. Use of this buffer as opposed to TBE results in increased sharpness of high molecular weight bands (over 3 kb) but fuzzier low molecular weight bands (Pemberton, 1989). The tracking dye/loading buffer is described in the previous section on restriction enzymes.

Procedure (Figure 6-4)

1. The ends of the gel casting tray are sealed with autoclave tape (unless casting gates are available) and the unit is placed in the tray holder on a leveling table.
2. The agarose mixture, at no greater than 50°C is poured into the tray giving a gel thickness of 3- to 5-mm. The comb is then inserted, with the teeth remaining at least 0.5 mm above the gel plastic support. The agarose mixture should be agitated during the cooling period prior to pouring. This should prevent the development of gradient pockets in the gel, which can cause band distortion and shifts.
3. When the gel has set (within approximately 40 min of pouring), the surface is flooded with buffer and the comb gently removed.
4. The casting tray is removed from its holder, the tape is removed, and the tray is submerged in the electrophoresis tank.
5. Size standards, controls, and individual samples are carefully added through the buffer into the wells. The specimens sink because of the dense sucrose

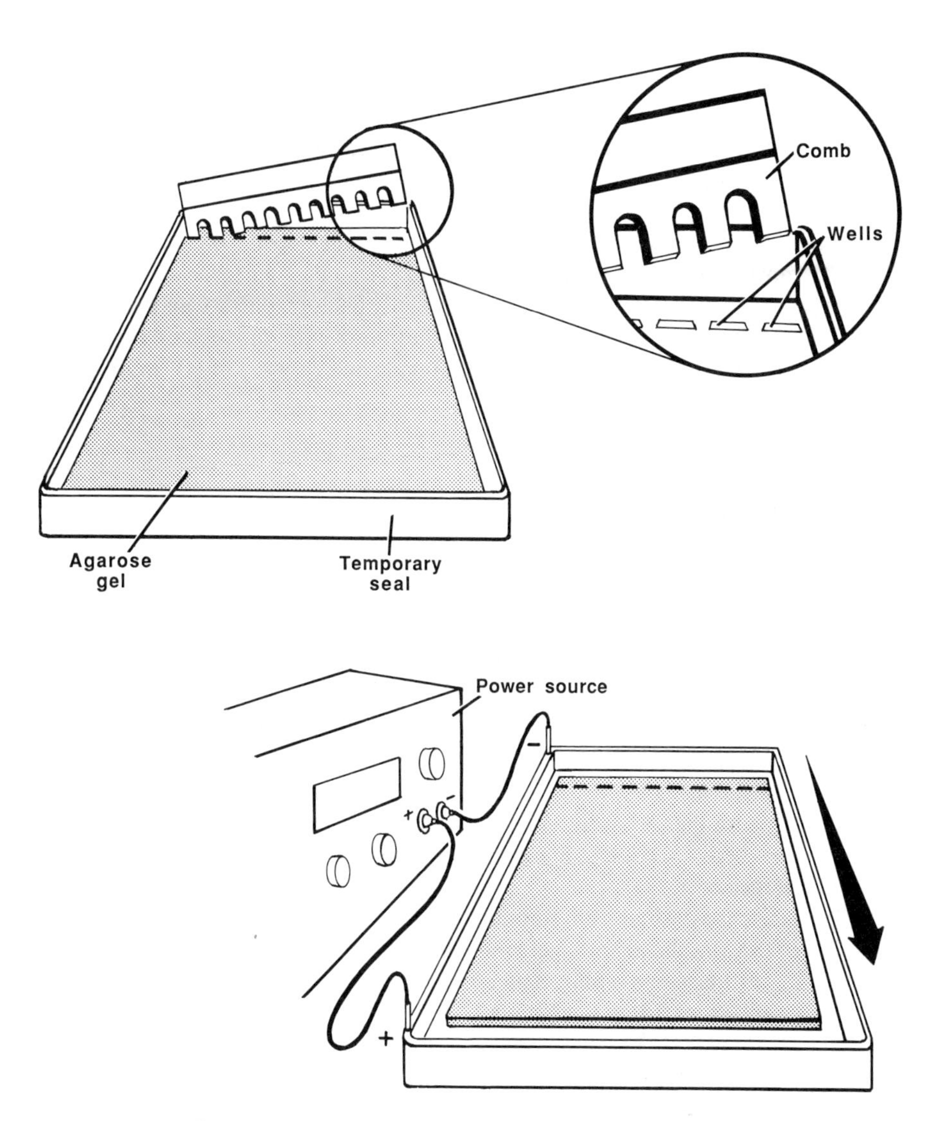

Figure 6-4. Agarose gel and electrophoresis setup.

loading buffer. Some analysts prefer to add the specimens to the wells before flooding the gel with buffer. This technique apparently reduces the possibility of well cross-contamination.

6. The tank cover is set in place, and the electrical connectors are attached. The sample migrates from negative to positive.
7. The voltage is adjusted as required for mini-gels (approximately 100V) or maxi-gels (usually 30–33 V).

PHOTOGRAPHY

Photos of gels are invaluable as records of size-marker locations, DNA digestion, comparisons of DNA concentration, and migration distances. A convenient apparatus includes a Polaroid Land camera station and a UV transilluminator (Figure 6-5). Thirty-five mm negatives are also suitable. A filter on the camera is required. The specifications for the filter unit will depend on the film and transilluminator suggested guidelines. Ethidium bromide-stained DNA fluoresces under UV irradiation and can, therefore, be readily photographed. Caution should be exercised during this process to minimize the time of exposure because DNA can be damaged when stained with ethidium bromide and exposed to UV irradiation (Cariello 1988).

DNA RECOVERY

It is sometimes necessary to isolate restriction-digested DNA fragments from gels. This procedure is used, for example, to obtain plasmid inserts to be used as probes. A number of techniques are available; however, enzyme inhibitors are also often extracted and the efficiency of recovery may be less than 25% for fragments over 20 kb. One relatively simple isolation technique follows:

1. Pour a low melting point 1.0 to 1.5% contaminant-free agarose gel containing 0.5 μg/ml ethidium bromide. (Low melting point agarose melts at 65°C and solidifies at 30°C.)
2. Dispense the DNA digest and separate by electrophoresis as previously described. (Caution should be exercised to ensure that the gel is not overheated.)
3. Excise the desired band from the gel and carry out (*i*) or (*ii*).
   (*i*) (*a*) Mix the agarose gel with water in the ratio of 3 ml/g of gel.
   (*b*) Place in a boiling water bath for approximately 7 min.
   (*c*) Store frozen in aliquots.
   (*d*) When required for labeling, reboil for 2 min and place in a 37°C water bath for at least 10 min.
   (*ii*) (*a*) Add 5 volumes of 20 mM Tris HCl, 1 mM EDTA pH 8.0 to the agarose block.
   (*b*) Heat for 5 min at above 65°C.
   (*c*) Extract with phenol, chloroform, and precipitate with ethanol as described in Chapter 4.
   (*d*) Label the resuspended DNA pellet as required.

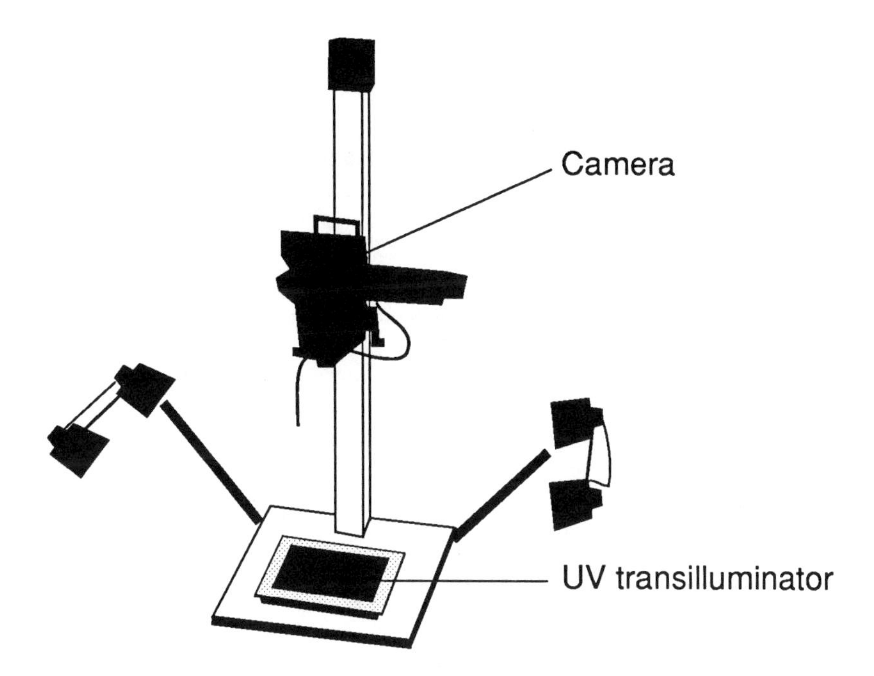

Figure 6-5. Camera station for photographing ethidium bromide-stained gels.

MEMBRANE TRANSFER

Southern Blots

The separation of DNA fragments of different sizes usually can be efficiently accomplished by agarose gel electrophoresis. These gels are, however, easily damaged and are difficult to process in terms of hybridization, washing, and autoradiography. In addition, gel bands diffuse after electrophoresis is completed. These problems can be largely overcome by transferring the DNA to nylon membrane supports such as Amersham's Hybond-N or N+. Caution must be exercised when using nylon membranes because of possible variation in performance between different lots. Each new lot should be pretested prior to service use.

The transfer technique was first described by E. Southern (1975) and was later modified to immobilize and cross-link single-stranded DNA fragments on the membrane by the action of UV irradiation. Nylon membranes can be reprobed repeatedly, provided the original radioactive tag has sufficiently decayed or the blots have been stripped of probe from the previous hybridization.

The Southern blot procedure for capillary transfer to a nylon membrane under neutral conditions is as follows:

1. After electrophoresis, place an orientation mark in the upper right-hand corner of the gel and flood the gel with denaturing solution (1.5 M NaCl, 0.5 M NaOH) for at least 30 min. If large fragments must be transferred, such as digested mammalian genomic DNA, transfer efficiency can be increased by first placing the gel in 0.25 M HCl for 15 min. This depurination process shears the DNA into smaller pieces. Caution must be exercised if the depurination process is carried out because probe binding sites on the DNA templates may be damaged.
2. Replace the denaturing solution with neutralizer (1.5 M NaCl, 0.5 M Tris HCl pH 7.2, 0.001 M Na_2 EDTA) for 30 min or longer.
3. Set up the blotting apparatus and fill the reservoir with 20 × SSC (3 M NaCl, 0.3 M Na_3 citrate) transfer solution.
4. Place the gel, DNA side down, on the prewetted filter paper wick and press out all air bubbles.
5. Place the nylon membrane on the gel and again ensure that all air bubbles are removed.
6. Cover the membrane with three sheets of 2 × SSC°C wetted Whatman 3 mm filter paper. Add absorbent papers such as paper towels and weights.
7. Allow the capillary transfer process to proceed for 5 to 16 h. The 20 × SSC solution transfers the DNA fragments from the gel directly to the membrane.
8. After the transfer is completed, remove the membrane, rinse in 2 × SSC to remove residual agarose, air-dry, and wrap in Saran Wrap.
9. Fix the DNA on the blot by irradiating with UV light for 3 min. This step is not required with Hybond-N+.

The Southern blot procedure for capillary transfer to a nylon membrane under alkaline conditions follows. (This procedure is similar to that followed in the FBI protocol—Budowle 1989.) The ability of positively charged nylon membranes to bind denatured genomic DNA under alkaline conditions appears advantageous when hybridizing with repeat sequence probes.

1. After electrophoresis, flood the agarose gel with denaturing solution (0.4 M NaOH) and gently shake for 30 min. Place a mark in the upper right-hand corner of the gel to provide for orientation.
2. Immerse the nylon membrane (precut to the gel size) in 0.4 M NaOH for 15 min.
3. Set up the blotting apparatus and fill the reservoir (sponge) with 0.4 M NaOH (Figure 6-6).
4. Place the gel, DNA side down, on the prewetted blot pad. (The original gel top should be face down on the pad.) Press out all air bubbles.
5. Place the nylon membrane on the gel and roll out all air bubbles. This can be accomplished by rolling a pipette over the membrane.

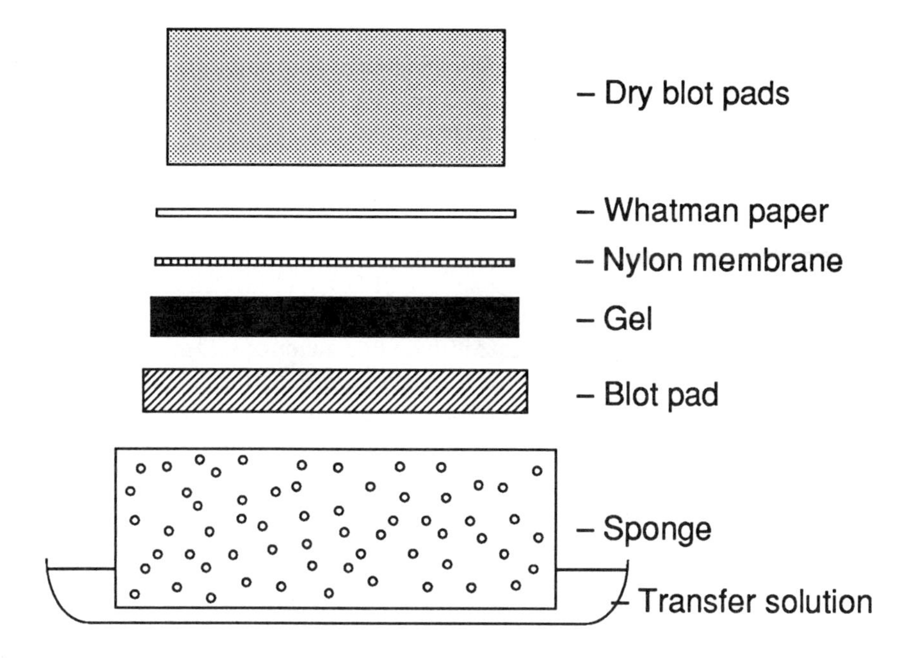

Figure 6-6. Blotting apparatus used by the FBI. The blot pads consist of precut pads of blotting paper that can be purchased commercially. An apparatus can also be set up as diagrammed in Figures 6-1 and 6-2.

6. Cover the membrane with Whatman 3 mm paper presoaked in 0.4 M NaOH. Roll out all air bubbles.
7. Place nine dry blot pads on the Whatman paper and place a glass plate over the pads to complete the sandwich.
8. Allow the transfer to proceed for at least 6 h at RT.
9. Remove the membrane, wash with gentle shaking in 0.2 M Tris pH 7.5 and 2 × SSC for 15 min to remove residual agarose, and blot on Whatman paper.
10. Fix the DNA on the blot by placing in a vacuum oven at 80°C for 30 min or by exposing the blot to UV light for 3 min. Nylon membranes produced by some manufacturers do not require a fixing step; follow the manufacturer's directions.

The efficiency of the Southern transfer process should be determined. This can be accomplished by immersing the gel in the electrophoresis buffer or a water solution containing 0.5 µg ethidium bromide per ml for approximately 45 min. If destaining is required to remove unbound ethidium bromide, the gel can be transferred to distilled water or to a 1 mM $MgSO_4$ solution for 1 h.

Blots should not be allowed to dry out after hybridization because this can give rise to increased background on the autoradiograms. Blot storage is best accomplished by wrapping in Saran Wrap and placing in a –20°C freezer. This is especially important for long-term storage in order to prevent DNA degradation.

Dot- (Slot-)Blots

Certain DNA testing, such as sex determination and detection of PCR-amplified products, may only require direct application of DNA, or labeled probe, to blot membranes, thus, bypassing electrophoresis and fragment transfer. Both dot-blot and reverse dot-blot procedures are used in the hybridization of SSO probes to PCR-amplified products. The reverse dot-blot technique differs from the standard dot-blot format in that a number of oligonucleotide probes are immobilized on a single membrane strip then PCR products (labeled during amplification) are hybridized to the probes (Figures 6-7, 6-8). This approach results in considerable savings in terms of number of blots, because in the dot-blot process, a separate blot is required for each allele detected. Because of the high specific allele copy number, nonradioisotope detection procedures are sufficiently sensitive for use with these blot techniques. The dot- or slot-blot procedure can be performed as follows:

1. Heat the DNA solution to 95°C for 5 min (or denature in NaOH) and place on ice for 2 min.
2. Apply the DNA in a slot-blot apparatus such as Schleicher and Schwell's Manifold II (Figure 6-9), or simply pipette approximately 2 μl aliquots directly on a nylon membrane. Allow to dry and reapply.
3. Wet the membrane in denaturing solution (1.5 M NaCl, 0.5 M NaOH) for 1 min. This process facilitates a rupture of the hydrogen bonds linking the complementary DNA strands.
4. Transfer the membrane to neutralizing solution (1.5 M NaCl, 0.5 M Tris HCl pH 7.2, 0.001 M Na_2 EDTA) for 1 min, blot, and air-dry.
5. Wrap with Saran Wrap, then irradiate with UV light for 3 min.

RADIOISOTOPE PROBE LABELING

The oligonucleotide primer method is an efficient procedure to radio-label DNA probes for hybridization. The technique, developed by Feinberg and Vogelstein (1984), is based on the annealing of random sequence hexanucleotides as primers for DNA synthesis at numerous sites on single-stranded templates (Figure 6-10). The DNA polymerase I Klenow fragment catalyzes the synthesis of DNA from deoxyribonucleotide triphosphates. Since this enzyme lacks 5'-3' exonuclease activity, newly incorporated nucleotides are not removed at a later stage. One (or more) of the deoxyribonucleotide triphosphates is labeled with ^{32}P to provide a radioactive tracer tag.

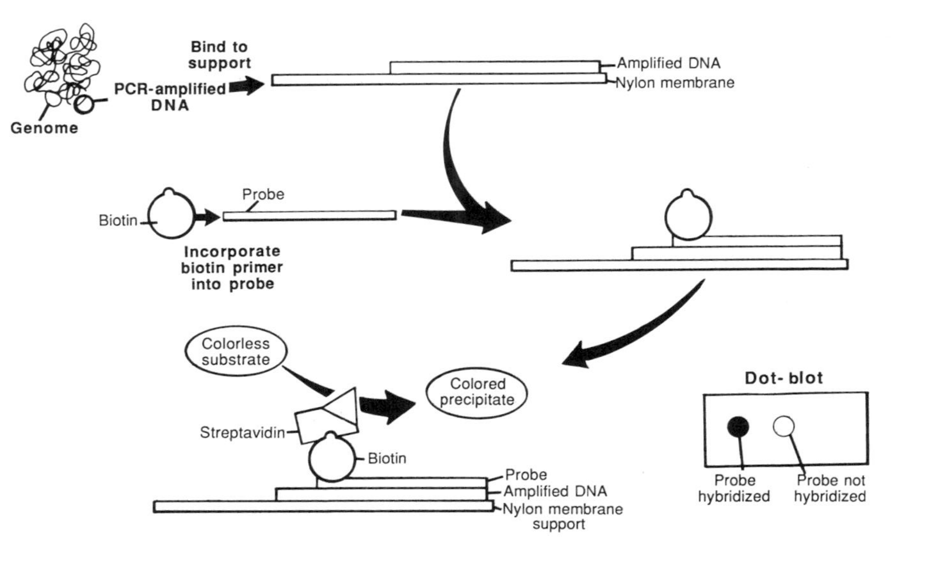

Figure 6-7. Schematic illustration of the dot-blot procedure. Amplified target DNA is immobilized on a nylon membrane, and a biotinylated probe is hybridized to the target, provided there are no nucleotide mismatches. An avidin-horseradish peroxidase (HPR) conjugate is added that binds to the probe. Lastly, the HPR converts a colorless substrate to a colored product.

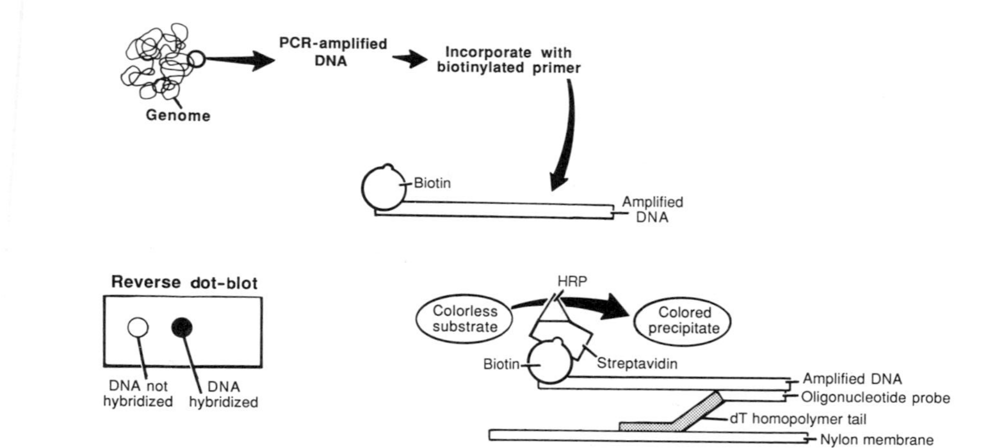

Figure 6-8. Schematic illustration of the reverse dot-blot procedure. A series of specific oligonucleotide probes with dT homopolymer tails are immobilized on a nylon membrane strip. The amplified PCR target DNA, with incorporated biotin, hybridizes only to a probe with the complementary sequence (no mismatch is tolerated). An avidin-horseradish peroxidase (HPR) conjugate is added that binds to the biotinylated DNA. Finally, the HPR converts a colorless substrate to a colored product.

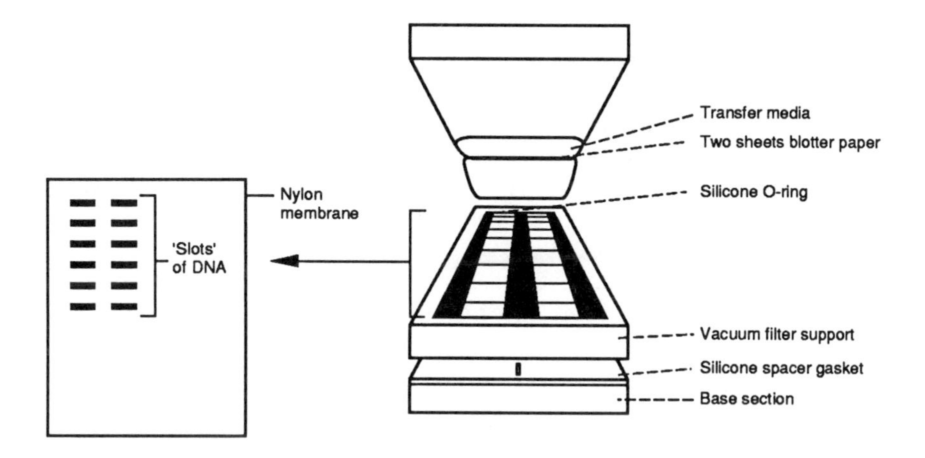

Figure 6-9. Slot-blot apparatus.

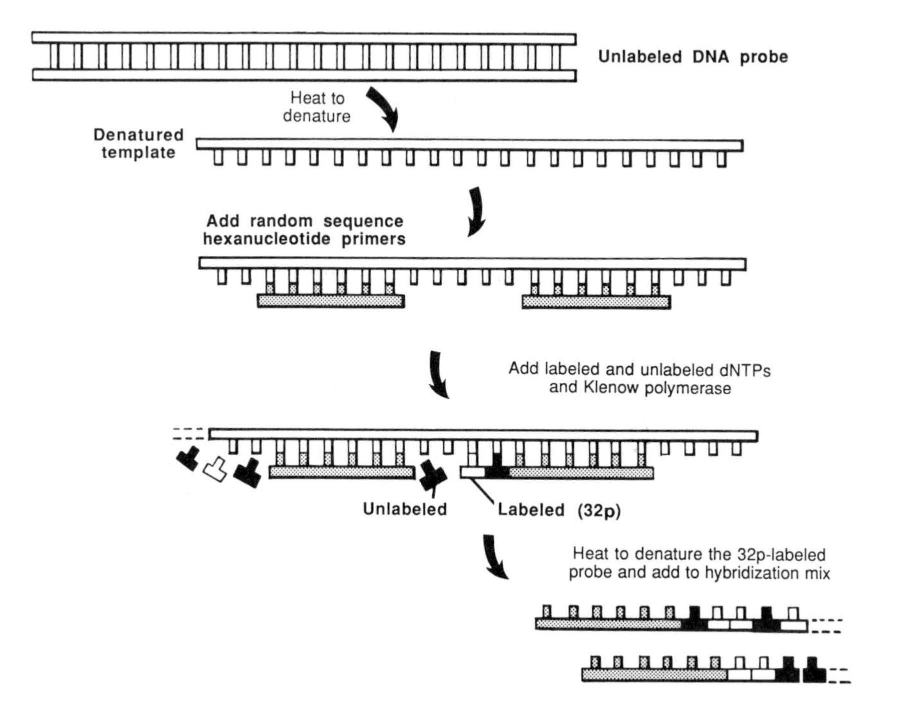

Figure 6-10. Oligonucleotide primer method for labeling DNA probes.

Labeling systems in kit form are available from a number of biological supply companies. These materials are produced under stringent quality control conditions and are fully guaranteed. The Amersham kit, for example, consists of (1) multiprime buffer with the nucleotides dATP, dGTP, and dTTP (the nuclide compound (α-^{32}P)dCTP is purchased separately); (2) a primer solution of random hexanucleotides; (3) the enzyme DNA polymerase I Klenow fragment; and (4) control plus carrier DNA. Approximately 25 ng of DNA are sufficient for labeling with 50 µCi (5 µl) of (α-^{32}P)dCTP at 3×10^3 Ci/mmol. If a greater or lesser quantity of DNA must be labeled, the reaction components should be adjusted accordingly. The protocol outlined usually results in a specific activity (SA) of over 1×10^9 cpm/µg of probe.

Mix the following on ice:

1. 25 ng (1-10 µl) of denatured DNA template
   This is prepared by heating the double-stranded DNA to 100°C for 2 min, then chilling on ice. If the probe was isolated in low melting point agarose gel, the boiled material should be placed in a 37°C water bath for 10 min before labeling. If the reaction mix gels during incubation, this need not be of concern because the reaction will not be affected.
2. 10 µl buffer with nucleotides
3. 5 µl primer solution
4. Distilled water to give a final reaction volume of 50 µl
5. 5 µl (α-^{32}P)dCTP
6. 2 µl polymerase

The reaction is incubated at RT for 5 h to overnight. Then 50 µl of LTE buffer (1 mM EDTA, 10 mM Tris-HCl pH 8.0) are added to give a final volume of 100 µl. The probe, after denaturation, is ready for hybridization.

Unincorporated nuclide materials can be separated from labeled probe before denaturation by passing the reaction mix through a DNA Sephadex G50 (medium) column (Figure 6-11). The procedure is as follows:

1. Prepare a column by filling an Isolab Quik-Sep tube (or a 1 ml sterile plastic syringe with glass wool as a plug) with Sephadex G50 (medium) kept swollen at 4°C in LTE buffer. Centrifuge at 3,000 rpm in a swinging bucket centrifuge for 5 min. The filling and centrifuging processes are repeated until the column is approximately 80% filled. (Premade columns can also be purchased from suppliers such as Pharmacia.)
2. Wash the packed columns by applying 100 µl of buffer and centrifuging.
3. Apply the 100 µl of radioactive reaction mix to the column and collect the resultant labeled probe by centrifuging. (Unincorporated nucleotides are retained on the column.)

The above gel filtration step may be unnecessary if there has been at least 50% nuclide incorporation during the labeling reaction.

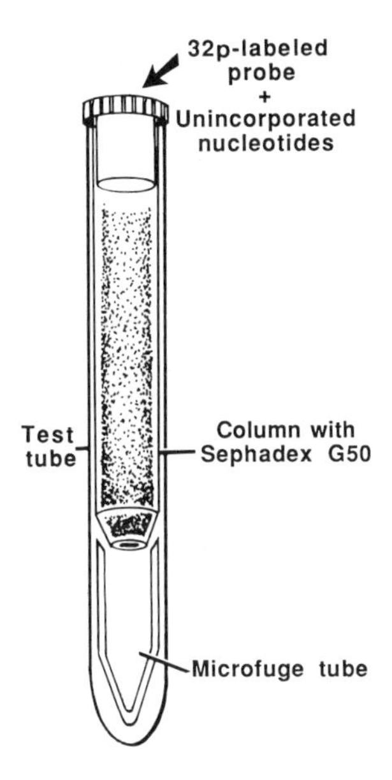

Figure 6-11. Sephadex G50 (medium) spin column for separating ^{32}P-labeled DNA probe from unincorporated nucleotides. The radioactive probe is collected in a 1.5 ml microfuge tube during centrifugation of the column.

The approximate SA of the gel-filtered probe can be determined by adding 1 or 2 μl of the 100 μl column eluent to a water-compatible scintillation cocktail and counting in a ß-counter. If 1 μl (0.25 ng probe) gives 0.5×10^6 cpm, the SA is 2×10^9 cpm/μg DNA probe. This calculation gives an approximate value because it is often difficult to accurately quantitate small DNA amounts such as 25 ng. If net DNA synthesis has occurred during the multiprime reaction, the calculation as outlined will be an overestimate.

NONRADIOISOTOPE PROBE LABELING

There is need of a technically simple, sensitive, reproducible, nonradioisotope technique for labeling probes for mammalian genomic DNA analysis. The system must also facilitate rehybridization of the blot membranes. This is especially important in forensic cases where the quantity of specimen is often limited. Prelabeled probes with a shelf life of months, as opposed to a week for ^{32}P, would then be commercially available. DNA analysis, as a kit technique, would make the technology available to a much broader cross section of laboratories and should provide improved quality control.

Although nonradioisotope labeling systems with streptavidin-biotin-alkaline phosphatase conjugate detection have been available since the early 1980s, sensitivity adequate to detect single-copy mammalian sequences has been lacking. Carrano et al. (1989) developed a fluorescence-based, semiautomated method for DNA fingerprinting. Dykes (1988) indicated that after appropriate modifications (including blocking conditions, type of blot membrane, and labeling conditions), a nonradioactive methodology was suitable for routine RFLP analysis. Bugawan et al. (1988), of Cetus Corporation, successfully used nonradioactive specific allele oligonucleotide probes in a dot-blot format. This group analyzed PCR-amplified DNA in the prenatal diagnosis of ß-globin gene defects and forensic HLA typing. Baum and McKee (1989) of Lifecodes Corporation are developing a sensitive chemiluminescent DNA detection system for paternity and forensic specimen analysis. Alkaline phosphatase-labeled synthetic probes are used and the light emitted during cleavage of the chemiluminescent substrate is detected on X-ray film. One μg of genomic DNA is sufficient to generate a signal within 1 h, and 10 ng (from 1 μl of blood stain) sufficient if the X-ray film is exposed overnight. Depending on the quantity of DNA available for analysis and the probe length, the time for both hybridization and detection can be as little as 4 h. Lastly, Urdea et al. (1988) compared a number of nonradioisotopic methods and concluded that enzyme-modified oligonucleotides were significantly better labeling materials than fluorescent or chemiluminescent derivatives. The sensitivity was comparable to ^{32}P-labeled probes.

The general framework for use of the horseradish peroxidase (HRP) system is as follows:

1. Probes are prepared by covalently linking oligomers containing a free sulfhydryl group to their 5' ends with derivatized HRP.
2. The HRP-oligonucleotide is purified by HPLC and tested to ensure that the enzyme activity remains unchanged. In appropriate buffer, these conjugates remain stable for at least one year.

HYBRIDIZATION

Hybridization involves the annealing of single-stranded labeled probes with complementary specimen DNA sequences immobilized on blot membranes. Numerous protocols using radioactive and nonradioactive probes have been developed (Hames 1985, Sambrook 1989). The general principles governing hybridization and subsequent washing, together with four protocols, are outlined in this section. The protocols include one for nonrepetitive sequence ^{32}P-labeled probes, two for highly conserved repetitive sequence ^{32}P-labeled probes, and a fourth for a horseradish peroxidase (HRP) labeling system. Each probe has its own characteristic "personality." This may necessitate modifications of the posthybridization wash conditions and perhaps the hybridization protocol to ensure optimum autoradiogram band resolution.

Principles

The maximal rate of hybridization for probes over 150 bases in solution at 1 MNaCl ionic strength occurs at 25°C below the T_m for the duplex. (T_m is the temperature at which half of the hybrids are dissociated.) T_m is affected by buffer ionic strength, pH, probe length, base composition, and helix-destabilizing agents such as formamide. The optimal temperature decreases for shorter probes. If 50% formamide is added, the hybridization can be carried out at approximately 42°C; otherwise, temperatures of about 63°C are necessary. The presence of formamide decreases the rate of hybridization. Dextran sulfate or other polymers, at approximately 10% final concentration, can be added to increase the rate of DNA reassociation. It is believed that this agent acts by increasing the effective probe concentration due to exclusion from part of the hybridization mix. There is little influence on oligonucleotides and, in practice, dextran sulfate is usually not used even for larger probes.

Genomic blots are hybridized with, on average, 25 ng of probe having an SA of at least 1×10^9 cpm/µg DNA. It is advisable to trial test different probes to determine the optimum quantity. The SA should always be the maximum possible. The solution volume should be kept as low as feasible to increase the probe concentration. The filters must be covered and all air bubbles eliminated. Fifteen ml of mix should be sufficient for up to 3 blots in a 20 × 25 cm hybridization bag. Provided the blots are covered with solution, agitation is usually unnecessary.

One of the following must be added to the hybridization solution to reduce nonspecific probe binding to the filter:

1. Denhardt's solution consisting of Ficoll, PVP (polyvinylpyrrolidene) and BSA (bovine serum albumin)
2. Non-homologous DNA, for example, sheared single-stranded salmon sperm
3. Other agents such as powdered milk

Care must be exercised when adding exogenous DNA as a nonspecific blocking agent because highly conserved sequences may bind and decrease probe association and tag signal (Vassart 1987). This creates a special problem when working with many of the repetitive sequence probes. The addition of sodium dodecyl (lauryl) sulfate (SDS) aids in removing nonspecifically-bound probe. This is also a component of many wash solutions.

The wash procedure facilitates removal of loosely bound probe. This in turn reduces background on the filter. Wash stringency increases as the temperature is increased and the buffer salt concentration is decreased. Base pair mismatches between membrane-bound specimen DNA and probe can be tolerated with low-stringency washes. As the wash stringency increases, greater amounts of mismatched probe are removed until, at very high stringency, only completely complementary strands remain bound, at least when short oligonucleotide probes are hybridized. The T_m for probes greater than approximately 100 bases decreases by 1°C for each one percent of mismatched base pairs.

Both low- and high-stringency protocols are used in DNA profiling. Single-locus multiallele hybridization systems are usually followed by high-stringency washes, whereas, low-stringency conditions are generally utilized with the multi-locus multiallele approach. Because of the low-stringency conditions, probes may cross-react with nonhuman contaminating DNA. This is a particular disadvantage of the multilocus system.

Procedures

Nonrepetitive sequence probes. The following protocol is a modification of the Amersham Hybond membrane technique. Denatured, sonicated, nonhomologous DNA is added to the hybridization solution; dextran sulfate may or may not be included, and formamide is not used.

A prehybridization/hybridization solution (at least 15 ml per hybridization container) is prepared consisting of 6 × SSC (5 × SSPE may be used in place of SSC), 5 × Denhardt's solution (0.1% BSA, 0.1% Ficoll, 0.1% PVP), and 0.5% SDS. 0.2 ml of 50% dextran sulfate may be added per ml of mix. Sonicated, denatured, non-homologous DNA, such as salmon sperm at 20 µg/ml, is added. (Sonicated sperm is heated for 5 min in a boiling water bath, immediately chilled on ice, then added.)

Prehybridization and hybridization are carried out in heat-sealed plastic bags, in plastic containers, in an apparatus such as an OmniBlot system or a Robbins tube incubator, or in Jeffreys' containers (Figure 6-12). Prehybridization is undertaken for a minimum of 1 h at 65°C. Excess mix is removed and 15 ml of fresh solution containing approximately 25 ng of denatured labeled probe added. Extreme care must be exercised in removing all air bubbles from the bag or other containers before resealing. The blots are hybridized at 65°C for at least 12 h.

The following wash conditions are recommended in the Hybond protocol.

(1) Low stringency: After hybridization the blots are removed from their containers and incubated in 50 ml of 2 × SSC for 15 min at 65°C. This solution is replaced with 50 ml of 2 × SSC containing 0.1% SDS and the incubation is continued for an additional 30 min.
(2) High stringency: The wash procedure in (1) is continued at 65°C for 10 min with 50 ml of 0.1 × SSC.

The blots can be rewashed if background remains a problem. To determine the optimum wash conditions, a series of trial washes should be tested for each probe.

Repetitive sequence probes containing highly conserved sequences (Method 1). The following prehybridization/hybridization solution is similar to that outlined by P. Gill (Central Research Establishment, Home Office Forensic Science Service, U.K., personal communication). A solution is prepared consisting of 1.5 × SSPE, 0.5% dried milk (Marvel or equivalent—see Johnson 1984), 1% SDS, and 6% PEG.

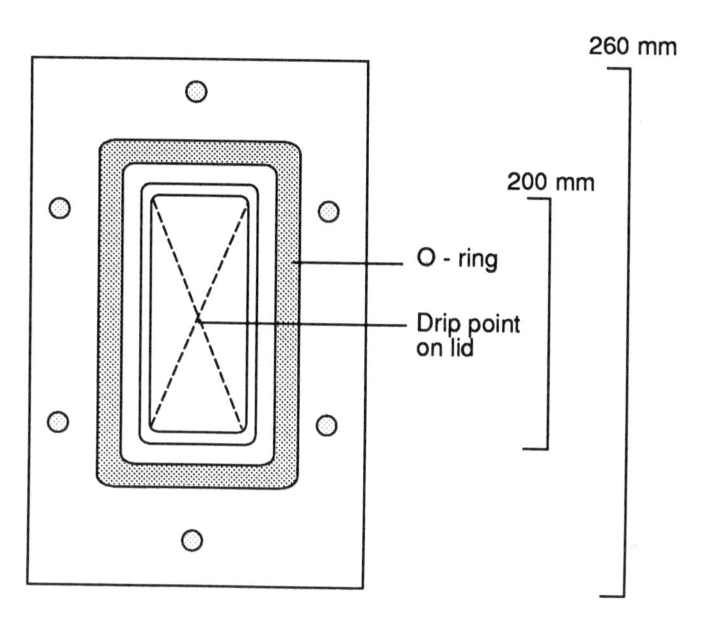

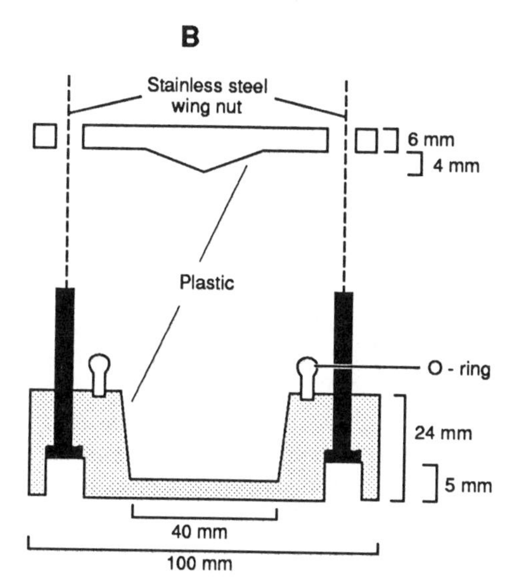

Figure 6-12. Jeffreys' hybridization unit. This container can be constructed with different dimensions to accommodate different size blots.

Up to three blots are prehybridized for at least 1 h at 61°C in approximately 15 ml of the solution. Excess mix is removed and an additional 15 ml containing approximately 25 ng of labeled denatured probe are added. All air bubbles must be removed before the bag is sealed. Hybridization is carried out at 61°C for approximately 12 h.

A low-stringency wash is used with the above dried milk hybridization procedure to facilitate detection of multiple bands when multiallele multilocus probes are hybridized. The stringency can be gradually increased by simply decreasing the SSC concentration. Blots are first placed in approximately 100 ml of 3 × SSC with 0.1% SDS and preheated to 61°C. This is repeated with new wash solution each minute until the waste liquid is at a background level of approximately 50 cpm. The time interval can be increased if desired. The blots are then incubated at 61°C for two washes of 15 min each in 1 × SSC, 0.1% SDS.

Repetitive sequence probes containing highly conserved sequences (Method 2). The following protocol is based on a procedure used by the Federal Bureau of Investigation (Budowle 1989) when single-locus multiallele probes are hybridized.

A prehybridization step is not required because the high concentration of SDS appears to block the unoccupied membrane sites. Dextran sulfate is replaced with less expensive PEG. The simple hybridization solution and protocol described here have resulted in increased sensitivity of target DNA detection, reduced membrane background, and lower reagent costs. Approximately 100 ml of hybridization solution are prepared by combining 34 ml of sterile distilled water, 20 ml of 50% PEG, 7.5 ml of 20 × SSPE, 35 ml of 20% SDS, and 3.3 ml of probe denaturation solution (0.95 ml of 0.2 M NaOH, 2.35 ml of herring sperm DNA stock solution). The radiolabeled probe and size marker are added to the probe denaturation solution and the mixture is boiled for 5 min prior to addition of the mixture to the hybridization solution, already in the hybridization bag (or plastic box) with the blots. The container is sealed and agitated to ensure that the solution is uniformly distributed over the blots. The hybridization is carried out at 65°C overnight with constant shaking.

At the end of the hybridization period, the blots are removed from the bag and washed under high-stringency conditions as follows: (1) two washes for 15 min each at RT in approximately 100 ml of 2 × SSC, 0.1% SDS solution, and (2) one wash for 10 to 30 min (depending on the probe used) at 65°C in approximately 100 ml of preheated 0.1 × SSC, 0.1% SDS solution.

For both Methods 1 and 2, it is important to maintain the wash temperatures as indicated. This can be accomplished by preheating the wash solutions and performing the procedures in closed containers. The radioactive counts can be monitored with a Geiger counter, which can also be used to scan the blot lanes to determine the degree of probe hybridization to specific specimen fragments.

HRP-labeled probes. Prehybridization and hybridization buffer consists of 5 × SSPE, 5 × Denhardt's solution, and 0.5% Triton X-100. Prehybridization is performed at 42°C for 15 min and hybridization for 1 h at 42°C. For hybridization, one to two moles of probe are added per 2 ml of buffer (Bugawan 1988).

The wash procedure for the HRP-label (nonradioisotope) hybridization procedure includes an initial wash with 1 × SSPE, 0.1% Triton X-100 at 42°C for 10 min. To remove excess probe, an additional 10 min wash is carried out in 100 mM NaCl, 1 M urea, 5% Triton X-100, and 1% dextran sulfate.

PCR Products

Because of the large number of specific sequence DNA copies produced using the PCR technique, relatively simple methods can be used to detect the amplified material. Sequence polymorphisms such as those found in the HLA region of nuclear DNA or the D-loop region of mitochondrial DNA can be detected using sequence-specific oligonucleotide (SSO) probes. Under appropriate hybridization and wash conditions, the SSO probes will hybridize only to perfectly matched complementary sequences. A different probe must be produced for each different sequence allele at a marker locus; this requires that the sequence of all alleles be known. The PCR product can be immobilized on a membrane and the SSO probes hybridized, or the probes can be immobilized and the PCR product hybridized.

VNTR alleles at a locus can be detected by size fractionation of the PCR-amplified alleles. Knowledge of the allele sequence is not required in this approach because only sufficient sequence data flanking the repeat regions are necessary for the production of primers for the PCR reaction (see Chapter 5).

PRINT DETECTION

Autoradiography is used to detect ^{32}P-labeled hybridization probes on the blot membranes. After the wash process is completed, the moist blots are covered with Saran Wrap, taken to a darkroom, placed with film in an X-ray cassette holder, and the cassettes placed in a –80°C freezer. The cassettes are lined with calcium tungstate intensifying screens (such as Dupont Cronex Lightening Plus) rather than the rare earth type. These screens together with the low temperature provide up to a tenfold increase in ^{32}P sensitivity.

More than one X-ray film, such as Kodak X-Omat XRP-1, or XAR5, can be taped to each blot. This technique facilitates removal of one film as a guide for determining the length of exposure required, as well as providing autoradiograms with different band intensities. Exposure time usually ranges from a few hours to days, depending on the probe SA, the degree of hybridization, and the quantity of target DNA bound to the membrane.

The films are developed in an automated processor, such as a Kodak M35A X-Omat, or by manually passing each film through develop, stop, fix, wash, and dry procedures. The film should be processed as rapidly as possible after it is removed from the freezer; otherwise, band resolution may be reduced owing to the accumulation of condensation.

HRP-labeled blots can be developed after the final wash step. The colorless TBM chromogen is enzymatically oxidized during the development process to form an insoluble blue precipitate. The blots are first prewetted for 5 min with 100 mM sodium citrate pH 5.0 containing 0.1 mg/ml 3,3'5,5'-tetramethyl benzidine (TMB). They are then placed in a solution of 100 mM sodium citrate, 0.1 mg/ml TMB and 0.4 mM H_2O_2 for 5 to 10 min. A final soaking for color development is performed for 30 to 60 min in 100 mM sodium citrate.

BLOT STRIPPING AND REPROBING

Nylon blots can be stripped and reprobed. The previously hybridized probe must be removed from the blot (the probe-target DNA hybrids denatured) or, provided a different probe is to be used, the radioisotope must have sufficiently decayed to ensure that excessive background will not develop. A protocol followed by the FBI (Budowle 1990) involves placing the membrane in a solution containing 55% formamide, 2× SSPE, and 1% SDS at 65°C for 45 to 90 min. The membrane is then rinsed for 1 min at RT in a solution containing 0.1× SSC and 0.1% SDS. According to the *Promega GenePrint DNA Typing Technical Manual* (Promega Corporation), over 80% of the initial hybridization blot signal is obtained when the blot is hybridized after five strippings. The blots should be stored in plastic wrap at freezer temperature.

DATA PROCESSING

The identification of autoradiogram allele fragments by size, the processing of the resultant data for print comparisons, and the storing of data for future requirements are key aspects of the DNA analysis process. Computer-directed gel scanners with data-handling programs are essential for enabling DNA profile service facilities to reduce the subjective aspects of band measurement and to streamline the tedious operation of profile comparisons.

Allele Fragment Size Determination

Recognition of a match in DNA identity analysis is technically demanding because of the difficulty of resolving the large number of alleles associated with the hypervariable loci profiled. Visual matching alone is not appropriate because of similar-size fragments and lack of an objective measurement base for accuracy, precision, and probability calculations.

When determining identity with two sample profiles, the FBI first uses a visual match which is later checked by computer-assisted image analysis after compensation for measurement imprecision. Factors affecting measurement imprecisions include band width, electrophoresis resolution, and electrophoretic mobility. It is anticipated that the effect of these will be considerably reduced when an analysis system that detects invariant bands is instituted. Because these bands are common to most individuals, they will serve as internal standards (Budowle 1990).

A camera scanner linked to a computer is usually used to digitize the data for band measurement. Allele sizes in a population can be determined by electrophoresis of identical-size fragments in different gel lanes and measurement of the variation in distance moved. Baird et al. (1986) determined that the error of these measurements, in their system, had a standard deviation (SD) of 0.6% of the fragment size. Matching rules can then be established. For example, two fragments are said to match if their positions do not differ by more than 3 SD in either direction.

The relation between the base pair (bp) lengths of double-stranded DNA fragments and their mobilities through an agarose gel can be used to estimate fragment size when the mobility distance through the gel has been measured (Southern 1979; Elder 1983, 1983*a*). The accuracy of this system depends on the mobility-length relation calculation used and the method of distance measurement.

The MW of double-stranded DNA fragments is inversely proportional to the mobility of the fragments through agarose gels, provided electrophoresis is performed at low voltage. Since MW is proportional to the DNA bp length, a straight line can be plotted for length (L) versus one/mobility ($1/m$) or calculated using linear regression. If the voltage is too high, curvature will develop. This can be corrected by including a factor (m_0) so that a plot of L versus $1/(m-m_0)$ is linear. The calculation of (m_0) is based on three standards:

L_1, L_2, and L_3, with mobilities m_1, m_2, and m_3.

$$m_0 = \frac{m_3 - m_1 A}{1 - A} \quad \text{where } A = \frac{L_1 - L_2}{L_2 - L_3} \times \frac{m_3 - m_2}{m_2 - m_1}$$

Accurate results are obtained when the standards span the range of unknowns to be measured. Plots can be constructed using the standards with the calculations described above. Alternatively, L or m can be calculated directly from the relation:

$$L = \frac{K_1}{(m-m_0)} + K_2 \text{ where } K_1 = \text{slope} = \frac{L_1 - L_2}{\dfrac{1}{(m_1-m_0)} - \dfrac{1}{(m_2-m_0)}}$$

$$\text{and } K_2 = \text{line intercept} = L_1 - \frac{K_1}{(m_1-m_0)}$$

A maximum bp length measurement error of 0.1% can be achieved when four standards are used in two sets of three and the length of the unknown is derived as a mean. The unknown is located between standards 2 and 3 in the ranked series 1, 2, 3, 4, where 1, 2, and 3 constitute one set and 2, 3, and 4 constitute the second set. This 0.1% error applies only when both standards and unknown have the same base composition and sequence. This is not usually the case and errors of up to 3% can then result. Better accuracy can be achieved when a few standards similar in size to the unknown are used, rather than a larger number of a considerably different size.

A number of methods are available for the physical measurement of DNA fragment mobility in agarose gels. The simplest, but least accurate, method is by metric ruler. The ruler is laid on the gel or the gel photograph and measurements made from the application slots to the bands. A digitizing tablet consisting of a cursor and a coordinate system will resolve distances to 0.1 mm with errors of up to 0.6% in DNA length. This system is superior to a ruler; however, the percent error may be unacceptable when fragments of similar size must be resolved.

The most accurate measurement method involves use of a computer-directed digital microdensitometer, whereby optical density profiles of the gel lanes containing the bands are produced and the peaks analyzed (Figure 6-13). This system is ideal for a computerized laboratory where the data can be readily processed for profile comparisons.

Digital Analysis

A DNA typing operation will be considerably streamlined if an automated optical scanner is incorporated into the system for analysis of autoradiograms (Sulston 1989). The scanner consists of a precision video camera mated to a microcomputer. This should result in faster and more accurate quantitative, objective matching of genotype patterns. The microcomputer should be capable of communicating with other computers for high-speed transfer of raw and processed images. The software is written to control the digitization, display images in multiple windows, and file, store, and retrieve data. Pattern-matching algorithms including statistical analysis are important in determining similarities and differences in banding structure on autoradiograms for different samples. A distortion correction system is also important to facilitate the comparison of patterns from different gels.

The FBI System

The sequence of operations performed by the system developed by the FBI (Monson 1988) includes generation of digital versions of the autoradiograms, location of lane boundaries, production of integrated intensity profiles, location of bands from intensity peaks, geometric correction for nonuniform fragment migration across the gel width, and calculation of fragment sizes. The generation of a digital version of the profiles, including size standards and controls, on an autoradiogram is prerequisite to computerized image analysis. This can be achieved with a video camera. In the FBI system, the video image is digitized at a geometric resolution of 480 lines by 512 columns and a gray scale resolution of 256 levels. An IBM PC can be mated to the camera for a total system cost of less than $12,000.

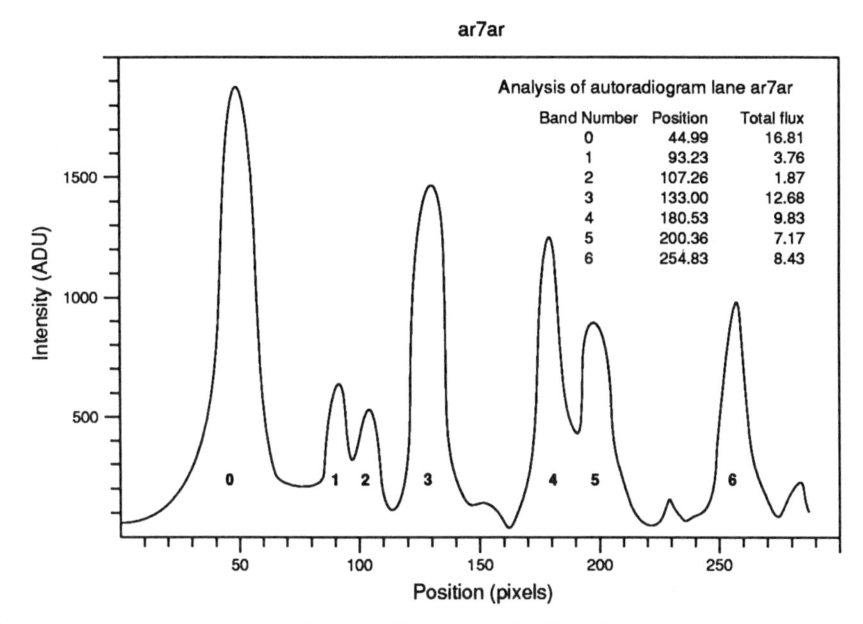

Figure 6-13. Optical density profile of a DNA fingerprint. Both band position and intensity are recorded.

A data base is necessary to manage the information resulting from image analysis of the autoradiogram profiles. The FBI has produced a data base of allele frequencies for different racial groups. (A different base is necessary for each probe hybridized.) These data can be used to estimate probabilities such as the probability that two DNA profiles match.

Interpretation of Profile Matches

The ability to declare a match between two profiles, such as DNA from an evidentiary semen stain and DNA from a defendant's blood specimen, can be influenced by many factors. These factors primarily revolve around the causes of band shifting and the corrections applied to compensate for the shifts. Band shifting is the phenomenon where DNA fragments in one lane of an electrophoresis gel migrate across the gel more rapidly than identical fragments in a second lane (Figure 6-14). Possible causes of the shifts are outlined in Table 6-2. It has been estimated that band shifting can occur in up to 30% of forensic cases. A shift can be determined by hybridizing target DNA with probes that detect monomorphic fragments of known size (as opposed to VNTR fragments). These constant-size fragments, such as the 2,731-bp Hae III fragment at locus D7Z2, appear to be present in every individual. If the constant bands are displaced, then band shifting has occurred.

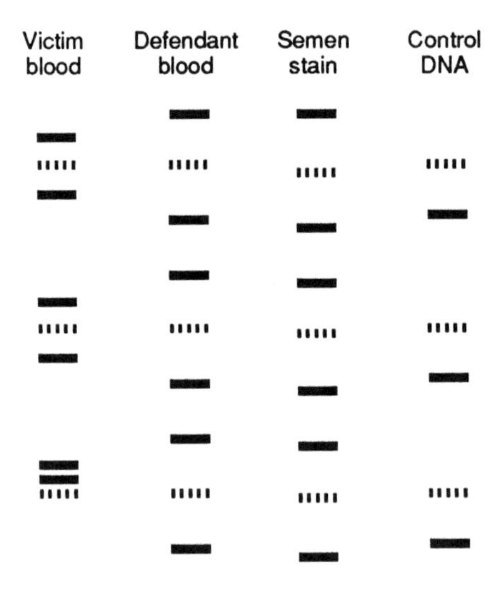

::::: Monomorphic control bands

▬ Profile VNTR bands

Figure 6-14. Diagrammatic composite DNA profiles for a forensic case. Note the shifting of bands (both monomorphic and VNTR) in the semen stain relative to the position of those in the defendant's blood specimen. The monomorphic bands in the stain profile have also shifted downward relative to those in the victim and control profiles.

Can a match be declared if two profile patterns are the same, but one profile is displaced in one direction? For example, all of the bands in an evidentiary stain profile appear slightly larger than the equivalent bands in a suspect's blood sample profile. Bruce Budowle of the FBI has suggested that if band shifting moves bands outside of the acceptable limits for declaring a match, then the forensic scientist must declare that the samples do not match, or that the evidence is inconclusive (Norman 1989). Lifecodes Corporation has attempted to apply a correction factor based on the shift of a "control" constant-size monomorphic band found in most individuals. A match is declared if after correction, the profile bands fall within the threshold required to declare a match.

Controversy has developed as to whether sufficient experimental work has been undertaken and peer reviewed to justify the use of band-shift correction factors in forensic cases (Anderson 1989, Norman 1989). Most experts agree that the degree of band displacement varies throughout the gel and that different correction factors

Table 6-2. Trouble Shooting Guide*

PROBLEM	*CAUSES*	*SOLUTION*
Barbell shape autoradiogram fragment bands	Loading solution adheres to the edges of the agarose wells	Ensure the wells are filled with loading solution.
Curved (◡) autoradiogram fragment bands	Wells overloaded with DNA	Reduce the quantity of DNA.
	Ethidium bromide stain in the gel	Stain the gels after electrophoresis.
	Excessive voltage during electrophoresis	Reduce the voltage.
	Electrophoresis buffer exhausted	Recirculate the buffer.
	Faulty electrophoresis apparatus	Replace or repair the unit.
	Gel overheating	Use refrigeration.
	EtOH in the DNA sample	Increase the sample drying time.
Diffuse autoradiogram bands	Excessive DNA loaded on gels	Reduce the DNA quantity. Ensure equal amounts are added per lane.
	Excessive gel thickness	Reduce thickness to 3–5 mm.
	Comb too thick	Use a thinner comb.
	Excessive voltage during electrophoresis	Reduce to 30–33 V.
	Excessive time from electrophoresis completion to start of Southern transfer.	Reduce the time to less than 30 min.
	Loose contact between the blot/X-ray sheet and intensifying screens	Clamp the cassettes to ensure close contact.
	Excessive exposure time of the blot to the X-ray sheet	Reduce the exposure time. Include "test" X-ray strips in the cassette.

PROBLEM	*CAUSES*	*SOLUTION*
	Electrophoresis buffer	For fragments above 3 kb use TAE buffer.
	Wash stringency too low	Increase wash temperature and decrease the salt concentration.
	Over exposure of some bands in multiband profiles	Use 2 or more X-ray sheets/blot. Remove the sheets at different times.
	Excessive probe genomic DNA mismatch; therefore, excessive cross reactivty	Use a more specific probe.
	EtBr incorporated into the gel	Stain with EtBr after electrophoresis.
DNA recovery poor	Cell membranes disrupted, DNAase released, low nucleated cell count	Request a fresh repeat specimen; do a cell count.
DNA insoluble in buffer	Excessive dehydration	Heat to 50°C for 10 or more min.
	EtOH in the DNA sample	Increase the sample drying time.
DNA spectrophotometric scan atypical	Phenol, protein, or RNA contamination	Reextract DNA, add DNAase-free RNAase.
	Visible spectrum cuvette used	Use a UV spectrum cuvette.
DNA appears degraded on gel	Old or improperly stored specimen	Request a repeat specimen.
	Nuclease contamination of DNA, buffers, other source	Prepare new reagents.
	An excessive amount of RNA present	Digest the sample with RNAase.
DNA gel lanes wavy	Excessive heat during electrophoresis	Refrigerate during electrophoresis or reduce voltage.
	Poor quality agarose	Replace agarose.

PROBLEM	*CAUSES*	*SOLUTION*
	Gradient pockets in the agarose gel	Agitate the agarose mixture while cooling prior to pouring.
	Contaminants such as salts in the samples	Dialyze the DNA.
DNA not digested	Faulty buffer or restriction enzyme	Check enzyme and buffer with a λDNA digest.
	Impurities in DNA	Reextract DNA with phenol/chloroform and reprecipitate in EtOH.
DNA band shifts	Quantity of DNA on the electrophoresis gel	Add equal amounts of DNA to each gel lane.
	DNA degraded	Obtain fresh specimens if possible. Consider PCR procedures to amplify short target sequences.
	Gradient pockets in the agarose gel	Agitate the agarose mixture while cooling prior to pouring the gel.
	EtBr incorporated in the agarose gel	Stain the gel by overlaying with EtBr when electrophoresis is completed.
	Electrophoresis buffer ion depletion	Reduce the gel running time. Recirculate or replace the buffer.
	Contaminants bound to the DNA	Reextract the DNA. Obtain a fresh specimen.
Incomplete DNA transfer	No Saran Wrap between nylon membrane filter, or transfer buffer ran dry	Repeat transfer, increase time, increase volume of transfer buffer.
	Poor quality agarose	Use only high-quality agarose.
	Air trapped between gel and filter	Press filter to the gel.
	DNA fragments too large	Depurinate DNA on the gel.

PROBLEM	*CAUSES*	*SOLUTION*
	Poor quality nylon membrane	Test membranes from a different lot number or supplier.
High background signal in DNA lanes	Posthybridization wash stringency too low	Increase stringency of final washes.
	Insufficient blocking agent in the hybridization solution	Ensure sufficient blocking agent is present.
High background over filters	Inadequate prehybridization	Repeat stringency washes, remove probe, and rehybridize. Use a different hybridization procedure. Wash new filters before hybridization in 1 × SSC, 0.1% SDS, RT, 15 min.
	Drying of filters	Ensure solution covers filter, and rotate or shake filter during hybridization. Keep filters moist after hybridization.
	Handling filters without gloves	Wear gloves.
	Cassette or film contamination	Test cassette and X-ray film; wear gloves.
	Insufficient blocking agent	Ensure denatured blocking DNA is added to the hybridization solution.
	Improper probe preparation	Ensure unincorporated nucleotides are removed from the labeled probe solution.
No bands detected after autoradiography	DNA not binding to membrane	Include ^{32}P-labeled λ DNA on gel; if not detected, change membrane and prepare fresh SSC.
	Check DNA transfer	Stain gels after transfer.
	No or incorrect insert in plasmid probe	Analyze probe insert size.

PROBLEM	*CAUSES*	*SOLUTION*
	Inadequate sensitivity of detection	Electrophorese 20–100 pg of probe in one lane of gel; if signal is obtained, prepare and characterize new probe. Repurify probe.
	Incorrect hybridization or wash solutions	Remake solutions.
	Specific activity of probe too low	Repeat labeling with fresh ^{32}P dNTP or fresh enzyme.
	Probe not denatured before hybridization	Denature probe.
	Improper X-ray film-intensifying screen combination	Match film with the cassette screen.
	Insufficient exposure time for the blot/X-ray at –80°C	Extend exposure at –80°C.
Unknown DNA bands	Microorganism or other DNA contaminants, such as mixtures of evidentiary specimens	Trace the source of contamination. Repeat the analysis with a new specimen if possible.

*A number of the suggestions in this table were recommended by R. Fourney of the RCMP Central Forensic Laboratory, Ottawa.

are required for the different-size fragments. This necessitates using a number of control monomorphic fragments spanning the size range of the profile bands.

To address such problems, as well as numerous other issues concerning DNA fingerprinting, the National Academy of Sciences recently appointed a committee to prepare guidelines for DNA technology in forensic science. The report is to be completed by the end of 1990.

PROFILE REPORT

Profile data must be prepared and presented in a form acceptable as evidence in a court of law (see Chapter 10). The following should be considered when preparing a report.

Test objectives (e.g., human forensic, parentage, poaching investigation):
Chain of custody (continuity):

(i) Details by date from sampling to final result
(ii) Disposition of remaining samples and blots

Materials used:

(i) Probes
(ii) Enzymes
(iii) Controls
(iv) Size markers
(v) Population allele frequency data base

Analysis results:

(i) DNA quantity and quality
(ii) Allele sizes detected
(iii) Profile comparisons

Special comments: outline of unusual aspects of the analysis
Conclusions: outline of match probability and other conclusions regarding the objectives of the specific analysis
Signatures of those involved with the analysis, including titles and dates

If requested by the court, a copy of the complete laboratory procedure manual should also be made available. This manual includes the quality assurance program, the assay protocols, and the population allele frequency data base work-up.

DNA PROFILE DATA BASES

The development of DNA profile data bases has received considerable attention from law enforcement agencies and some state legislatures during the past two years. The FBI and the Home Office have revealed plans for DNA fingerprinting files. The states of California, Colorado, Nevada, Virginia, and Washington are formulating or have already passed legislation that allows for the collection of blood from a person convicted of a sex offense, to be used to develop DNA profile data bases (Kearney 1989, Sattaur 1989). Ethical concerns regarding privacy and data security have also been voiced by a number of groups (see Chapters 10 and 11).

A document, "The Combined DNA Index System (CODIS): A Theoretical Model," was completed recently by the Data bases Subcommittee of TWGDAM (Kearney 1989). The proposed system initially would include

1. Profiles produced from the examination of DNA from randomly selected (unrelated) individuals to provide a statistical DNA population data base

2. Open-case DNA profiles
3. Convicted violent offenders' DNA profiles
4. DNA profiles of missing persons and unidentified bodies

The considerations addressed in the model include

1. Privacy and security
2. Justification
3. System organization and functional responsibilities
4. Data acceptance, data entry, and reports
5. Performance requirements
6. Resource requirements

Because of the sensitive nature of many aspects of the CODIS document, the text has been reproduced in Appendix II. The public should get a balanced view regarding the benefits and hazards of DNA profile data bases. People should not be stampeded or misguided by demagoguery on these important subjects.

POTENTIAL PITFALLS

Unavailability of suspect specimens is one of the greatest potential pitfalls in establishing DNA forensic services. Refusal by a suspect to provide a blood sample could render crime sample analysis worthless. (If a victim's body fluid stain is found on a suspect, it could be very incriminating and a suspect blood sample would not be required.) In the United States, the taking of a blood specimen from a suspect is permitted with a court order (see Chapter 10). In Canada and the United Kingdom, a suspect has the right to refuse.

Potential problems in DNA testing over which the analyst may have little control include insufficient specimen, as well as specimen degradation, contamination, and mutation. Also, the use of an appropriate statistical approach and availability of an adequate population allele frequency base for the group under consideration may be of concern. Sufficient population data should be available provided laboratories use the same probe series and restriction enzymes.

At least 10 to 50 ng of DNA are required per digest for hybridization with single-locus probes. This should be sufficient for testing with at least four different probes but may be insufficient for additional analysis. Specimen availability may be critical if the initial testing fails or a sample is required by another laboratory to confirm a result. Amplification may be considered if insufficient DNA is available for VNTR analysis. Court admissibility of this form of evidence should, however, be carefully considered before embarking on the amplification process. This information may be determined from court transcripts of previous cases and from discussions with those involved.

DNA degradation, especially in rape cases, is not uncommon. On average, 50 percent of rape samples submitted to one commercial laboratory were deemed untestable. Vaginal swabs taken 1 h to 48 h postcoitus were apparently unanalyzable in 40% of the cases. Warm and moist conditions are not conducive to preserving intact DNA; dry, cool conditions are best. Amplification of highly degraded DNA (e.g., to fragments only a few hundred base pairs in length), can be very successful and may be the only alternative, provided the court acceptability caveat is considered.

Contamination, in general, should not present a problem; however, the analyst must be aware of this possibility. DNA from an extraneous source exists if more than two bands are detected with a single-locus probe and the restriction enzyme does not digest within an allele. Screening tests for microorganisms and other nonhuman DNA can be performed, and techniques are available to separate sperm DNA from that of vaginal epithelial cells.

Mutation of an allele is of concern in DNA transmitted to an offspring because all hereditary material in the offspring will carry the new size fragment. The average mutation rate at any locus in the human has been estimated at 1 in 100,000 per generation. This is insignificant in terms of DNA typing for parentage testing; however, it must be remembered that this value is an average and a mutation rate of 5% has, for example, been recorded at the D1S7 locus. A probe that recognizes alleles at this locus may be valuable in forensic analysis, as is MSI, but it would be disastrous for use in paternity testing.

Although a number of pitfalls must be considered when preparing DNA identity profiles, the possibility of obtaining a false positive result (a match when none in fact exists) because of insufficient DNA quantity or quality is negligible. "No result" is the outcome in these situations. A false negative result could arise if a DNA contaminant is present in the probe and only one of the specimen pairs. Careful control must be maintained for technical factors, such as ensuring that the full range of small to large DNA fragments are transferred and are resolvable, and that a large quantity of DNA in one lane does not contaminate an adjacent lane containing only a small amount of specimen.

It is critical to ensure that bands match when comparing crime and suspect profiles. This assurance, as previously discussed, may best be achieved by analyzing "control" monomorphic bands . The analysis of DNA mixtures, such as crime and suspect, is discouraged. The analyst should be aware of the possibility that irreversibly bound contaminants on DNA from evidentiary stains and other factors, such as DNA concentration, could alter electrophoretic migration when mixed with pristine DNA from a suspect. Even though the DNA samples were derived from the same individual, the profiles may not be superimposable.

Small populations, especially isolates, may present a problem in terms of skewed allele frequencies. The statistical power of DNA identity analysis is based on the low probability that different individuals share rare alleles at a number of loci by chance. This is valid for groups such as the nonisolated Canadian population

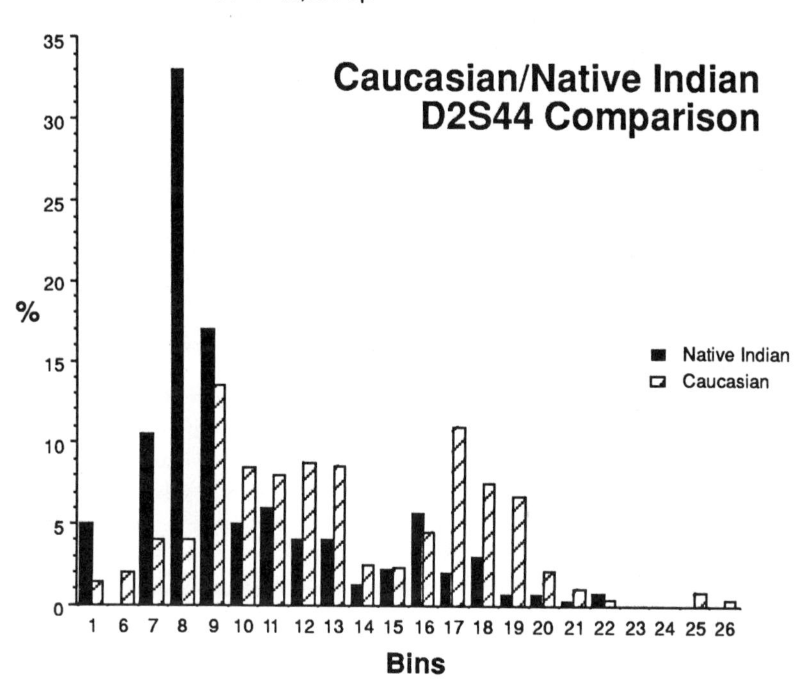

Figure 6-15. Caucasian/Native Indian allele frequencies for the D2S44 locus. The alleles were grouped into bins defined by specific size markers. (Courtesy of R. Fourney, RCMP Central Forensic Laboratory, Ottawa.)

with a low coefficient of inbreeding at 0.00004 to 0.0007. Isolated groups, however, may have relatively high coefficients, such as the Hutterites and Pennsylvania Dunkers in the United States at 0.02 and 0.03 respectively, with resultant fewer alleles but at higher frequencies. An example of allele frequency distribution for the D2S44 locus in two Canadian populations is outlined in Figure 6-15. DNA samples from 700 individuals in the nonisolated Canadian Caucasian population and from 250 individuals in an isolated Canadian Native Indian population were digested with Hae III and hybridized with the probe YNH24. The data bases for these populations are considerably different so that the groups must be considered separately in terms of allele frequency calculations (Fourney 1990). Determination of test loci allele frequencies in all populations of concern should provide the long-term solution to this quandary. In the interim, use of conservatively high frequency figures and testing of a larger number of loci will be required.

TROUBLESHOOTING GUIDE

The ability of an analyst to effectively troubleshoot technical problems is key to the efficient operation of a DNA laboratory. Typical problem areas, together with causes and solutions, are outlined in Table 6-2.

ANALYSIS COST

The cost of sample processing and analysis varies with the type of specimen and the number of specimens simultaneously processed. Laboratory overhead, including equipment depreciation, accounts for up to 50% of the costs. If a privately funded laboratory is in the business of DNA profiling, a profit is also necessary to provide the investors a satisfactory return on capital.

The largest number of samples feasible should be processed to achieve maximum efficiency. An automated DNA extractor can accommodate up to eight specimens per cycle with two to three cycles per 24 h. At least 15 DNA samples can be processed on each electrophoresis agarose gel and subsequently Southern blotted, hybridized, and autoradiographed. The following is a range of fees (1990) charged by larger, established U.S. corporations:

Processing (sample recording, extraction, and determination of suitability for analysis)—$40 to $75 per specimen
Paternity (whole blood) testing—$120 to $200 per specimen
Forensic (tissue or dried stains) analysis—approximately $300 per specimen
Expert witness fees—over $1000 per day plus expenses.

ANALYSIS REAGENTS

Denhardt's solution (10×):

5.0 g	BSA (Sigma fraction V)
5.0 g	Polyvinylpyrrolidone (40,000 MW)
5.0 g	Ficoll (MW 400,000)

Dissolve these components in 500 ml of distilled H_2O. (This solution should be prepared fresh.)

Herring sperm stock (denatured) (10 mg/ml)

10 mg	herring sperm

Dissolve to a volume of 1 ml in sterile distilled water. Herring sperm DNA must be denatured prior to use. This can be accomplished by boiling in a water bath for 5 min.

PEG (50%)

50 g	Polyethylene glycol (MW 8,000)

Dissolve to a volume of 100 ml in distilled water.

SDS (20%):

20 g Sodium dodecyl sulfate

Dissolve to a volume of 100 ml in distilled water. (Heat to approximately 65°C to dissolve.)

SSPE (20×):

210.4 g NaCl (3.6 M)

27.6 g $NaH_2PO_4 \cdot H_2O$ (0.2 M)

7.4 g $Na_2EDTA \cdot 2H_2O$ (0.02 M)

Dissolve and titrate to pH 7.0 with NaOH. Bring to 1.0 liter with distilled H_2O.

SSC (20×) (autoclave):

175.3 g NaCl (3.0 M)

88.2 g $Na_3citrate \cdot 2H_2O$ (0.3 M)

Dissolve in 800 ml distilled H_2O. Adjust to pH 7.0 with HCl. Bring to 1.0 liter with distilled H_2O.

TBE (10×):

109 g Tris (0.9 M)

9.31 g EDTA (25 mM)

55.65 g Boric Acid (0.9 M)

Dissolve and adjust to pH 8.2 with concentrated HCl. Bring to 1.0 liter with distilled H_2O.

TAE (50×) (autoclave):

242 g Tris-base (2.0 M)

57.1 ml Glacial acetic acid

100.1 ml EDTA (0.5 M) pH 8.0

Adjust the volume to 1.0 liter with distilled H_2O.

TE buffer (autoclave):

1.21 g Tris-base (10 mM)

0.037 g $Na_2EDTA \cdot 2H_2O$ (0.1 mM)

Dissolve the Tris in 800 ml distilled H_2O and adjust the pH to 7.5. Add the EDTA and adjust the volume to 1.0 liter with distilled H_2O. Recheck the pH.

REFERENCES

Anderson A. 1989. DNA fingerprinting on trial. *Nature* 342:844.

Baird M, Balazs I, Giusti A, Miyazaki L, Nicholas L, Wexler K, Kanter E, Glassberg J, Allen F, Rubinstein P, and Sussman L. 1986. Allele frequency distribution of two highly polymorphic DNA sequences in three ethnic groups and its application to the determination of paternity. *Am. J. Hum. Genet.* 39:489–501.

Baum HJ and McKee R. 1989. The use of a sensitive chemiluminescent DNA detection system for paternity and forensic identifications. *Am. J. Hum. Genet.* 45(Suppl):0677.

Berger SL and Kimmel AR, eds. 1987. Guide to Molecular Cloning Techniques. In *Methods in Enzymology*. Vol 152. Academic Press, New York.

Budowle B and Baechtel FS. 1990. Modifications to improve the effectiveness of restriction fragment length polymorphism typing. *Appl. Theoret. Electro.* (in press).

Budowle B and Monson KL. 1989*a*. A statistical approach for VNTR analysis. In *Proceedings—DNA Symposium.* International Symposium on the Forensic Aspects of DNA Analysis. Government Printing Office, Washington, D.C. (in press).

Bugawan TL, Saiki RK, Levenson CH, Watson RM, and Erlich HA. 1988. The use of nonradioactive oligonucleotide probes to analyze enzymatically amplified DNA for prenatal diagnosis and forensic HLA typing. *BioTechnology* 6:943–947.

Cariello NF, Keohavong P, Sanderson BJS, and Thilly WG. 1988. DNA damage produced by ethidium bromide staining and exposure to ultraviolet light. *Nucleic Acids Res.* 16:4157.

Carrano AV, Lamerdin J, Ashworth LK, Wilkins B, Branscomb E, Slezak T, Raff M, deJong PJ, Keith D, McBride L, Meister S, and Kronick M. 1989. A high-resolution, fluorescence-based, semi-automated method for DNA fingerprinting. *Genomics* 4:129–136.

Cawood AH. 1989. DNA fingerprinting. *Clin. Chem.* 35:1832–1837.

Chambers GK. 1990. Running the perfect gel. *Fingerprint News* 2:4–6.

Dykes DD. 1988. The use of biotinylated DNA probes in parentage testing: Nonisotopic labeling and non-toxic extraction. *Electrophoresis* 9:359–368.

Elder JK, Amos A, Southern EM, and Shippey GA. 1983. Measurement of DNA length by gel electrophoresis. I. Improved accuracy of mobility measurements using a digital microdensitometer and computer processing. *Anal. Biochem.* 128:223–226.

Elder JK and Southern EM. 1983*a*. Measurement of DNA length by gel electrophoresis II: Comparison of methods for relating mobility to fragment length. *Anal. Biochem.* 128:227–231.

Feinberg AP and Vogelstein B. 1984. A technique for radiolabeling DNA restriction endonuclease fragments to high specific activity. *Anal. Biochem.* 137:266–267.

Fourney RM, Shutler GG, Monteith N, Bishop L, Gaudette B, and Waye JS. 1989. DNA typing in the Royal Canadian Mounted Police. In *Proceedings of an International Symposium on the Forensic Aspects of DNA Analysis.*

Fowler JCS, Skinner JD, Burgoyne LA, and McInnes JL. 1988. Improved separation of multi-locus hypervariable DNA restriction fragments by field inversion gel electrophoresis and fragment detection using biotinylated probe. *Appl. Theoret. Electro.* 1:23–28.

Hames D and Higgins S, eds. 1985. Nucleic Acid Hybridization. *IRL Press*, Oxford.

Johnson DA, Gautsch JW, Sportman JR, and Elder JH. 1984. Impoved technique utilizing nonfat dry milk for analysis of proteins and nucleic acids transferred to nitrocellulose. *Gene Anal. Techn.* 1:3–8.

Kearney JJ, Baechtel FS, Budowle B, Forsen GE, Guerrieri RA, Kahn MR, Konzak KC, Monson KL, Wanlass SA, and Waye JS. 1989. A combined DNA index system (CODIS): A theoretical model. An unpublished document prepared by the Database Subcommittee of TWGDAM.

McGourty C. 1989. Profiles bank on the way. *Nature* 339:327.

Monson, KL. 1988. Semiautomated analysis of DNA autoradiograms. In *Crime Laboratory Digest*, U.S. Department of Justice, Federal Bureau of Investigation 15:104–105.

Norman C. 1989. Maine case deals blow to DNA fingerprinting. *Science* 246:1556–1558.

Pemberton J. 1989. Running a straight gel. *Fingerprint News* 3:5.

Sambrook J, Fritsch EF, and Maniatis T. 1989. *Molecular Cloning: A Laboratory Manual.* Vols 1, 2 and 3. 2nd ed. Cold-Spring Harbor, Cold-Spring Harbor, NY.

Sattaur O. 1989. Home Office reveals plans for DNA fingerprint files. *New Scientist* 122:32.

Signer E, Kuenzle CC, Thomann PE, and Hubscher U. 1988. Modified gel electrophoresis for higher resolution of DNA fingerprints. *Nucleic Acids Res.* 16:7739.

Southern EM. 1975. Detection of specific sequences among DNA fragments separated by gel electrophoresis. *J. Mol. Biol.* 98:503–527.

Southern EM. 1979. Measurement of DNA length by gel electrophoresis. *Anal. Biochem.* 100:319–323.

Sulston J, Mallett F, Durbin R, and Horsnell T. 1989. Image analysis of restriction enzyme fingerprint autoradiograms. *Comput. Applic. Biosci.* 5:101–106.

Uitterlinden AG, Slagboom PE, Knook DL, and Vijg J. 1989. Two-dimensional DNA fingerprinting of human individuals. *Proc. Natl. Acad. Sci. USA* 86:2742–2746.

Urdea MS, Warner BD, Running JA, Stempien M, Clyne J, and Horn T. 1988. A comparison of non-radioisotopic hybridization assay methods using fluorescent chemiluminescent and enzyme labeled synthetic oligodeoxyribonucleotide probes. *Nucleic Acids Res.* 16:4937.

Vassart G, Georges M, Monsieur R, Brocas H, Lequarre AS, and Christophe D. 1987. A sequence in M13 phage detects hypervariable minisatellites in human and animal DNA. *Science* 235:683–684.

Waye JS, and Fourney RM. 1990. Agarose gel electrophoresis of linear genomic DNA in the presence of ethidium bromide: Band shifting and implications for forensic identity testing. *Appl. Theoret. Electro.* (in press).

CHAPTER 7

Probes, Allele Mutations, and Restriction Enzymes

Positive identification is the ultimate objective of forensic analysis of blood and other tissue specimens. Nucleotide probes can be very effective tools for detecting genetic markers in this identification process. The genetic markers should be highly polymorphic; allelic variants should be easily and readily detectable; if amplification is required, the alleles should be efficiently amplified using PCR technology; and a statistically sound estimate of the population allele and genotype frequencies should be available.

Probes are single-stranded fragments of DNA or RNA containing the complementary code for a specific sequence of genome bases. Probes available for DNA profile analysis will, no doubt, eventually number in the hundreds. Currently, the most valuable detect tandem repetitive sequence fragments either at a specific locus under high-stringency analysis conditions or at numerous loci under low-stringency conditions. Each locus consists of many possible alleles with frequencies that vary depending on the specific population. Other factors also enter into the selection of probes, including ease of amplification, stability, cross-reactivity, and general availability.

Rate of allele mutation is also a prime consideration in probe selection. Mutation can be considered at two levels: as the basis for the large number of tandem repeat (VNTR) alleles formed during evolution and as a possible reason for spurious unassignable bands in typing analysis. Although highly unlikely, somatic mutations may be of concern in forensic testing if DNA from different tissues, such as blood and hair roots, are being matched. Germ line (gamete) mutations must be considered when parentage analyses are undertaken. These situations could give rise to false negative results and, therefore, false exclusions. Different considerations also apply for single versus multilocus probes. If a band that is not seen in the putative father is detected in an offspring, the man could incorrectly be excluded if the single-locus probe approach is used. This situation would necessitate testing with more than the usual four or five probes. Similarly, a somatic mutation in one body tissue could be interpreted as the suspect and crime samples not deriving from the same person. One unassignable band on a multilocus print would be very suggestive of a mutation and not of a true biological difference in the specimen source. This would also not be a valid indication of nonpaternity.

Lastly, a number of different restriction enzymes can be used, provided the enzyme does not cleave within the tandem repeat units and is not blocked by recognition site methylation. Other considerations when choosing an endonuclease include stability, availability, and price. Service laboratories, especially those involved in human forensic analysis, should narrow the number of enzymes and probes used. This will simplify not only the technical procedures, but also facilitate data exchange between analysis centers. One enzyme such as Hae III and a series of six probes comparable to α-globin 3'HVR and the VNTRs produced in R. White's laboratory (Nakamura 1987) should suffice.

PROBES

Polymorphism

DNA polymorphisms that arise due to the presence or absence of a restriction enzyme recognition site, that is, a polymorphic site, provide useful codominant markers in genetics. These RFLPs have low heterozygosity relative to those formed by variation in the number of tandem repeat (VNTR) units between fixed (constant) restriction enzyme sites. Heterozygosity, because of the polymorphic enzyme site between two constant sites, reaches a maximum of only 50% in the absence of selection. (If the frequency of alleles p and q are each 0.5, according to the Hardy-Weinberg law, $p^2 + 2pq + q^2 = 0.25 + 0.50 + 0.25$. The frequency, $2pq$, of heterozygotes is 0.5.) Approximately 1 per 150 base pair sites in noncoding DNA is hypothesized to be polymorphic; however, the overall variability in human DNA is low with a mean heterozygosity per base pair of about 0.2 percent. Probes to detect these markers, although useful in pedigree analysis for certain genetic diseases, are not very informative for DNA profiling. Hypervariable sites, on the other hand, because of the large number of different alleles and high heterozygosity, may be extremely informative.

Hypervariable regions (HVR), also referred to as minisatellites or variable number of tandem repeats (VNTR), consist of core tandem repeat sequences where hypervariability results from changes in the number of repeat units (Figures 7-1, 7-2). From observations to date, it appears that hundreds if not thousands of these loci are scattered throughout each animal genome.

Tandem Repeat Single-Locus versus Multilocus Probes

Depending on the probe specificity and stringency of the analysis conditions, some probes used in profiling will hybridize to only a single locus or many loci simultaneously. Under stringent wash conditions, the α-globin 3'HVR probe, consisting of tandem 17-bp repeat units, hybridizes to a single locus on chromosome 16, 3' to the α-globin complex. Under low stringency and with deletion of carrier DNA from the prehybridization and hybridization solutions, this probe will loosely hybridize with alleles at many autosomal loci. The result is a profile with an average of more than 15 detectable fragment bands in the 4- to 23-kb size range. A maximum of 2 bands are detected under high-stringency conditions. The sequences at the different loci are closely related but are not identical; therefore, under low-stringency conditions, some base pair mismatch with the 3'HVR probe is tolerated, resulting in multiband profiles (Fowler 1988, Higgs 1986, Jarman 1986). Probes such as pYNH24, produced in R. White's laboratory, are mainly used as single-locus markers. Many of the loci detected are highly polymorphic with more than 30 alleles and 95 percent heterozygosity.

The minisatellite system with probes 33.15 and 33.6, developed by Alex Jeffreys (1987), detects alleles at multiple autosomal loci. The allele repeat units, 16 to 64 bp long, have in common 11- to 16-bp cores. The original tandem repeat of a 16-bp core and 17-bp noncore was isolated from a myoglobin intron. The probes were constructed of only core sequence repeats. Mismatch between the probes and alleles is the rule. In many laboratories, this has resulted in assay difficulties in terms of band detection and reproducibility. Because the possible alleles at any specific minisatellite locus differ mainly in the number of tandem repeats, it is possible to develop a different specific probe for each locus (Wong 1986, 1987). These probes can then be used in single-locus analysis.

Although the probes described have been used primarily for human analysis, the minisatellite, 3'HVR, a 'per' gene sequence (Georges 1987), and a bacteriophage M13 sequence (Medeiros 1988, Vassart 1987, Westneat 1988) also cross-react with genomes from other animals. Good quality profiles using 33.15 and 33.6 can be produced for dogs, cats, birds, and fish. These probes are, however, not suitable for artiodactyls (such as sheep, goats, pigs, and cows). When used as a multilocus probe, α-globin 3'HVR will detect DNA fragments in cows, horses, and dogs, as

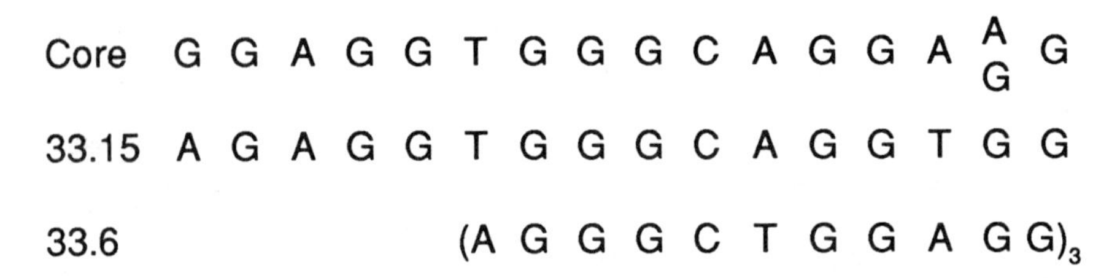

Figure 7-1. Sequence of Jeffreys' 33.15 and 33.6 minisatellite probes.

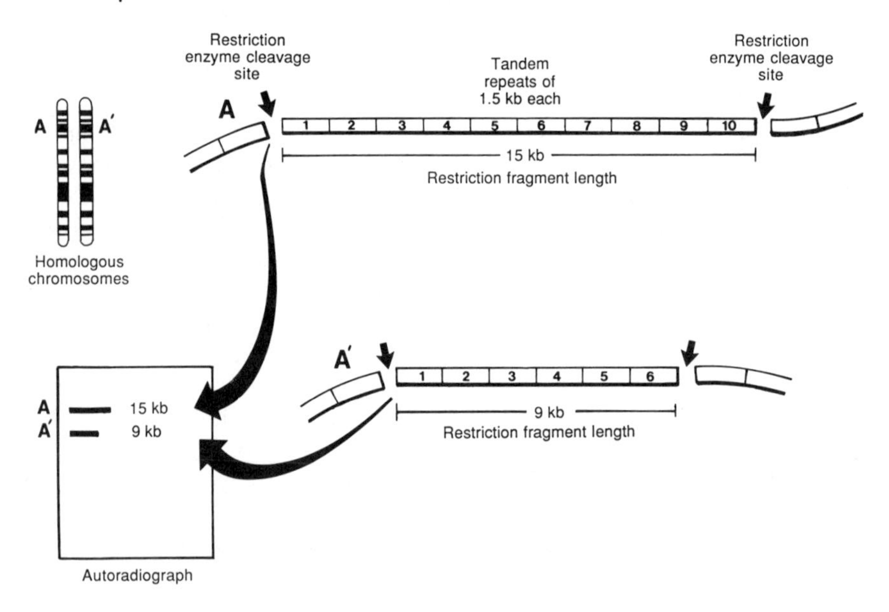

Figure 7-2. Alleles (A and A') at a specific locus on a homologous pair of chromosomes. Allele A consists of 10 tandem repeats and allele A' consists of 6 tandem repeats. The restriction enzyme cleaves the DNA at fixed sites in the flanking regions of the alleles to generate 15-kb and 9-kb fragments.

well as in big game animals such as deer, moose, and elk. Wild-type bacteriophage M13 contains a hybridizing sequence consisting of two clusters of 15-bp repeats in the protein II gene. These sequences complement with DNA from human, equine, bovine, canine, deer, moose, and elk, although, as with 3'HVR, the multiple banding pattern may be faint and overall service value remains to be determined.

The M13 tandem repeat has recently been used to screen a Charon 40 human chromosome 16 library. A 4.5-kb fragment from one of over 70 clones isolated has been very useful as a probe (pV47-2) in hybridizations with Hae III-digested genomic DNA from a cross section of wildlife and domesticated animal species

(Longmire 1990 and personal communication). This probe is more sensitive than the wild-type bacteriophage M13, resulting in improved band resolution and greater usefulness in animal studies (such as determining relatedness in endangered species). Mitochondrial probes can be useful for analyzing maternal inheritance in humans and other animals. These probes are also valuable for species identification (Avise 1989).

There are advantages and disadvantages to the single- and multilocus probe systems (Table 7-1). The main power of the multilocus approach resides in the very large number of fragments profiled, whereas, that of the single-locus system resides in the analyst's ability to determine allele frequencies in different populations. Human forensic analysis involves almost exclusive use of single-locus probes—at least in North America. Twin zygosity testing, parentage analysis, and animal population studies lend themselves to both the single-locus and multilocus techniques. The main disadvantages of multilocus probing are band resolution and the resultant profile interpretation difficulties and the lack of population allele frequency data bases.

Oligonucleotide Probes

Under specific temperature and other hybridization and wash conditions, oligonucleotides (17-mers or greater) will anneal to genomic sequences, provided there is a perfect match (Ali 1986, Schafer 1988, 1988*a*). A single base pair mismatch will be destabilizing and prevent hybridization. When used in conjunction with nonradioactive labeling systems and amplified DNA, these probes may provide a powerful tool in forensic analysis, even for genomic material degraded to 50–150-bp lengths.

Sequence-specific oligonucleotide (SSO) probes have been successfully used in dot-blot and reverse dot-blot formats with PCR-amplified allelic variants from the second exon of the human HLA-DQ α locus (von Beroldingen 1989). In situations where an oligonucleotide probe specifies a specific allele (ASO), that single probe is sufficient to identify the allele. If, however, the probe recognizes the identical sequence in more than one allelic variant, a second or third specific sequence probe will be required to uniquely identify the alleles. In this situation, patterns of probes define the different alleles.

Amplification and Stability

Some repetitive sequence probes are unstable and difficult to amplify. It is, therefore, important to ensure that the probe being used is the original as constructed (see Chapter 9). Amplification problems may be due to mutation or to probe degradation. Unless a laboratory is prepared to troubleshoot such problems, it is most efficient to purchase, if possible, the required materials from a supplier who is well-recognized and operates under a stringent quality assurance program.

Table 7-1. Comparison of Single-locus and Multilocus Probes

Single-locus	vs	*Multilocus*
Target DNA consists of VNTR (minisatellite) alleles.		Target DNA consists of VNTR (minisatellite) alleles.
At least 10–50 ng DNA/digest.		At least 0.5–1.0 μg DNA/digest. Insufficient DNA is available for many forensic specimens.
One or two fragment bands/ hybridization; therefore, usually clear band resolution.		Over 15 bands/hybridization; therefore, band resolution may be reduced.
Bands can be interpreted to 500-bp size. Probes can be targeted to smaller size—for example, 0.5- to 4-kb fragments. Some DNA degradation can, therefore, be tolerated with continued good band resolution.		Bands can be interpreted to about 4-kb size. DNA degradation cannot be tolerated without loss of resolution.
Mixed samples should be obvious since more than two bands are usually detected.		Mixed samples may be difficult to interpret.
Probability calculations are based on population allele frequency data.		Probability calculations are not based on population allele frequency data because the loci are not defined.
Probes are usually species-specific.		Probes will often cross-react with a wide range of species. This may be advantageous for animal studies; however, interpretation problems could arise with contaminated forensic specimens.
Usually must hybridize with four or more different probes for well-defined non-linked loci.		Hybridize with only one or two probes for undefined loci that may be linked.
Use blocking DNA and high stringency hybridization and wash conditions. These conditions result in the hybridization of only closely matched probe and target DNA.		Blocking DNA is not used and low-stringency hybridization and wash conditions are required. These factors facilitate a probe-target DNA mismatch rate of 30–40%.

Availability

Probes for research purposes are generally available. However, barriers, such as patent restrictions, have been encountered when requests were made for some of the same probes for service use.

There are obvious disadvantages and perhaps some advantages to restricting probe availability. The ingenuity of researchers has been tapped in efforts to circumvent such obstacles by developing alternative materials at a more rapid pace. There are now many excellent probes for profile analysis. These are usually available with the payment of royalties.

A laboratory about to embark on service genotyping is well-advised to select fully characterized probes with long-term availability at reasonable prices. Characterization includes chromosome locus location (as recorded by the Yale Gene Library or the International Human Genetic Mapping Workshops), size, rate of mutation, number and size distribution of alleles detected, degree of heterozygosity, and cross-reactivity. The probes should be in widespread use and under an external quality control program. A population data base of allele frequencies for each probe should be available to all laboratories.

Probes Used in Forensic Analysis

A summary of probes used currently in forensic analysis is outlined in Table 7-2. The final probes used by the RCMP in specimen analysis include those for loci D4S139, DYZ1 (specific to the Y chromosome), and D7Z2 (Fourney 1989). These probes have high sensitivity and remain useful even if DNA is lost from the membranes during stripping. Male and female control DNA samples are included on every blot. These provide marker lanes with a specific pattern for each probe; they also provide controls for sex. Locus D7Z2 provides an internal monomorphic marker band (2,731-bp fragment with an Hae III digest) for evaluation of electrophoretic mobility of the DNA across the agarose gel. Monomorphic markers are also useful in determining the accuracy and precision of molecular size calculations.

In the future, probes for loci such as D11S129 and D20S15 that detect fragments in the 0.8- to 2.3-kb range could be very useful when used in conjunction with high-resolution techniques employing rehydratable polyacrylamide silver stained gels (see chapter 5).

Table 7-2. Examples of Probes Used in Forensic Analysis

Probe	*Locus*	*Source*
MS1	D1S7	Cellmark
YNH24	D2S44	Promega/GenMark
pH30	D4S139	Genelex/BRL
3'HVR	D16S85	Collaborative
VI	D17S79	Lifecodes

MUTATIONS

Allele Formation

As noted previously, the raw materials of genotyping are for the most part the VNTR alleles located at numerous loci throughout the genome. The different alleles located at a specific locus may have formed during meiosis by unequal crossing over or replication slippage (Chandley 1988, Jeffreys 1988, Royle 1988). Because newly formed alleles are inherited in a Mendelian fashion, they become part of the hereditary material if transmitted to offspring. The identical or almost identical repeat units of an allele can number in the thousands. These alleles have no known function in terms of body chemistry. If there is a function, the addition or deletion of a few repeat units certainly appears to be without effect. This perhaps accounts for the ability of the genome to accommodate the large number of alleles and the relatively high mutation rate at some loci. More than 70 alleles are present at the D2S44 (probe pYNH24) locus, and a mutation rate of at least 5 percent per gamete exists at the D1S7 locus (probe MS1).

Genotype Analysis

Probes that hybridize at relatively unstable loci should not be used. Mutation rate often increases with the degree of heterozygosity at a locus. As heterozygosity approaches 100 percent, the analyst must be more cognizant of possible fragment length changes. Fortunately, the mean mutation rate as observed in multilocus systems is approximately 0.004 per fragment per gamete; thus, most loci are relatively stable.

Somatic mutations have been observed in two situations. The restriction enzyme sites for Sau 3A and Hinf I were blocked by methylation in one tissue but not in another from the same individual and tumors will often display numerous mutated sites. In the latter case, it may be possible to trace cancer development by studying these changes.

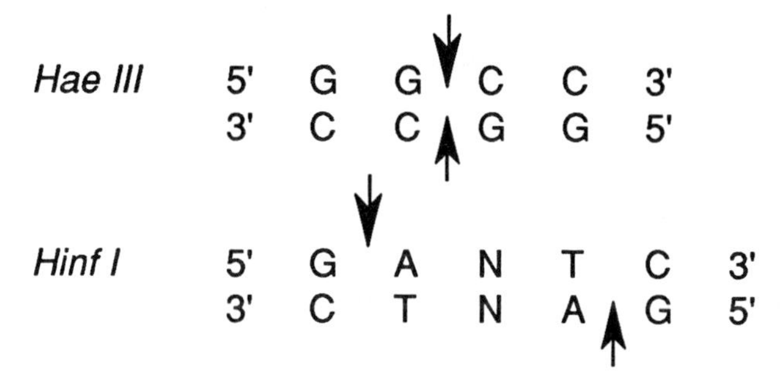

Figure 7-3. Restriction enzymes Hae III and Hinf I recognition sequences.

RESTRICTION ENZYMES

Recognition Site Features

It is imperative that the enzyme or enzymes chosen in conjunction with specific probes not have recognition sites within the tandem repeat units. Cleavage sites within the alleles may result in the production of short uninterpretable fragment patterns. Endonucleases with recognition sequences flanking the alleles and generating fragments 2 to 15 kb or less in length are best. Two enzymes, Hae III and Hinf I (Figure 7-3), fit this criterion for many of the VNTR probes; however, Hinf I sites can be blocked by methylation (Figure 7-4) resulting in incomplete DNA digestion. Since DNA methylation is tissue specific, artifactitious differences could also arise if the enzymes used are sensitive to this phenomenon.

Fragment Resolution

Fragments should be produced that are easily resolvable by gel electrophoresis. Due to the nature of electrophoresis, small size differences in the fragments are more easily resolved using smaller DNA fragments. Enzymes such as Hae III with 4-bp recognition sites cleave DNA more often (1/256 bp) than enzymes such as Pst I and Pvu II with 6-bp recognition sites (1/46,656 bp) (Figure 7-5).

Sensitivity to Inhibitors and Stability

Certain restriction enzymes such as Eco R1 are sensitive to inhibitors. Others such as Pvu II may be quite labile and require exceptional care in storage and handling. Detailed descriptions of endonucleases are usually available in supply company catalogs; these should be consulted when in doubt.

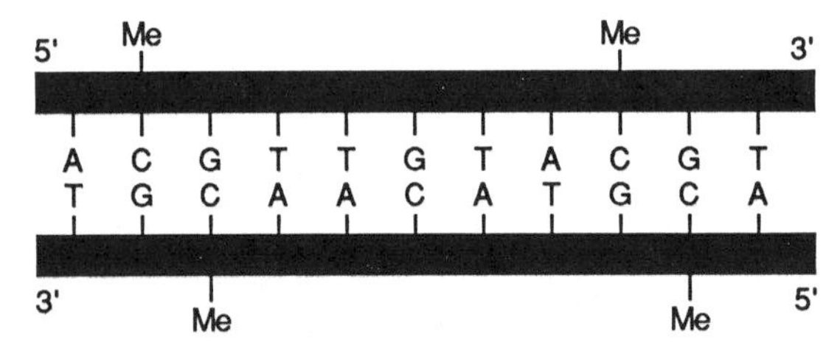

Figure 7-4. DNA methylation. In mammals, methylation of the 5' cytosine that immediately precedes guanine is the most frequent modification identified in DNA. This CpG dinucleotide is not present in the Hae III recognition sequence but may be present at sites of enzymes such as Hinf I.

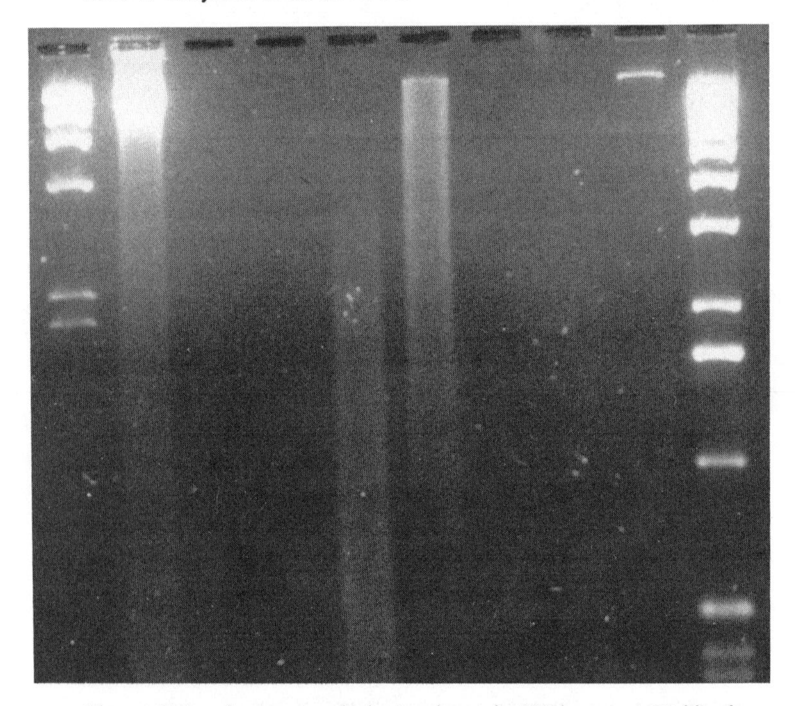

Figure 7-5. Agarose gel electrophoresis (16 h run at 30 V) of moose genomic DNA stained with ethidium bromide and viewed under UV irradiation. The lanes from left to right are (1) λDNA Hind III digest size marker, (2) 5 μg of undigested DNA, (3) blank, (4) blank, (5) 1.5 μg of DNA digested with Hae III (a four-cutter), (6) 1.5 μg of DNA digested with Pvu II (a six-cutter), (7) blank, (8) 5 μg of RNA, (9) 300 ng of undigested DNA, (10) BRL 1-kb size marker. The marker sizes are outlined in Figure 6-3. Note the broad band of undigested DNA (greater than 20 kb) in lane (2), the smears of DNA in lanes (5) and (6), and the diffuse spot of RNA at the bottom of lane (8).

Price and Availability

Enzyme availability is generally good and many biological supply companies stock a broad range of both common and uncommon restriction endonucleases. There is little sense in choosing an expensive enzyme when a much cheaper one will suffice.

The Enzymes Selected for Forensic Analysis

The FBI and the RCMP (Budowle 1990) have selected Hae III as the restriction enzyme of choice. This enzyme is compatible with the VNTR loci D2S44, D1S7, D4S139, D16S85, D17S74, D17S79, D14S13, D20S15, and many others. Because Hae III has a 4-bp recognition site (compared to other enzymes, such as Pst I, which has a 6-bp site), smaller and, therefore, more easily resolvable fragments are produced. Methylation does not occur in mammalian DNA Hae III recognition sites. The enzyme is robust. Human DNA can be digested in low to high ionic strength buffers, temperatures to 56°C are tolerated, and DNA contaminated with protein and other impurities is readily digested. Hae III is available commercially at competitive prices.

The Home Office Forensic Science Service in the U.K. uses Hinf I in conjunction with the multilocus probe 33.15. This enzyme provides the largest number of resolvable fragments of the endonucleases tested when used in conjunction with the 33.15 probe.

REFERENCES

Ali S, Muller CR, and Epplen JT. 1986. DNA fingerprinting by oligonucleotide probes specific for simple repeats. *Hum. Genet.* 74:239–243.

Avise JC, Bowen BW, and Lamb T. 1989. DNA fingerprints from hypervariable mitochondrial genotypes. *Mol. Biol. and Evol.* 6:258–269.

Budowle B, Waye JS, Shutler GG, and Baechtel FS. 1990. Hae III—A suitable restriction fragment length polymorphism analysis of biological evidence samples. *J. Forensic Sci.* (in press).

Chandley AC and Mitchell AR. 1988. Hypervariable minisatellite regions are sites for crossing-over at meiosis in man. *Cytogenet. Cell Genet.* 48:152–155.

Fourney RM, Shutler GG, Monteith N, Bishop L, Gaudette B, and Waye JS. 1989. DNA typing in the Royal Canadian Mounted Police. In *Proceedings DNA Symposium.* International Symposium on the Forensic Aspects of DNA Analysis. Government Printing Office, Washington, D.C. (in press).

Fowler SJ, Gill P, Werrett DJ, and Higgs DR. 1988. Individual specific DNA fingerprints from a hypervariable region probe: Alpha-globin 3' HVR. *Hum. Genet.* 79:142–146.

Georges M, Cochaux P, Lequarre AS, Young MW, and Vassart G. 1987. DNA fingerprinting in man using a mouse probe related to part of the Drosophilia 'Per' gene. *Nucleic Acids Res.* 15:7193.

Higgs DR, Wainscoat JS, Flint J, Hill AVS, Thein SL, Nicholls RD, Teal H, Ayyub H, Peto TEA, Falusi AG, Jarman AP, Clegg JB, and Weatherall DJ. 1986. Analysis of the human α-globin gene cluster reveals a highly informative genetic locus. *Proc. Natl. Acad. Sci. USA* 83:5165–5169.

Jarman AP, Nichols RD, Weatherall DJ, Clegg JB, and Higgs DR. 1986. Molecular characterization of a hypervariable region downstream of the human α-globin gene cluster. *EMBO J.* 5:1857–1863.

Jeffreys AJ. 1987. Highly variable minisatellites and DNA fingerprints. *Biochem. Soc. Trans.* 15:309–317.

Jeffreys AJ, Royle NJ, Wilson V, and Wong Z. 1988. Spontaneous mutation rates to new length alleles at tandem-repetitive hypervariable loci in human DNA. *Nature* 332:278–281.

Longmire JL, Kraemer PM, Brown NC, Hardekopf LC, and Deaven LL. 1990. A new multi-locus DNA fingerprinting probe: pV47-2. *Nucleic Acids Res.* (in press).

Medeiros AC, Macedo AM, and Pena SDJ. 1988. A simple non-isotopic method for DNA fingerprinting with M13 phage. *Nucleic Acids Res.* 16:10394.

Nakamura Y, Leppert M, O'Connell P, Wolff R, Holm T, Culver M, Martin C, Fujimoto E, Hoff M, Kumlin E, and White R. 1987. Variable number of tandem repeat (VNTR) markers for human gene mapping. *Science* 235:1616–1622.

Royle NJ, Clarkson RE, Wong Z, and Jeffreys A. 1988. Clustering of hypervariable minisatellites in the proterminal regions of human autosomes. *Genomics* 3:352–360.

Schafer R, Zischler H, and Epplen JT. 1988. $(CAC)_5$, a very informative oligonucleotide probe for DNA fingerprinting. *Nucleic Acids Res.* 16:5196.

Schafer R, Zischler H, Birsner U, Becker A, and Epplen JT. 1988*a*. Optimized oligonucleotide probes for DNA fingerprinting. *Electrophoresis* 9:369–374.

Vassart G, Georges M, Monsieur R, Brocas H, Lequarre AS, and Christophe D. 1987. A sequence in M13 phage detects hypervariable minisatellites in human and animal DNA. *Science* 235:683–684.

von Beroldingen CH, Blake ET, Higuchi R, Sensabaugh GF, and Erlich E. 1989. Applications of PCR to the analysis of biological evidence. In *PCR Technology Principles and Applications for DNA Amplification*, 209–223. Erlich HA, ed. Stockton Press, New York.

Westneat DF, Noon WA, Reeve HK, and Aquadro DF. 1988. Improved hybridization conditions for DNA 'fingerprints' probed with M13. *Nucleic Acids Res.* 16:4161.

Wong Z, Wilson V, Jeffreys AJ, and Thein SL. 1986. Cloning a selected fragment from a human DNA fingerprint: Isolation of an extremely polymorphic minisatellite. *Nucleic Acids Res.* 14:4605–4616.

Wong Z, Wilson V, Patel I, Povey S, and Jeffreys AJ. 1987. Characterization of a panel of highly variable minisatellites cloned from human DNA. *Ann. Hum. Genet.* 51:269–288.

CHAPTER 8

Probability and Statistical Analysis*

Determination of the probability of specimen match and estimation of population allele frequency distributions are two key areas of DNA profiling requiring probabilistic and statistical analyses. Statistical calculations can be tedious and slight changes in the wording of probability statements can result in vastly different meanings. It may, therefore, be prudent for the analyst to seek the advice of a qualified statistician before assembling population frequency data or submitting probability statements in a court of law.

Statistical methods provide powerful tools to assist with decision-making. Because of easy access to powerful statistical software on personal computers, these methods are easy to use. However, for this same reason they can also often be misused and questionable data presented in a favorable light. One must always be aware of those skilled in the misleading use of statistics or those who simply make incorrect statistical statements even though the quantity of base data may be considerable and the quality good.

The objective of this chapter is to review methods of probability and statistics relevant to DNA fingerprint analysis.

---

* This chapter was prepared in collaboration with Professor Martin L. Puterman, Faculty of Commerce, The University of British Columbia, Vancouver, B.C., Canada.

STATISTICAL METHODOLOGY

Random Sampling

Basic statistical data are usually derived from samples drawn from large populations. A population is a collection of individuals having stated features in common, such as all Orientals in the United States. A simple random sample is a subset of individuals selected from a population using a random choice mechanism (such as a random number table) which guarantees that all members of the population are equally likely to be chosen. Usually the sample size is denoted by n. Values in a sample are called observations and denoted by $X_1, X_2, \ldots, X_n$.

The features of interest in a population such as the true average IQ, or the true proportion of a specific allele at a given locus, are called parameters, while statistics are numerical summaries of data such as means or standard deviations. The science of statistics is concerned with inferring information about population parameters based on sample statistics.

Data Presentation

Data are best summarized in either graphical or tabular form. Common and useful graphical presentations include frequency distributions or histograms. When data can be classified into discrete categories, such as the identity of alleles at a specific locus, histograms (Figure 8-1) provide convenient representations of the allele frequencies. Figure 8-2 illustrates several commonly observed shapes of distributions, including symmetric unimodal, bimodal, skewed right, and skewed left.

Tabular presentations should always include appropriate measures of central tendency (means or medians) and dispersion (standard deviations or percentiles) as described in the following.

Summary Statistics

Measures of central tendency. The mean, median, and mode are measures of the center of the data. The (arithmetic) mean is the most common measure; it is computed by summing the data and dividing by the number of observations n. The Greek letter μ denotes a population mean and $\overline{X}$ a sample mean. When the population is finite, μ is the average over all items in the population; when it is infinite, it is a weighted average computed by integration. The quantity $\overline{X}$ is the average over all observations in the sample. It is computed by

$$\overline{X} = \frac{1}{n}\sum_{i=1}^{n} X_i$$

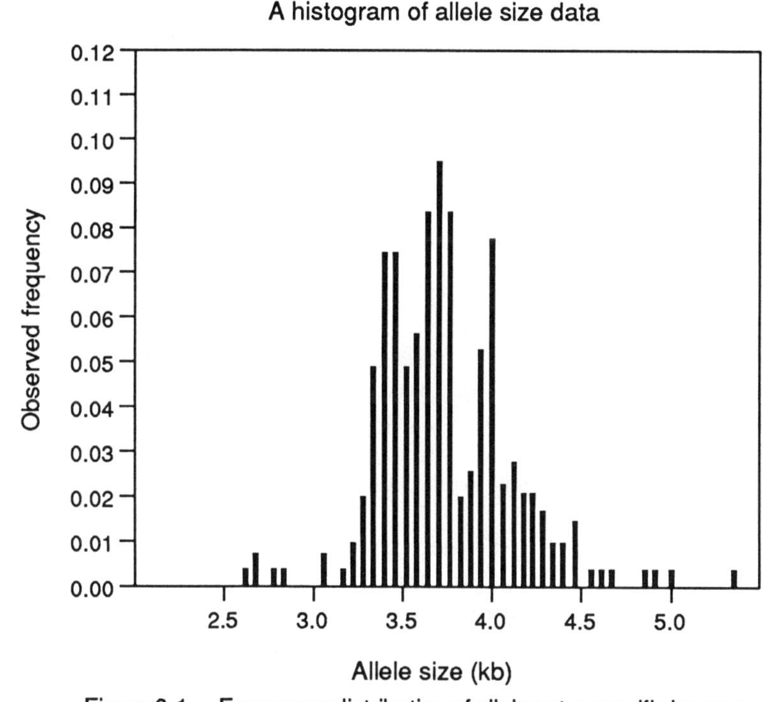

Figure 8-1. Frequency distribution of alleles at a specific human locus for a specific population. The frequency is the number of alleles of that size divided by the total number of alleles in the sample.

The median is the middle value among the sample observations in the population. Fifty percent of the values are above and 50% are below this point. If the number of data points is odd, the median is the middle value; if even, it is the average of the two central values. The median is found by arranging the data in ascending order and choosing the central value or the average of the two central values.

The mode is the value in a sample or population that occurs most frequently. Some distributions may be bimodal or multimodal. In fact, the mode has little practical significance. For data that are symmetric, the mean and median are good indicators of the center of the data. For skewed distributions, the median is preferable. When data are categorical, important summary statisitics include the proportion p of observations in each category.

Measures of dispersion. It is important to calculate the dispersion within a distribution as well as the mean or median because many distributions might have the same mean but the variation of observations around the mean might differ markedly. Measures of dispersion include the range, percentiles, standard deviation *(SD)*, and the coefficient of variation *(CV)*.

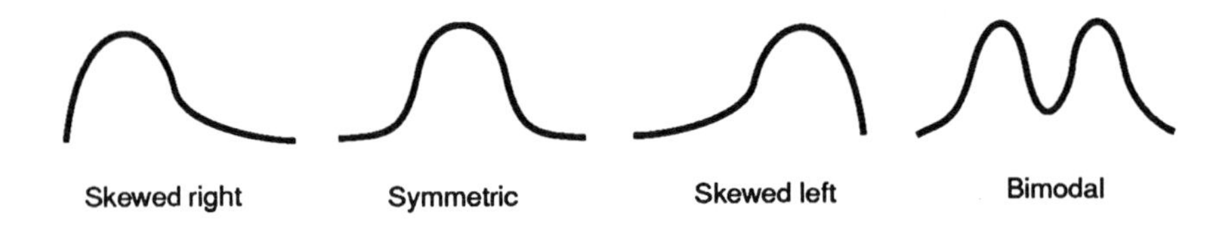

Figure 8-2. Commonly observed shapes of frequency distributions.

The range gives the smallest and largest numbers in a sample or population. The range provides a measure of spread but is overly sensitive to disparate measurements because it is based on only the two extreme values.

A percentile is the point in a sample or population below which a specific percent of values lie and above which the complementary percentage lies. The 95th percentile is defined to be the value which lies above 95% of the observations and below 5% of the observations.

The standard deviation is the most important measure of dispersion in a population or sample. It is the square root of the average of the sum of squared deviations from the mean. The population standard deviation (in a finite population) is defined by

$$\sigma = \sqrt{\frac{\sum_{i=1}^{n}(X_i - \mu)^2}{n}}$$

and the *sample standard deviation* by

$$s = \sqrt{\frac{\sum_{i=1}^{n}(X_i - \overline{X})^2}{n - 1}}$$

In a symmetric population (or sample), roughly 95% of the observations lie within two standard devations of the mean, and in any population (or sample), 75% of the observations lie within two standard deviations of the mean.

The coefficient of variation *(CV)* is a relative measure of dispersion expressed as a percentage. The sample coefficient of variation is given by

$$CV = \frac{s}{\overline{X}} \times 100$$

The other measures of variability, such as the *SD*, have magnitudes dependent on the scale of the data, whereas the *CV* is a ratio and is dimensionless.

Statistical Inference

The purpose of almost all of statistics is to infer something about a population based on observations in a sample. This is carried out through point and interval estimation, and hypothesis testing.

Point estimation. Sample quantities such as sample means, sample standard deviations, and sample proportions are used as point estimates of the analogous population quantities. They are called point estimates because they are single values with no associated measures of precision. Desirable properties of such estimators are that they use the data efficiently and are certain to be sufficiently close to the true parameter value if a sufficiently large sample is selected.

Interval estimation. Interval estimates or confidence intervals are useful for determining the precision of a point estimate. They are determined so that if an experiment or sampling procedure were repeated many times, the true parameter value would lie in the interval a prespecified proportion of the time. For example, a 95% confidence interval would be computed from a rule that ensured that in 95% of the samples, the interval would contain the actual parameter value.

When computing confidence intervals for population means, a key quantity is the standard error of the mean. It is often referred to as the standard error or *SEM* and is usually computed using the formula $s/\sqrt{n}$. It provides an idea of how variable a sample mean is as an estimate of the population mean. This quantity should not be confused with the sample standard deviation, which describes the variability of a single observation. The standard error is an important component in computing a $(1-\alpha) \times 100\%$ confidence interval. When the sample size is large enough (say greater than 50), confidence intervals are given by the formula

$$X - Z_{\alpha/2}\frac{s}{\sqrt{n}} \leq \mu \leq \overline{X} + z_{\alpha/2}\frac{s}{\sqrt{n}}$$

In this expression, the quantity $z_{\alpha/2}$ is the value a standard normal random variable exceeds with probability $\alpha/2$ (Table 8-1). For 90% confidence intervals, $z_{.05} = 1.645$, and for 95% confidence intervals, $z_{.025} = 1.96$. A 95% confidence interval for the population mean is approximately the sample mean plus or minus two standard errors.

Confidence intervals for a population proportion π are obtained from the sample proportion p using the formula

$$p - z_{\alpha/2}\sqrt{\frac{p(1-p)}{n}} \leq \pi \leq p + z_{\alpha/2}\sqrt{\frac{p(1-p)}{n}}$$

This formula applies only when the sample is sufficiently large so that np and n(1–p) both exceed five. If p were .02 and n were 100 so that $np = 2$, the interval would not be guaranteed to contain the true population frequency in $(1-\alpha) \times 100\%$

of the samples. As a rule of thumb, such confidence intervals are invalid when one wishes to estimate low frequencies. Methods based on the Poisson distribution or exact procedures are then preferred.

Calculation of allele frequency distributions. Allele frequencies for the specific population in question must be determined before the proportion of the population that are potential contributors of an evidentiary specimen can be calculated. Natural populations are often characterized by differences in allele and genotype frequencies in various geographic regions. Probability calculations, as discussed under the Hardy-Weinberg and Wahlund principles, must take account of this fact; otherwise, the resultant predictions will not be valid.

Depending on the confidence limits chosen, hundreds or even thousands of control tissue samples may have to be analyzed to determine the frequency of rare alleles. These specimens usually constitute only small samples from the specific populations; thus, an indicator of the confidence one has in the data estimated proportions must also be determined. A 95% confidence interval based on a rare allele frequency of $p = 1/50 = .02$ and $n = 1{,}000$ specimens is

$$.02 - 1.96\sqrt{\frac{(.02 \times .98)}{1000}} \leq \pi \leq .02 + 1.96\sqrt{\frac{(.02 \times .98)}{1000}}$$

or equivalently, 0.02 ± 0.0086, or 0.0114 to 0.0286. In this population, the true frequency of this rare allele could range from 114/10,000 to 286/10,000. If the sample size n were increased to 5,000, this range would be reduced to 161/10,000 to 239/10,000.

Sample size determination. The confidence interval formulae can be used to determine the sample size required to attain a prespecified degree of precision. If one desires an estimate that is accurate to plus or minus L with $(1-\alpha) \times 100\%$ confidence and the population proportion is expected to be p' then the required sample size is

$$n = \left(\frac{z_{\alpha/2}}{L}\right)^2 p'(1-p')$$

Suppose the allele frequency is expected to be (0.01), a 95% confidence interval is sought, and a precision of plus or minus 0.001 is desired. The number of analysis specimens (n) required is $(1.96/0.001)^2 (0.01 \times 0.99) = 38{,}031$.

Simultaneous confidence intervals. Often the analyst is interested in estimating more than one allele frequency. For example, if several alleles are present at one locus, then estimates of both frequency and precision are required. If individual confidence intervals, such as those described above, were obtained for each estimate, the overall confidence statement (that pertaining to several allele frequencies), would not hold at the predetermined confidence level. For example, with two independent 95% confidence intervals, the probability that both contain the true allele frequencies would be $(0.95)^2 = 0.9025$. If instead we wished to determine five allele frequencies, the probability that all five individual 95% confidence intervals would contain the true frequencies is $(0.95)^5 = 0.7738$.

To ensure that the probability that all confidence intervals contain the true parameter values is as specified, for example 95%, then each individual interval must be made wider. Bonferroni's method is the easiest to use and applies in many situations (Goodman 1965). If K quantities, such as allele frequencies, are to be estimated and the goal is to have an overall $(1-\alpha) \times 100\%$ confidence interval, then choosing each interval to be a $(1-\alpha/K) \times 100\%$ interval achieves this objective. For example, if we wish to obtain an overall 95% confidence interval, and two allele frequencies are being estimated, then each should be estimated with confidence $(1-.05/2) \times 100\%$ or 97.5%. If five frequencies are being estimated, then using a 99% confidence interval, $(1- 0.05/5) \times 100\%$, for each ensures an overall 95% confidence interval.

Suppose we estimate K frequencies, say $\pi_1, \ldots, \pi_K$, by $p_1, \ldots, p_K$ based on samples of size $n_1, \ldots, n_K$. Then the following individual confidence intervals will give an overall $(1-\alpha) \times 100\%$ confidence interval.

$$p_i - z_{\alpha/2K} \frac{\sqrt{p_i(1-p_i)}}{n_i} \leq \pi_i \leq p_i + z_{\alpha/2K} \frac{\sqrt{p_i(1-p_i)}}{n_i}$$

Some refinements of the above intervals for multiple alleles at a single locus, based on the multinomial distribution, have been provided by Angers (1989).

Simultaneous allele frequency determination. Suppose in samples of size 100, we observe the frequency of two alleles to be 0.05 and 0.10 and we seek confidence intervals that will contain both true frequencies at a confidence level of 95%. As described above, we compute 97.5% confidence intervals for each. In the above formula, $\alpha = .05$ and $K = 2$, so we obtain $z_{.0125} = 2.24$ from Table 8-1. These intervals are

$$.05 - 2.24 \frac{\sqrt{(0.05)(0.95)}}{100} \leq \pi_1 \leq .05 + 2.24 \frac{\sqrt{(0.05)(0.95)}}{100}$$

and

$$.10 - 2.24 \frac{\sqrt{(0.10)(0.90)}}{100} \leq \pi_2 \leq .10 + 2.24 \frac{\sqrt{(0.10)(0.90)}}{100}$$

This gives the intervals 0.05 ± 0.049 for the first allele and 0.10 ± 0.067 for the second allele.

Table 8-1. Areas Under the Standard Normal Distribution

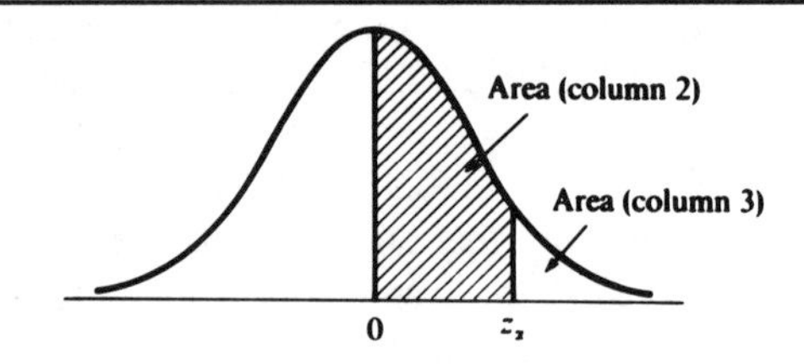

(1) z_x	(2) Area between Mean and z_x	(3) Area beyond z_x	(1) z_x	(2) Area between Mean and z_x	(3) Area beyond z_x	(1) z_α	(2) Area between Mean and z_α	(3) Area beyond z_α
0.00	0.0000	0.5000	0.30	0.1179	0.3821	0.60	0.2257	0.2743
0.01	0.0040	0.4960	0.31	0.1217	0.3783	0.61	0.2291	0.2709
0.02	0.0080	0.4920	0.32	0.1255	0.3745	0.62	0.2324	0.2676
0.03	0.0120	0.4880	0.33	0.1293	0.3707	0.63	0.2357	0.2643
0.04	0.0160	0.4840	0.34	0.1331	0.3669	0.64	0.2389	0.2611
0.05	0.0199	0.4801	0.35	0.1368	0.3632	0.65	0.2422	0.2578
0.06	0.0239	0.4761	0.36	0.1406	0.3594	0.66	0.2454	0.2546
0.07	0.0279	0.4721	0.37	0.1443	0.3557	0.67	0.2486	0.2514
0.08	0.0319	0.4681	0.38	0.1480	0.3520	0.68	0.2517	0.2483
0.09	0.0359	0.4641	0.39	0.1517	0.3483	0.69	0.2549	0.2451
0.10	0.0398	0.4602	0.40	0.1554	0.3446	0.70	0.2580	0.2420
0.11	0.0438	0.4562	0.41	0.1591	0.3409	0.71	0.2611	0.2389
0.12	0.0478	0.4522	0.42	0.1628	0.3372	0.72	0.2642	0.2358
0.13	0.0517	0.4483	0.43	0.1664	0.3336	0.73	0.2673	0.2327
0.14	0.0557	0.4443	0.44	0.1700	0.3300	0.74	0.2704	0.2296
0.15	0.0596	0.4404	0.45	0.1736	0.3264	0.75	0.2734	0.2266
0.16	0.0636	0.4364	0.46	0.1772	0.3228	0.76	0.2764	0.2236
0.17	0.0675	0.4325	0.47	0.1808	0.3192	0.77	0.2794	0.2206
0.18	0.0714	0.4286	0.48	0.1844	0.3156	0.78	0.2823	0.2177
0.19	0.0753	0.4247	0.49	0.1879	0.3121	0.79	0.2852	0.2148
0.20	0.0793	0.4207	0.50	0.1915	0.3085	0.80	0.2881	0.2119
0.21	0.0832	0.4168	0.51	0.1950	0.3050	0.81	0.2910	0.2090
0.22	0.0871	0.4129	0.52	0.1985	0.3015	0.82	0.2939	0.2061
0.23	0.0910	0.4090	0.53	0.2019	0.2981	0.83	0.2967	0.2033
0.24	0.0948	0.4052	0.54	0.2054	0.2946	0.84	0.2995	0.2005
0.25	0.0987	0.4013	0.55	0.2088	0.2912	0.85	0.3023	0.1977
0.26	0.1026	0.3974	0.56	0.2123	0.2877	0.86	0.3051	0.1949
0.27	0.1064	0.3936	0.57	0.2157	0.2843	0.87	0.3078	0.1922
0.28	0.1103	0.3897	0.58	0.2190	0.2810	0.88	0.3106	0.1894
0.29	0.1141	0.3859	0.59	0.2224	0.2776	0.89	0.3133	0.1867

Table 8-I is abridged from Table II of Fisher and Yates: *Statistical Tables for Biological, Agricultural and Medical Research* published by Longman Group UK Ltd. London (previously published by Oliver and Boyd Ltd., Edinburgh) and by permission of the authors and publishers.

(1) z_x	(2) Area between Mean and z_x	(3) Area beyond z_x	(1) z_x	(2) Area between Mean and z_x	(3) Area beyond z_x	(1) z_x	(2) Area between Mean and z_x	(3) Area beyond z_x
0.90	0.3159	0.1841	1.35	0.4115	0.0885	1.80	0.4641	0.0359
0.91	0.3186	0.1814	1.36	0.4131	0.0869	1.81	0.4649	0.0351
0.92	0.3212	0.1788	1.37	0.4147	0.0853	1.82	0.4656	0.0344
0.93	0.3238	0.1762	1.38	0.4162	0.0838	1.83	0.4664	0.0336
0.94	0.3264	0.1736	1.39	0.4177	0.0823	1.84	0.4671	0.0329
0.95	0.3289	0.1711	1.40	0.4192	0.0808	1.85	0.4678	0.0322
0.96	0.3315	0.1685	1.41	0.4207	0.0793	1.86	0.4686	0.0314
0.97	0.3340	0.1660	1.42	0.4222	0.0778	1.87	0.4693	0.0307
0.98	0.3365	0.1635	1.43	0.4236	0.0764	1.88	0.4699	0.0301
0.99	0.3389	0.1611	1.44	0.4251	0.0749	1.89	0.4706	0.0294
1.00	0.3413	0.1587	1.45	0.4265	0.0735	1.90	0.4713	0.0287
1.01	0.3438	0.1562	1.46	0.4279	0.0721	1.91	0.4719	0.0281
1.02	0.3461	0.1539	1.47	0.4292	0.0708	1.92	0.4726	0.0274
1.03	0.3485	0.1515	1.48	0.4306	0.0694	1.93	0.4732	0.0268
1.04	0.3508	0.1492	1.49	0.4319	0.0681	1.94	0.4738	0.0262
1.05	0.3531	0.1469	1.50	0.4332	0.0668	1.95	0.4744	0.0256
1.06	0.3554	0.1446	1.51	0.4345	0.0655	1.96	0.4750	0.0250
1.07	0.3577	0.1423	1.52	0.4357	0.0643	1.97	0.4756	0.0244
1.08	0.3599	0.1401	1.53	0.4370	0.0630	1.98	0.4761	0.0239
1.09	0.3621	0.1379	1.54	0.4382	0.0618	1.99	0.4767	0.0233
1.10	0.3643	0.1357	1.55	0.4394	0.0606	2.00	0.4772	0.0228
1.11	0.3665	0.1335	1.56	0.4406	0.0594	2.01	0.4778	0.0222
1.12	0.3686	0.1314	1.57	0.4418	0.0582	2.02	0.4783	0.0217
1.13	0.3708	0.1292	1.58	0.4429	0.0571	2.03	0.4788	0.0212
1.14	0.3729	0.1271	1.59	0.4441	0.0559	2.04	0.4793	0.0207
1.15	0.3749	0.1251	1.60	0.4452	0.0548	2.05	0.4798	0.0202
1.16	0.3770	0.1230	1.61	0.4463	0.0537	2.06	0.4803	0.0197
1.17	0.3790	0.1210	1.62	0.4474	0.0526	2.07	0.4808	0.0192
1.18	0.3810	0.1190	1.63	0.4484	0.0516	2.08	0.4812	0.0188
1.19	0.3830	0.1170	1.64	0.4495	0.0505	2.09	0.4817	0.0183
			1.645	0.4500	0.0500			
1.20	0.3849	0.1151	1.65	0.4505	0.0495	2.10	0.4821	0.0179
1.21	0.3869	0.1131	1.66	0.4515	0.0485	2.11	0.4826	0.0174
1.22	0.3888	0.1112	1.67	0.4525	0.0475	2.12	0.4830	0.0170
1.23	0.3907	0.1093	1.68	0.4535	0.0465	2.13	0.4834	0.0166
1.24	0.3925	0.1075	1.69	0.4545	0.0455	2.14	0.4838	0.0162
1.25	0.3944	0.1056	1.70	0.4554	0.0446	2.15	0.4842	0.0158
1.26	0.3962	0.1038	1.71	0.4564	0.0436	2.16	0.4846	0.0154
1.27	0.3980	0.1020	1.72	0.4573	0.0427	2.17	0.4850	0.0150
1.28	0.3997	0.1003	1.73	0.4582	0.0418	2.18	0.4854	0.0146
1.29	0.4015	0.0985	1.74	0.4591	0.0409	2.19	0.4857	0.0143
1.30	0.4032	0.0968	1.75	0.4599	0.0401	2.20	0.4861	0.0139
1.31	0.4049	0.0951	1.76	0.4608	0.0392	2.21	0.4864	0.0136
1.32	0.4066	0.0934	1.77	0.4616	0.0384	2.22	0.4868	0.0132
1.33	0.4082	0.0918	1.78	0.4625	0.0375	2.23	0.4871	0.0129
1.34	0.4099	0.0901	1.79	0.4633	0.0367	2.24	0.4875	0.0125

(1) z_x	(2) Area between Mean and z_x	(3) Area beyond z_x	(1) z_x	(2) Area between Mean and z_x	(3) Area beyond z_x	(1) z_x	(2) Area between Mean and z_x	(3) Area beyond z_x
2.25	0.4878	0.0122	2.64	0.4959	0.0041	3.00	0.4987	0.0013
2.26	0.4881	0.0119	2.65	0.4960	0.0040	3.01	0.4987	0.0013
2.27	0.4884	0.0116	2.66	0.4961	0.0039	3.02	0.4987	0.0013
2.28	0.4887	0.0113	2.67	0.4962	0.0038	3.03	0.4988	0.0012
2.29	0.4890	0.0110	2.68	0.4963	0.0037	3.04	0.4988	0.0012
2.30	0.4893	0.0107	2.69	0.4964	0.0036	3.05	0.4989	0.0011
2.31	0.4896	0.0104	2.70	0.4965	0.0035	3.06	0.4989	0.0011
2.32	0.4898	0.0102	2.71	0.4966	0.0034	3.07	0.4989	0.0011
2.33	0.4901	0.0099	2.72	0.4967	0.0033	3.08	0.4990	0.0010
2.34	0.4904	0.0096	2.73	0.4968	0.0032	3.09	0.4990	0.0010
2.35	0.4906	0.0094	2.74	0.4969	0.0031	3.10	0.4990	0.0010
2.36	0.4909	0.0091	2.75	0.4970	0.0030	3.11	0.4991	0.0009
2.37	0.4911	0.0089	2.76	0.4971	0.0029	3.12	0.4991	0.0009
2.38	0.4913	0.0087	2.77	0.4972	0.0028	3.13	0.4991	0.0009
2.39	0.4916	0.0084	2.78	0.4973	0.0027	3.14	0.4992	0.0008
2.40	0.4918	0.0082	2.79	0.4974	0.0026	3.15	0.4992	0.0008
2.41	0.4920	0.0080	2.80	0.4974	0.0026	3.16	0.4992	0.0008
2.42	0.4922	0.0078	2.81	0.4975	0.0025	3.17	0.4992	0.0008
2.43	0.4925	0.0075	2.82	0.4976	0.0024	3.18	0.4993	0.0007
2.44	0.4927	0.0073	2.83	0.4977	0.0023	3.19	0.4993	0.0007
2.45	0.4929	0.0071	2.84	0.4977	0.0023	3.20	0.4993	0.0007
2.46	0.4931	0.0069	2.85	0.4978	0.0022	3.21	0.4993	0.0007
2.47	0.4932	0.0068	2.86	0.4979	0.0021	3.22	0.4994	0.0006
2.48	0.4934	0.0066	2.87	0.4979	0.0021	3.23	0.4994	0.0006
2.49	0.4936	0.0064	2.88	0.4980	0.0020	3.24	0.4994	0.0006
2.50	0.4938	0.0062	2.89	0.4981	0.0019	3.25	0.4994	0.0006
2.51	0.4940	0.0060	2.90	0.4981	0.0019	3.30	0.4995	0.0005
2.52	0.4941	0.0059	2.91	0.4982	0.0018	3.35	0.4996	0.0004
2.53	0.4943	0.0057	2.92	0.4982	0.0018	3.40	0.4997	0.0003
2.54	0.4945	0.0055	2.93	0.4983	0.0017	3.45	0.4997	0.0003
2.55	0.4946	0.0054	2.94	0.4984	0.0016	3.50	0.4998	0.0002
2.56	0.4948	0.0052	2.95	0.4984	0.0016	3.60	0.4998	0.0002
2.57	0.4949	0.0051	2.96	0.4985	0.0015	3.70	0.4999	0.0001
2.576	0.4950	0.0050	2.97	0.4985	0.0015	3.80	0.4999	0.0001
2.58	0.4951	0.0049	2.98	0.4986	0.0014	3.90	0.49995	0.00005
2.59	0.4952	0.0048	2.99	0.4986	0.0014	4.00	0.49997	0.00003
2.60	0.4953	0.0047						
2.61	0.4955	0.0045						
2.62	0.4956	0.0044						
2.63	0.4957	0.0043						

Binning. A panel of VNTR probes, with each probe recognizing an independent and hypervariable locus, is used to produce a composite DNA profile unique to each individual. This approach currently provides the best route for characterization of body tissues for excluding an individual falsely associated with an evidentiary sample. For many profiles, the analyst must determine the proportion of the population that could have contributed the specimen. This calculation must take into consideration the limitations of the technology and the available population data.

The resolution of alleles that differ by one or a few repeat units is not possible by present gel electrophoresis and autoradiography techniques. This is especially true when an allele is large and the tandem repeat units are small. The true number of alleles associated with a hypervariable locus may be extremely difficult to ascertain, especially if a number of the alleles are infrequent in the population. The D1S7 locus alleles range from approximately 1 to 20 kb or 110 to 2,220 tandem repeats. Theoretically, over 2,100 alleles could exist.

Binning is used by the Federal Bureau of Investigation to determine population allele frequencies and in turn to determine the frequency of an allele in an evidentiary sample (Budowle 1989). The approach compensates for limited fragment resolution. The system is conservative; consequently, it is unlikely that an allele from a sample specimen would be assessed a frequency of occurrence that is lower than the true frequency in the population of unrelated individuals.

Bins are designed with boundaries defined by size standards, such as restriction digests of viral DNA. The difference in the sizes of the two fragments that define each bin must be greater than the measurement imprecision of the analytical system. Sample population alleles, according to size, are placed in the bins and a frequency of occurrence for each bin is calculated (Table 8-2). An allele from a suspect or crime sample is assigned the frequency of the bin the allele falls within.

Allele's are said to match only if they are of the same size not if they fall in the same bin. If an allele overlaps two bins (i.e., it could be placed in either of two adjacent bins), the bin with the larger frequency is chosen. It should be noted that different-size alleles reside within the same bin, and although they do not match, each will be assigned the same frequency. Lastly, how can low frequency alleles be used in the binning system? This is an important consideration because it is these alleles that can provide match probabilities measuring in the billionths; however, due to subpopulation differences and sampling errors, there may be very low precision associated with the probabilities. Until adequate population data are available, Budowle and Monson (1989) recommend that bins be combined to contain a minimum of five alleles each.

A floating bin approach (as opposed to fixed bins as described above) has been used by Balasz et al. (1989) whereby the size assigned to an evidentiary sample allele is assumed to be the mean size. Because, with current assay methodology, it cannot be demonstrated that the value assigned is the true mean, the Federal Bureau of Investigation avoids using this approach.

Table 8-2. Black Population Data for D14S13

Bin	Range (base pairs)		Allele count	95% LCL	Point Est	95% UCL
1	0	871	7	.006	.018	.052
2	872	963	9	.008	.023	.059
3	964	1077	7	.006	.018	.052
4	1078	1196	5	.003	.013	.044
5	1197	1352	22	.029	.055	.102
6	1353	1507	27	.038	.068	.117
7	1508	1637	20	.026	.050	.095
8	1638	1788	32	.047	.080	.132
9	1789	1924	18	.022	.045	.089
10	1925	2088	31	.045	.078	.129
11	2089	2351	36	.055	.090	.144
12	2352	2522	27	.038	.068	.117
13	2523	2692	23	.031	.058	.105
14	2693	2862	12	.013	.030	.069
15	2863	3033	17	.021	.043	.086
16	3034	3329	14	.016	.035	.076
17	3330	3674	36	.055	.090	.144
18	3675	3979	7	.006	.018	.052
19	3980	4323	26	.036	.065	.114
20	4324	4821	7	.006	.018	.052
21	4822	6368	6	.005	.014	.048
22	6369		11	.011	.028	.066

N.B.: One individual displayed a three-band pattern which is not included in the above tabulation.

Low-frequency bins must have a minimum of five events
LCL lower confidence limit
UCL upper confidence limit
LCL and UCL calculated according to Goodman (1965)

(This table is reproduced, with permission, from Budowle B and Monson KL, 1989.)

Hypothesis Testing

Hypothesis-testing ideas are quite involved. For more details, the reader should refer to basic statistics texts, such as those in the references at the end of this Chapter. A brief description relevant particularly to forensic science appears below.

Testing for proportions. The basic purpose of hypothesis testing is to determine whether or not there is sufficient evidence in the data to reject a hypothesis about the value of the true population proportion in favor of an alternative hypothesis. In this setting, the null hypothesis, denoted by H_0, is a statement that the value of the true population proportion π equals a specified value π_0. This is written H_0: $\pi = \pi_0$. The alternative hypothesis might be that π does not equal π_0. A test or procedure is said to have significance level α if it rejects the null hypothesis when it is true with probability α. For this problem, the following rule gives a test with significance level α.

Decision rule: Reject H_0 if π_0 is *not* contained in a (1-α) x 100% confidence interval for π. Otherwise, do not reject H_0.

This rule has the desired rejection probability because a (1-α) x 100% confidence interval will contain the true value of π with probability 1-α. Note that the same decision rule applies for hypotheses about the population mean.

The above hypothesis test is called two-sided because it rejects H_0 when π is either too large or too small. The *power* of this decision rule is the probability it fails to reject H_0 when it is false, that is, when π does not equal π_0. Note that tests with large power and low significance are desirable. Prespecified levels of power and significance are achieved by adjusting the sample size.

Another important concept related to hypothesis testing is that of a *p*-value. A *p*-value gives the probability of observing a proportion at least as extreme as that observed when the null hypothesis is true. Small *p*-values lead to rejection of H_0 while large *p*-values favor the null hypothesis. A rule of thumb that applies in all hypothesis-testing problems is to reject H_0 at significance level α whenever the *p*-value is less than or equal to α.

Example. Theoretical considerations suggest that the frequency of a specific allele is 0.04. In a sample of 1,000 specimens, the observed frequency is .02. Do the data agree with the theoretical model?

This is formulated as a hypothesis testing problem as follows. The null hypothesis is H_0: $\pi = 0.04$. The above calculations indicate that a 95% confidence interval for π is (0.0114, 0.0286). Because this interval does not contain 0.04, reject H_0 at the .05 level. This suggests that the theoretical model might not be suitable. However, with such a large sample, we will reject almost any hypothesis that deviates from the observed value.

Chi squared goodness of fit test. The chi squared goodness of fit test is useful for comparing a set of observed sample data with the data expected under some natural law. For example, an investigator might wish to determine whether an experimentally determined genotype frequency distribution differs from that derived from theoretical considerations. The null hypothesis being tested is H_0: The observed frequency distribution agrees with the theoretical frequency distribution. The alternative hypothesis is that the two distributions differ.

Each observation falls into exactly one of M mutually exclusive categories. Let O_i be the number of observations in the ith category and let F_i be the expected number of observations in the ith category. This hypothesis is tested at significance level α by computing the following test statistic

$$\chi^2 = \sum_{i=1}^{M} \frac{(O_i - E_i)^2}{E_i}$$

and rejecting H_0 if it exceeds the reference value $\chi^2_{\alpha,M-1}$ obtained from a table or chart of critical values from a χ^2 distribution with M-1 degrees of freedom (Figure 8-3). This procedure requires that the expected number of observations in each category is five or greater.

Example
Suppose we observe 40 males and 10 females with a specific trait in a given population. Do these data differ from an expected 1:1 male to female ratio?

Solution.

Group	Observed	Expected	$(O_i-E_i)^2/E_i$
Male	40	25	9
Female	10	25	9
Total	50	50	18

The number of degrees of freedom in this example is 1; the number of categories (male + female) is 2 and 2 - 1 = 1. At the 0.05 significance level the critical value from the χ^2 chart (Figure 8-3) is $\chi^2_{.05,1} = 3.84$. Since the calculated value of the test quantity of 18 exceeds the critical value of 3.84, reject H_0 and conclude that the observed data differ from the expected 1:1 ratio. Note that this hypothesis is also rejected at the 0.001 probability level since the critical value at this significance level is 10.91. The p-value is .00002.

Example
The frequencies of alleles p and q in a population are 0.75 and 0.25, and the observed proportions in a sample of size 100 of genotypes pp, pq and qq are 0.50, 0.25, and 0.25. Do the observed frequencies differ from those expected according to the Hardy-Weinberg law $p^2 : 2pq : q^2$ for the population?

Solution.

Grouping	Observed	Expected	$(O_i-E_i)^2/E_i$
pp	50	56.25	0.70
pq	25	37.50	4.17
qq	25	6.25	56.25
Total	100	100.00	61.12

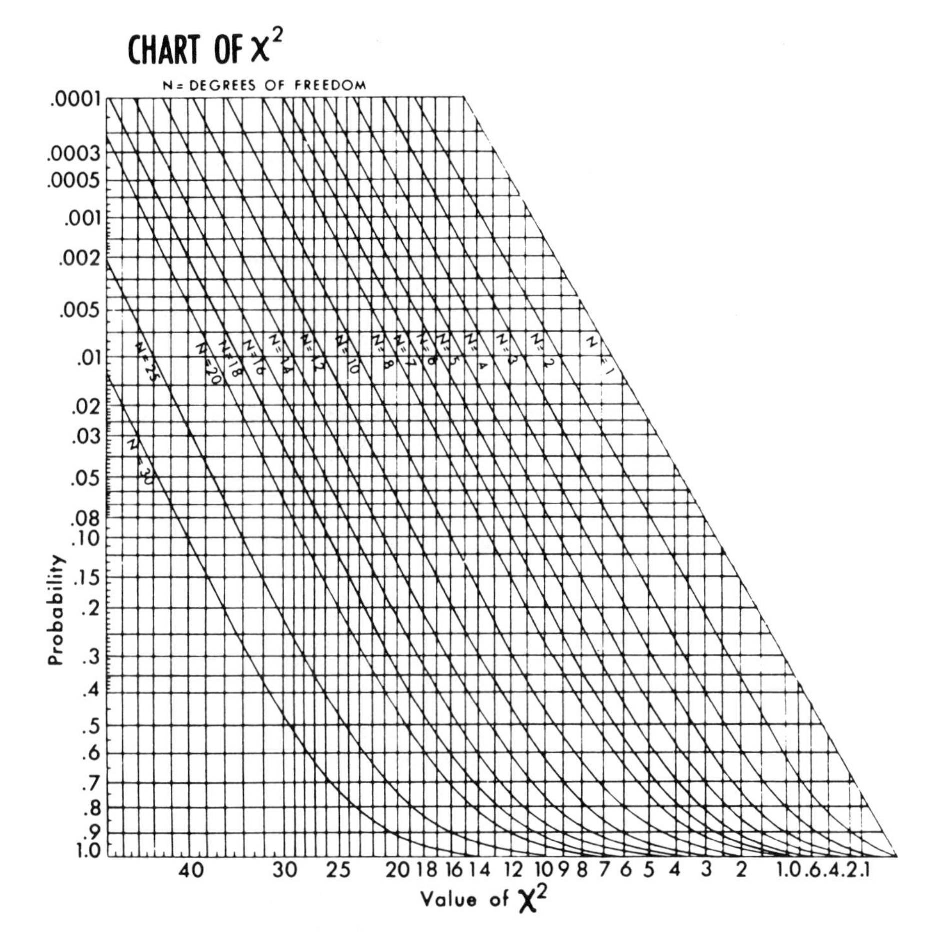

Figure 8-3. Chart of χ^2. Reprinted with permission of Macmillan Publishing Company from *A Dictionary of Genetics,* 3rd edition, by R.C. King and W.D. Stansfield. (New York: Macmillan Publishing Company, 1985.)

The number of degrees of freedom is 3-1 or 2. From the χ^2 distribution chart, at the 0.05 level of significance and 2 degrees of freedom, the critical value is 5.99. This value is considerably less than the calculated value of 61.12; thus, we reject the hypothesis at significance level .05, that the population distribution of genotypes follows the Hardy-Weinberg rule. The main reason for rejection in this case is the discrepancy between the observed and expected values in the *qq* cell.

PROBABILITY

A critical question arises in forensic DNA profile analysis: What is the chance or probability that suspect and crime samples have the same source given that they match? For paternity testing, a similar question must be asked: What is the probability that a putative father is the biological father when all of the child's alleles are found in the mother and putative father?

These are some of the questions that can be answered using probability theory. Probability is a means of quantifying uncertainty. Probabilities are numbers between 0 and 1 that indicate the degree of likelihood of various outcomes (events) of experiments. They can be determined either subjectively, objectively, or empirically.

Subjective probabilities are based on one's feelings, preferably based on experience, about the likelihood an event will or will not occur. If this subjective approach is based on solid experience (expert opinion), the conclusions could be sound. An example of a subjective probability calculation is a determination of the probability that it will rain more than two centimeters two weeks from now.

Objective probability calculations are relevant when an experiment has several equally likely outcomes such as occurs in games of chance. The probability that an event E occurs, $P(E)$ equals the number of outcomes corresponding to E, n_E, divided by the total number of possible oucomes of the experiment n_{TOT}. That is, $P(E) = n_E/n_{TOT}$. Clearly $0 \leq P(E) \leq 1$. For example, when rolling a die, there are six possible outcomes so the probability of each is 1/6. To compute the probability of the event E = "the outcome is odd," we note that if any of the three outcomes, 1, 3 or 5, occurs, then E occurs so that $n_E = 3$, $n_{TOT} = 6$, and $P(E) = 3/6 = 0.5$.

Empirical probability calculations are based on information acquired from data collection. For example, to determine the frequency of individuals posessing a certain allele in a population, choose a sample and define the probability of the allele to be the number of individuals exhibiting that allele divided by the number of individuals in the sample. If the sample size is large, the empirically determined frequency would be a very good approximation of the "true" frequency. Confidence intervals determine the precision of such calculations.

Probability of Combined Events: Addition and Multiplication Rules

Two events A and B are said to be *mutually exclusive* if the occurrence of one excludes the occurrence of the other. For example, the events "an individual is Caucasian" and "an individual is Oriental" are mutually exclusive, while the events "an individual is Caucasian" and "an individual is a male" are not. The probability that one of several mutually exclusive events occurs is the sum of the probability that the individual events occur. That is,

$$P(A \text{ or } B \text{ or } C \ldots) = P(A) + P(B) + P(C) + \ldots$$

In this calculation, it is essential that the events contain no items in common. If not, the calculation becomes more complex. In the case of two events,

$$P(A \text{ or } B) = P(A) + P(B) - P(A \text{ and } B)$$

The event "*A or B*" is sometimes referred to as the union of A and B, while the event "*A and B*" is referred to as the intersection of A and B.

The probability that both events A and B occur is computed by the formula

$$P(A \text{ and } B) = P(A/B)P(B) \quad \text{or} \quad P(A \text{ and } B) = P(B/A)P(A)$$

where $P(A/B)$ is the conditional probability that A occurs given that B occurred, while $P(B/A)$ is the conditional probability that B occurs given that A occurred. Conditional probabilities are probabilities that take evidence into account. Two events A and B are *independent* if $P(B/A) = P(B)$ and $P(A/B) = P(A)$. Independence means that evidence that A occurred does not influence our estimate of $P(B)$. If two (or more events) are independent, the joint probability that they all occur is the product of the probability that each occurs. That is,

$$P(A \text{ and } B \text{ and } C \ldots) = P(A)P(B)P(C) \ldots$$

Examples of the addition and multiplication rules follow: If at locus M the probability of detecting allele A is 0.2, and at locus R the probability of detecting allele B is 0.3, and occurrence of alleles at the loci are independent, then the probability of detecting both A and B is $P(A \text{ and } B) = P(A)P(B) = 0.2 \times 0.3 = 0.06$. The probability of detecting either allele A or allele B or both is $P(A \text{ or } B) = P(A) + P(B) - P(A \text{ and } B) = 0.2 + 0.3 - 0.06 = 0.44$.

When matching individual genotypes to forensic samples, the phrases "probability of a match" and "probability of a match given one of the set to be matched" are quite different. If the frequency of a given allele, A, in a population is 1/100, the probability of independently drawing two individuals at random from the population each with an A is (1/100)(1/100) = 1/10,000. If a crime specimen is known to contain an A allele (given), the probability of drawing an individual at random from the population matching this allele is 1/100. The former, P(match), is a case of an unconditional probability, while the latter P(match|one allele is A), is a conditional probability.

Bayes' Theorem and Revision of Probabilities

Bayes' Theorem provides a method for revising probabilities based on acquired information or evidence. We begin with a prior estimate of the probability (prior probability) of event A and combine it with the known conditional probability of the evidence B, given A has occurred, to obtain the revised or posterior probability of A given the evidence B. Thus we are given the prior $P(A)$ and the conditional $P(B/A)$

and wish to obtain the posterior $P(A/B)$. Bayes' Theorem can be expressed by the formula

$$P(B|A) = \frac{P(A|B)\,P(B)}{P(A|B)\,P(B) + P(A|not\ B)\,P(not\ B)}$$

where *not B* is the event known as the complement of *B* and corresponds to all outcomes for which *B* does not occur. Note that $P(B/A) + P(not\ B/A) = 1$.

In determining whether an individual is a carrier for a specific allele, such as a female for an X-linked defect, one might wish to compute the probability that a woman, whose mother is a confirmed carrier, is herself a carrier, when it is known that she has two non-affected sons. Let *B* be the event the mother is a carrier and *A* be the event that she has two non-affected sons. Suppose $P(B) = P(not\ B) = 0.5$. The conditional probability $P(A/B)$ is the chance of having two non-affected sons if the mother is a carrier, which is (0.5)(0.5)=0.25, while $P(A|not\ B)$ is the chance of having two non-affected sons if she is not a carrier, which is 1. Then

$$P(B|A) = \frac{P(A|B)P(B)}{P(A|B)P(B) + P(A|not\ B)P(not\ B)} = \frac{.25\times.5}{.25\times.5 + 1\times.5} = 0.2$$

Similarly, $P(not\ B/A) = 0.8$, so that the probability that the mother of two non-affected sons is a carrier is 0.2, while the probability she is not is 0.8. Thus, the original assumption of 0.5 (since her mother was a confirmed carrier) has been modified to 0.2 based on the information that she has two non-affected sons. If her mother had not been a confirmed carrier, we would use the population frequency of carriers for the prior probability and repeat the above calculation.

In paternity cases, Bayes' Theorem can be applied as follows. Let *B* be the event that the putative father is the biological father. Then $P(B)$ is the prior probability that he is the biological father and $P(not\ B)$ is the probability he is not. Let *A* be the event that the putative father could transmit a specific allele (one present in the offspring). Then $P(A/B)$ is the probability he could transmit the specific allele given he is the father and $P(A\ not\ B)$ is the probability that the putative father could transmit this allele given he is a randomly selected individual. If the putative father is homozygous for that allele, the conditional probability $P(A/B)$ is 1 while the $P(A\ not\ B)$ is the frequency of that allele in the population. If the putative father is heterozygous, $P(A/B) = 0.5$ and $P(A\ not\ B)$ is the frequency of the allele multiplied by 0.5, since there is only a 50% chance the allele would be transmitted to an offspring. $P(B/A)$ is the probability that the man is the father given he transmitted the specified allele. It can be computed by direct substitution into the above formula. In paternity and forensic laboratory testing, the prior probabilities are often considered equal and cancel from the genetic marker analysis calculations because of the above formula. Inclusion of the prior estimates is the prerogative of the judge or jury at a latter stage of the proceedings.

Random Variables and Distributions

Random variables (denoted by capital letters such as X or Z) are numerical value outcomes of a random experiment. For example, the number of dots observed when rolling a die is a random variable, as is the height of an individual selected at random from some population. A probability distribution assigns a probability to each possible value of the random variable. If the set of outcomes is discrete, such as when rolling a die, the probability distribution can be given by a list: $P(X = 1) = 1/6$, $P(X = 2) = 1/6, \ldots, P(X = 6)= 1/6$. The possible values together with the probability distribution for the random variable can also be determined theoretically, or a frequency distribution of the actual observations can be used as an approximation. Important discrete probability distributions include the binomial, Poisson, and hypergeometric; they will not be discussed here.

Probability distributions of continuous random variables, such as height of an individual, are best represented graphically or through formulae. The most important continuous distribution in statistical analysis is the normal distribution (Figure 8-4). Many quantitites, such as IQ and male height, have distributions resembling the normal curve. Characteristics of the curve include a symmetric bell shape with the mean, median, and mode at the same central location, and the curve tails approaching the horizontal axis. Probabilities are found by computing areas under the curve. The total area under the curve is designated as 1.0 or 100 percent.

The normal distribution is characterized by two parameters: its mean μ and its standard deviation σ. The areas or probabilities corresponding to intervals $\mu \pm \sigma$, $\mu \pm 2\sigma$, and $\mu \pm 3\sigma$ are depicted in Figure 8-4. They are 68.24%, 95.45%, and 99.73% respectively. To calculate other probabilities, a general normal random variable is transformed to one with mean 0 and standard deviation 1. This transformation is given by

$$Z = \frac{X - \mu}{\sigma}$$

This is referred to as a *standard normal random variable* and is usually denoted by Z. This quantity measures departures from the mean in standard deviation units. Probabilities $P(0 \leq Z \leq z_\alpha)$ and $P(z_\alpha \leq Z)$ are tabulated (Table 8-1) and are the basis for all probability calculations for the normal distribution.

Example

Suppose one wished to calculate the probability that a randomly selected individual has an IQ between 110 and 125. IQ scores have a mean of 100 and standard deviation of 15. Let X represent the IQ of the randomly selected individual. The desired probability is $P(110 \leq X \leq 125)$ and it is calculated as follows:

$$P\left(\frac{110 - 100}{15} \leq \frac{X - 100}{15} \leq \frac{125 - 100}{15}\right) = P(0.67 \leq Z \leq 1.67)$$

$$= P(0 \leq Z \leq 1.67) - P(0 \leq Z \leq 0.67) = 0.4525 - 0.2486 = 0.2039$$

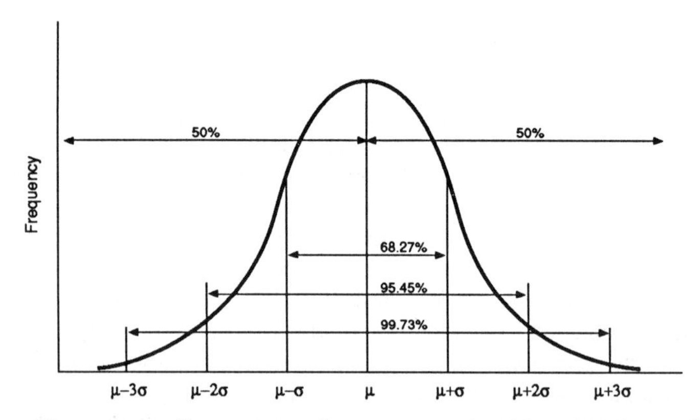

Figure 8-4. Percentage of scores contained in select intervals around the mean in a normal distribution.

Frequently, especially in statistical applications, it is necessary to determine a numerical value that Z exceeds with probability α, that is the $(1-\alpha)$th percentile of Z. This quantity is denoted by z_{α} and can also be obtained from Table 8-1. For example $z_{.05} = 1.645$ and $z_{.025} = 1.96$.

Genetic Applications of Probability

Hardy-Weinberg law. The Hardy-Weinberg law states that in a large random mating population, where no disturbances by outside influences such as mutation, migration, or selection exist, the relative proportions of the different genotypes remain constant from generation to generation. In a two-allele (A and a) system with the frequency of allele A equal to p and that of a equal to q, the genotype proportions of $AA : 2Aa : aa$, or $p^2 : 2pq : q^2$ are valid under the Hardy-Weinberg law. The frequency of the allele A plus the frequency of allele a must equal 1 ($p + q = 1$). If the frequency of A in a given population is 0.8 and that of a is 0.2, the genotype frequencies pp, $2pq$ and qq are $(0.8)(0.8) = 0.64$, $2(0.8)(0.2) = 0.32$, and $(0.2)(0.2) = 0.04$. In a three allele system, the expansion of $(p + q + r)^2$, or $p^2 + q^2 + r^2 + 2pq + 2pr + 2qr$ gives the relative frequencies.

Noting that $p + q = 1$ or $q = 1 - p$, the genotype proportions at one locus in a two-allele system in Hardy-Weinberg equilibrium can be expresed as

Genotype	Proportion
AA	$p^2 = (1 - q)^2$
Aa	$2pq = 2q - 2q^2$
aa	$q^2 = (1 - p)^2$

The frequency of allele a in the population is the sum of the frequencies of genotypes aa and Aa which, in this notation, equals $2q - q^2$. Similarly, the frequency of allele A is $1-q^2$.

Genotype frequencies for populations in Hardy-Weinberg equilibrium can be determined using a Punnett square. Examples for loci with two and three alleles are given in Table 8-3. In a two-allele system, the probability that an A-bearing sperm fertilizes an A-bearing egg is p^2. The probability that an a-bearing sperm fertilizes an a-bearing egg is q^2. The probability that an a-bearing sperm fertilizes an A-bearing egg is pq and the probability that an A-bearing sperm fertilizes an a-bearing egg is pq. The frequency of Aa is $pq + pq = 2pq$.

Valid use of the multiplication rule in probability calculations dictates that each event is random and independent of the others. In population genetics, the procedure of determining the combined power of identity from a series of individual powers is valid only if the loci profiled are not linked and Hardy-Weinberg equilibrium exists. If the observed genotype frequencies deviate significantly from those expected, then Hardy-Weinberg equilibrium has not been attained. The chi-square goodness of fit test described earlier can be used to compare observed and theoretical frquencies.

Non-random mating and migration can exert a considerable influence on Hardy-Weinberg equilibrium. A founder effect may occur in population isolates where a single gene spreads through the population in a few generations. In isolates, the phenomenon of random genetic drift can result in the fixation of one gene and loss of another in a single generation. The loss could occur if only a very few individuals have a specific gene that is not transmitted. This phenomenon could have occurred with human blood groups. The majority of North American Indians are group O; however, in the Blood and Blackfoot populations, type A is frequent. Perhaps this is due to immigrant ancestors carrying A, B, and O from Asia, but in a small isolate, A became common due to genetic drift.

Gene frequencies can be considerably affected by migration of population groups. Blood type O is frequent in American Indians but rare in Asiatics, even though there is good evidence that American Indians are descended from Asiatic ancestors who migrated across the Bering Straits several thousand years ago. Lastly, it has been possible to estimate gene flow between Caucasians and Blacks in the United States by determining blood group gene frequencies in these groups as well as Blacks in West Africa. The rate is from a low of 4% in the southern U.S. to a high of 25% in the north.

Wahlund's principle. The observation that the rate of homozygosity in a population consisting of distinct subgroups is greater than that expected under Hardy-Weinberg equilibrium is referred to as Wahlund's principle. The subgroups must be in Hardy-Weinberg equilibrium and have different allele frequencies. The pooled population allele will equal the average of those in the subpopulations.

Observed allele and genotype frequencies in two equal sized subpopulations of a specific population, together with the calculated genotype frequencies under Hardy-Weinberg equilibrium in each subpopulation and in a combined population with the average allele frequencies, are given below.

Table 8-3. Hardy-Weinberg Genotype Frequency Determination

Two alleles ($p = q = 0.5$)

		Sperm	
		A(p)	a(q)
Egg	A(p)	AA(p^2)	Aa(pq)
	a(q)	Aa(qp)	aa(q^2)

Offspring

Genotype	Frequency
AA	$p^2 = 0.25$
Aa	$2pq = 0.50$
aa	$q^2 = 0.25$
Total	1.00

Three alleles ($p = q = r = 0.333$)

		Sperm		
		A_1(p)	A_2(q)	A_3(r)
	A_1(p)	A_1A_1(p^2)	A_1A_2(pq)	A_1A_3(pr)
Egg	A_2(q)	A_2A_1(qp)	A_2A_2(q^2)	A_2A_3(qr)
	A_3(r)	A_3A_1(rp)	A_3A_2(rq)	A_3A_3(r^2)

Offspring

Genotype	Frequency	
A_1A_1	p^2	= .111
A_2A_2	q^2	= .111
A_3A_3	r^2	= .111
A_1A_2	2pq	= .222
A_1A_3	2pr	= .222
A_2A_3	2qr	= .222
Total *		1.000

* Note individual frequencies do not sum to 1.000 because of roundoff.

Hardy-Weinberg Equilibrium in Each Subpopulation

	Alleles		Genotypes		
	A	*a*	*AA*	*Aa*	*aa*
Subpopulation 1	.3	.7	.09	.42	.49
Subpopulation 2	.5	.5	.25	.50	.25
Mixed population (50 : 50)	.4	.6	.17	.46	.37

Hardy-Weinberg Equilibrium in the Mixed Population

	Alleles		Genotypes		
	A	*a*	*AA*	*Aa*	*aa*
Frequency	.4	.6	.16	.48	.36

The calculations show that because the population consists of distinct subgroups, the population composed of a 50 : 50 mixture of these two subgroups will have genotype frequencies 17 : 46 : 37. These differ from the Hardy-Weinberg frequencies of 16 : 48 : 36. Wahlund's principle is the observation that in this (or any) mixed population, the frequency of homozygotes (in the example, 17 + 37 = 54%) exceeds that in a population in Hardy-Weinberg equilibrium (in the example, 52%). If the subgroups randomly mate, the population will achieve Hardy-Weinberg proportions in one generation.

Wahlund's principle can be represented graphically using a De Finetti diagram (Hartl 1989).

Forensic probability determination. As mentioned earlier, a fundamental question in forensic science concerns the probability that a specific crime tissue specimen is derived from a specific suspect. If a multilocus multiallele DNA analysis approach is used, as with good quality fingerprints, a match of two sets of DNA fingerprints normally implies that they are from the same source because each person's DNA is unique. If the prints derive from the same source, the probability of a match is 1; if they derive from different sources, the probability of a match, from experience to date, is zero. Population studies can be carried out to estimate the probability that any specific fragment in a multiband profile present in individual *A* is also present in another individual *B* randomly selected from the population. Provided the loci analyzed are not linked, the probability that all fragments present in *A* are also present in *B* can be calculated using methods described and applied in the section "Estimation of probabilities for multilocus multiallele fingerprint systems" below.

If single-locus multiallele DNA analysis is used, this is comparable to but considerably more specific than conventional blood classification typing. When a match is observed, the probability that the match could have arisen by chance in the population must be calculated. Population allele frequencies and the frequencies of the loci genotypes must be known. Provided Hardy-Weinberg and linkage equilibria apply, the probabilities for the loci matched can be multiplied to determine the composite profile probability. The value of match evidence, in conjunction with other evidence, can be very incriminating especially if a number of loci are analyzed and rare alleles are present.

A match between a crime and suspect sample for a specific allele does not equate to the same specimen source. The probability of a match between suspect and crime specimens when the source is the same is 1. If the sources are different, the probability of a match, given the crime allele size is known, equals the specific population allele frequency. With a specific allele frequency of 0.01, the match probability would be 0.01. The probability of a profile match, with alleles at a number of loci, is the product of the individual locus genotype frequencies. For example, if five independent loci with genotype frequencies of 0.05, 0.1, 0.15, 0.3, and 0.4 are tested, the probability of a match is (0.05)(0.1)(0.15)(0.3)(0.4) = 0.00009.

The incriminating value *(IV)* of match evidence defined by

$$IV = \frac{\text{probability of match with the same source}}{\text{probability of match with the different source}}$$

is directly dependent on the probability of a match when the suspect and crime specimens are from different sources. If a genotype frequency is 0.01 and there is a match between the suspect and crime specimen for this genotype, then $IV = 1/(0.01) = 100$. If the genotype frequency were ten fold greater, then $IV = 1/(0.1) = 10$.

The FBI prefers to determine the percent of the population that could contribute the evidentiary sample for which the suspect cannot be excluded as a contributor. This approach is especially important when several mixed profiles or partial profiles must be considered.

Likelihood of paternity. During the past decade, statistical analysis of conventional blood-typing results applied to paternity testing has involved a degree of controversy. Representative journal articles include "Basic Fallacies in the Formulation of the Paternity Index" (Li 1985), and "No Fallacies in the Formulation of the Paternity Index" (Bauer 1986). The problem centers on the method of calculating and reporting the probability of paternity when exclusion of the alleged father is not possible.

Measures of the chance of paternity include statistical frequency, the paternity index, and the probability of paternity. (The chance of paternity refers to the probability of transmitting the offspring's paternal alleles.) The statistical frequency *(SF)* is the chance that a randomly selected male from the population is the father (δ). The paternity index *(PI)* or likelihood ratio *(LR)* is the ratio of the chance of paternity for the putative father (ß),assuming he is the biological father, to the chance of paternity for a randomly selected male (δ), $PI = ß/δ$. The probability of paternity *(PP)* expresses the paternity index as a percentage and equals $[ß/(ß + δ)] \times 100$.

These basic expressions, as applied in practice, employ only genetic markers identified in the mother, offspring, and putative father. The same calculation techniques can be applied when DNA markers are used but the system is simpler because only genotypes need be considered. Size measurement of the DNA marker bands (alleles) may create problems if the fragment sizes are similar. The simplest solution is to use only enzyme and probe systems where the alleles can be readily separated on gels and the sizes easily and accurately determined. A typical calculation proceeds as follows:

1. List the mother, child, and putative father genotypes.
2. Determine the alleles inherited from the father. (First define the maternal alleles; those remaining must be paternal.)
3. In (2) above, list the allele frequencies for the population in question. These data are used to calculate the statistical frequency of the paternal allele profile in the population.
4. Using the data from (3) above, determine the frequency of the paternal allele profile in the population. This value is the statistical frequency δ.
5. Calculate *PI* and *PP*.

This procedure is illustrated below. Given an offspring DNA profile consisting of paternal alleles *X, Y, Z,* and *W* from non-linked loci *A, C, O,* and *P* with population frequency of these alleles equal to 1/25, 1/50, 1/100, and 1/125; and, provided the putative father is homozygous at each locus, the calculations proceed as follows. Since the putative father is homozygous at each locus, ß = 1.

$$SF = δ = (0.04)(0.02)(0.01)(0.008) = 6.4 \times 10^{-8}$$

$$PI = ß/δ = 1/(6.4\times10^{-8}) = 1.56 \times 10^{7}$$

$$PP = [1/(1+6.4\times10^{-8})] \times 100 \approx 100\%$$

If the putative father is heterozygous at any locus, the allele frequency at that locus must be multiplied by 0.5 since there is only a 50% chance the allele would be transmitted to an offspring. If the four loci in the above example were instead heterozygous,

$$SF = (0.02)(0.01)(0.005)(0.004) = 4 \times 10^{-9},$$

$$ß = (0.5)(0.5)(0.5)(0.5) = 6.25 \times 10^{-2}$$

$$PI = \frac{6.25 \times 10^{-2}}{4 \times 10^{-9}} = 1.56 \times 10^{7}$$

$$PP = \left[0.0625 / \left(0.0625 + 4 \times 10^{-9}\right)\right] \times 100 \approx 100\%$$

Note that the statistical frequency is effected by heterozygosity while *PI* and *PP* are not.

Estimation of probabilities for multilocus multiallele fingerprint systems. The following method was proposed by Jeffreys (1985, 1987*a*) and used by Georges (1988) for estimating the probability that a specific allele is present in individual *B* given that individual *A* is known to possess the allele. As described in the discussion of conditional probability, this is the probability that a DNA fingerprint fragment (allele) is present in the population. This calculation serves as the basis for estimating the probability that two individuals have identical DNA fingerprints as the result of applying a minisatellite probe.

Suppose q is the frequency of a specific allele in the population. Then the probability (x) that an individual selected at random from this population contains this allele is $2q - q^2$. (See the section on the Hardy-Weinberg law.) When q is sufficiently small so that q^2 is much smaller than q, x can be approximated by $2q$. The "heterozygosity" or proportion of individuals posessing a specific allele who are in the heterozygous state is given by $h = (2q - 2q^2)/(2q - q^2) = 2(1 - q)/(2 - q)$.

Georges and Jeffreys use the following approach for estimating x. A sample of individuals is selected, DNA is isolated, and profiles are prepared. From this sample of fingerprints, distinct resolvable bands are identified and the proportion of the sample possessing each is recorded. The average of these sample proportions is the estimate of the "true" average probability x. From this estimated average probability x, the allele frequency is estimated by solving $x = 2q - q^2$ (Georges 1988) or $x = 2q$ (Jeffreys 1985, 1987*a*).

This method is best illustrated through a simple example with a small number of individuals and a small number of resolvable bands. Suppose there are six individuals in the sample and five bands are resolvable. The data are presented in the following table with an "X" indicating that the band is present in the individual. The column "Proportion" indicates the fraction of individuals in the sample who possess that band in their "fingerprint."

	Individuals						
Bands	1	2	3	4	5	6	Proportion
2.3 kb	X		X	X			.5
2.9 kb		X		X	X	X	.67
4.1 kb	X		X				.33
8.2 kb	X	X	X		X	X	.83
11.7 kb	X		X	X			.5
Total number of bands	4	2	4	3	2	2	

The estimated average probability of a match, $\hat{x}$, is the average of the "Proportion" column, which equals 0.57. From this, the average allele frequency is estimated by solving $2\hat{q} - \hat{q}^2 = 0.57$ which gives $\hat{q} = 0.34$. The estimated heterozygosity is $h = 0.80$. Using the formulas from the section on descriptive statistics shows that the mean number of bands per individual, m, is $17/6 = 2.83$ and the standard deviation of the number of bands is 0.98.

Using the probe that produced the above result, the probability that an individual selected at random has an identical profile to a specific individual is estimated by the method of Jeffreys and Georges to be $\hat{x}^m = (0.57)^{2.83} = 0.20$.

Jeffreys et al. (1985, 1987) calculated that the mean level of band sharing between unrelated individuals is 0.25 in both the North European and the Indian subcontinent populations when their Hinf I endonuclease-digested DNA is hybridized with minisatellite probes 33.15 or 33.6. The Home Office Forensic Science Service in Britain uses a factor of 0.26 with probe 33.15 and Hinf I-digested DNA. This is not a mean value. It is the most conservative figure taken from the 4 kb to 6 kb region where bands are most common. Examination of over 700 profiles shows the figure to be conservative, that is, the frequency of bands at any given position does not exceed 0.26 (P. Gill, Central Research Establishment, Home Office Forensic Science Service, personal communication). If one person has a profile consisting of 10 resolvable bands, the probability of an unrelated person having the identical pattern is approximately $(1/4)^{10} = 1/1,048,576$. If the profile consisted of 18 bands, the probability would be $(1/4)^{18} = 1/68,719,475,200$. With a world population of 5.2 billion, and increasing at the rate of three people per second, there is little question that 1 in 69 billion is significantly small.

Band sharing is considerably more common in biologically related individuals. The probability that a band in sibling *A* is also present in sibling *B* is approximately 0.5 (siblings have one-half of their genomes in common). If 18 bands are resolvable in sibling *A* when the digested DNA is hybridized with a minisatellite probe, the probability that the same bands will be detected in the DNA profile from sibling *B* is $(1/2)^{18} = 1/262,144$ or approximately 4×10^{-6}.

A critical review together with suggested improvements of some methods of statistical analysis of data on DNA fingerprinting has been carried out by Cohen (1990). The approach used by Alex Jeffreys received detailed attention in this article. Cohen's paper exemplifies a number of possible pitfalls in terms of data collection and handling in DNA profiling. The points of contention include (1) the sampling procedures used, (2) the small number of well defined populations sampled, (3) the suggestion that some loci analyzed may not be independent, that is, they may be linked, (4) the assumptions that DNA fragments occur independently and with constant frequency within a size class, and (5) use of the geometric mean instead of the arithmetic mean in some calculations. The degree to which the points made are real or perceived in terms of the final results in forensic identification cases will, in all likelihood, be resolved only when the data bases are sufficiently expanded to provide large enough sample numbers to verify or reject the statistical approaches used.

REFERENCES

Aickin M. 1984. Some fallacies in the computation of paternity probabilites. *Am. J. Hum. Genet.* 36:904 - 915.

Angers C. 1989. Note on quick simultaneous confidence intervals for multinomial proportions. *American Statistician* 43:91.

Balazs I, Baird M, Clyne M, and Meade E. 1989. Human population genetic studies of five hypervariable DNA loci. *Am. J. Hum. Genet.* 44:182 - 198.

Baur MP, Elston RC, Gurtler H, Hennindsen K, Hummel K, Matsumoto H, Mayr W, Moris JW, Niejenhauis L, Polesky H, Salmon D, Valentin J, and Walker R. 1986. No fallacies in the formation of the paternity index. *Am. J. Hum. Genet.* 39:528 - 536.

Budowle B, and Monson KL. 1989. A statistical approach for VNTR analysis. In *Proceedings-DNA Symposium.* International Symposium on the Forensic Aspects of DNA Analysis. Government Publishing Office, Washington, D.C. (in press).

Cohen JE. 1990. DNA fingerprinting for forensic identification: Potential effects on data interpretation of subpopulation heterogeneity and band number variability. *Am. J. Hum. Genet.* 46: 358 - 368.

Evett IW. 1983. What is the probability that this blood came from that person? *J. Forensic Sci. Soc.* 23:35 - 39.

Feller W, 1968. *An Introduction to Probability Theory and its Applications.* Vol. I, 3rd ed. John Wiley and Son Limited, New York.

Freedman D, Pisani R, and Purves R. 1980. *Statistics.* WW Norton and Company, New York.

Georges M, Lequarre AS, Castelli M, Hanset R, and Vassart G. 1988. DNA fingerprinting in domestic animals using four different minisatellite probes. *Cytogenet. Cell Genet.* 47:127 - 131.

Gjertson DW, Mickey MR, Hopfield J, Takenouchi T, and Terasaki PI. 1988. Calculation of paternity using DNA sequences. *Am. J. Hum. Genet.* 43:860-869.

Goodman LA. 1965. On simultaneous confidence intervals for multinomial proportions. *Technometrics* 7:247 - 354.

Hartl DL and Clark AC. 1989. *Principles of Population Genetics*. 2nd ed. Sinauer Associates, Sunderland, MA.

Jeffreys AJ, Wilson V, and Thein SL. 1985. Individual-specific 'fingerprints' of human DNA. *Nature* 316:76 - 79.

Jeffreys AJ. 1987. Highly variable minisatellites and DNA fingerprints. *Biochem. Soc. Trans.* 15:309 - 317.

Jeffreys AJ and Morton DB. 1987*a*. DNA fingerprints of dogs and cats. *Anim. Genet.* 18:115.

Kirk RE. 1978. *Introductory Statistics*. Brooks/Cole Publishing Company, Monterey, CA.

Koopmans LH. 1987. *Introduction to Contemporary Statistical Methods*. 2nd ed. Duxbury Press, Boston.

Lander ES. 1989. DNA fingerprinting on trial. *Nature* 339:501 - 505.

Li CC and Chakravarti A. 1985. Basic fallacies in the formulation of the paternity index. *Am. J. Hum. Genet.* 37:809 - 818.

Walker RH, ed. 1983. *Inclusion Probabilities in Parentage Testing*. American Association of Blood Banks (AABB), Arlington, VA.

Weir BS. 1983. *Statistical Analysis of DNA Sequence Data*. Marcel Dekker Inc., New York.

Zar JH. 1984. *Biostatistical Analysis*. 2nd ed. Prentice-Hall, Englewood Cliffs, NJ.

CHAPTER 9

Quality Control

Quality control (QC) is one aspect of quality assurance (QA). Quality assurance also includes (1) periodic laboratory audits by external specialists, (2) keeping up-to-date clearly-written protocols, (3) preparation of QA reports, (4) troubleshooting, (5) equipment maintenance and calibration, (6) methodology development, (7) personnel training, (8) continuing education, and (9) laboratory safety. To be assured of ongoing quality performance, laboratory accreditation is mandatory. The above aspects have been discussed throughout this text; however, because DNA profiling often involves legal considerations, details of QC are presented in this chapter. An extremely important report, *Guidelines for a Quality Assurance Program for DNA Restriction Fragment Length Polymorphism Analysis,* prepared by the Quality Assurance Subcommittee of TWGDAM, is included in Appendix I (Mudd 1989).

Quality control, narrowly defined, is directly concerned with the accuracy and precision of laboratory results for specimens of verified origin. Implicit in this definition is assurance that correct and accurate population allele frequencies are used in probability calculations. The expert witness representing the laboratory must ensure that (1) the correct specimen is analyzed, (2) the DNA is not significantly degraded, (3) the specimens are not significantly contaminated with extraneous DNA, (4) the analysis procedure is well-controlled, (5) result interpretation is correct, and (6) no deviations from the authorized laboratory protocol have occurred.

The expert witness in DNA profiling should be a meticulous analyst with a solid grasp of genetic and biochemical principles combined with considerable experience in recombinant DNA processes. A reasonable understanding of the concepts and application of statistical techniques, especially probability, is a definite asset. The ability to articulate the results of an analysis and to respond under cross-examination to in-depth questions concerning principles, techniques, and chain of custody is mandatory. (See Wetli 1989 for a general overview on appearing as an expert witness.)

A set of objective QC criteria must be established and followed to ensure that neither false negative nor false positive DNA profile match results are released from a service laboratory. A false negative result equates to a conclusion that a match between two specimen profiles does not exist when in fact the specimens were derived from the same source. A false positive result equates to a conclusion that two specimen profiles match when the specimens were derived from different sources. An appreciation of the potential sources of error in a system is prerequisite to establishing good quality control. Every person associated with handling a specimen, the results, the reagents, and the equipment used must be cognizant of possible pitfalls.

DNA TEST SYSTEM STANDARDS

The American Association of Blood Banks (AABB 1989) has set standards for tests involving DNA polymorphisms. Although the standards were established specifically for parentage analysis, they are, for the most part, relevant to DNA identity profiling in general.

1. DNA loci used in parentage testing shall meet the following criteria prior to reporting results:
   (*a*) DNA loci shall be validated by family studies to demonstrate that the loci exhibit Mendelian inheritance and low frequency of mutation and/or recombination, less than 0.002 (2 per 1000).
   (*b*) The chromosomal location of the polymorphic loci used for parentage testing shall be recorded in the Yale Gene Library or by the International Human Gene Mapping Workshop.
   (*c*) Polymorphic loci shall be documented in the literature stating the restriction endonuclease and probes used to detect the polymorphism, the conditions of hybridization, and sizes of variable and constant fragments.

   (*d*) The type of polymorphism detected shall be known (i.e., single locus, multilocus, simple diallele, or hypervariable).
2. A method shall be available to assure complete endonuclease digestion of DNA for testing.
3. Size markers with discrete fragments of known size shall span and flank the entire range of the DNA loci being tested.
4. A human DNA control of known size shall be used on each electrophoretic run.
5. Autoradiographs or membranes shall be read independently by two or more individuals.
6. DNA reports shall contain at minimum the following information:
   (*a*) The name of the DNA locus tested as defined by the Nomenclature Committee of the International Human Gene Mapping Workshop
   (*b*) The probe used to detect the polymorphism
   (*c*) The restriction endonuclease used to cut the DNA
   (*d*) Reported allelic fragments listed by size or allelic description (alphanumeric).
7. Confirmatory testing by an independent laboratory shall be possible for all DNA loci. These laboratories shall meet AABB standards for tests involving DNA polymorphisms.

In addition, screens must be available for detecting nonhuman DNA contaminants; population allele frequency data must be statistically sound; and detailed protocols outlining every step in the chain of custody are required.

CORRECT SPECIMEN

Each specimen must be uniquely coded. In forensic and paternity cases, verification of sample origin must also be recorded (Figures 3-1, 3-2). If DNA is isolated from a number of specimens in a batch process, the technologist must take exceptional care to ensure that tubes are not interchanged during the many specimen transfers. If uncertainty arises concerning sample interchange, this must be recorded.

DNA QUALITY

There is no use continuing the analysis of a specimen when the DNA quality is poor unless specific fragments are to be amplified. In this situation, degraded DNA may be suitable, provided intact regions of the material to be reproduced are present.

DNA CONTAMINATION

With locus-specific multiallele systems, only one or two bands should be detected with each probe. If more than two bands are present, contamination should be suspected and a repeat sample obtained if possible. Contamination in a multilocus multiallele analysis may be difficult to detect unless considerably more bands than expected with the probe hybridized are present. If concern arises, a repeat sample should be sought.

Hybridization with a Y chromosome-specific probe is useful for confirming female DNA contaminated by a male source. Simple dot-blot techniques are often suitable for this type of analysis. A positive control, that is, a sample of confirmed male DNA, must be included on the blot to ensure proper functioning of the hybridization process.

Depending on the specific probe used, contamination with microorganism DNA may or may not present a problem. An uncharacteristic banding pattern may result if cross-reactivity occurs. To validate this possibility, hybridization of the suspected contaminants must be carried out with the probe. This can be readily achieved by placing a test membrane containing the suspected contaminants in the hybridization container with the sample blots.

Other than stressing to those in the field the importance of meticulous handling of specimens, there is little a laboratory analyst can do if a specimen arrives contaminated. If the collector believes more than one DNA source is present, this should be noted on the sample information card. Contamination within the laboratory also can occur; there is no excuse for sloppy techniques that result in such an event.

ANALYSIS CONTROL

The analysis procedure is best controlled by hybridizing with a probe that detects a few constant bands over the desired range and includes in-house and external QC samples on the gels (Figures 6-14, 9-1, 9-2). Mixtures of DNA may be profiled in paternity cases, especially if the individual genotypes appear to match. This is to ensure that the profiles do in fact superimpose. The analyst should, however, be aware of the possibility that factors such as irreversibly-bound contaminants and different concentrations of DNA can alter electrophoretic migration. The mixing of specimens in forensic cases is not recommended.

Hybridization with a probe that detects a range of constant band sizes facilitates control in determining the degree of DNA degradation. Large fragments will appear very faint or will be undetectable, and an excess of small fragments will be observed in highly degraded specimens. Differences in the rate of migration of DNA fragments in different gel lanes will also be obvious from measurements of the constant bands.

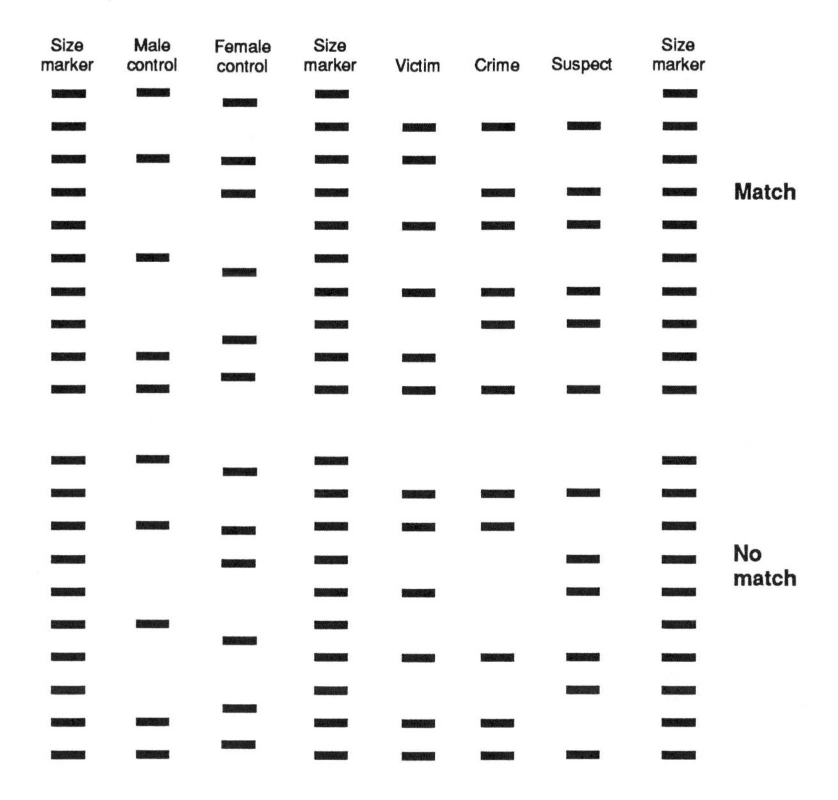

Figure 9-1. Diagrammatic composite DNA profiles for a forensic case.

Control materials must be universally available and should be of sufficient quantity to provide at least a one-year supply. Human lymphoblast culture cell lines, because of their "eternal" nature, are probably the best source of control DNA. Pool controls with fragment sizes spanning the range expected in unknowns should be used. Sizes are determined as described in Chapter 6. Mean ($\overline{X}$) $\pm$ standard deviation (*SD*) can be calculated from data collected over a period of assays (usually 20 or more) by placing controls in each batch analysis of unknowns. Once the means and standard deviations have been determined, limits are established using rules such as those described by Shewhart (1931) or Westgard (1981) to alert the analyst when an assay is not under control and action is required.

External quality control specimens facilitate interlaboratory comparisons. These controls are unknowns in the hands of each laboratory; thus, the consistency between laboratories in detecting specific alleles can be readily determined.

The terms accuracy and precision are often encountered in assay procedure descriptions. Accuracy refers to the degree an allele measurement agrees with the

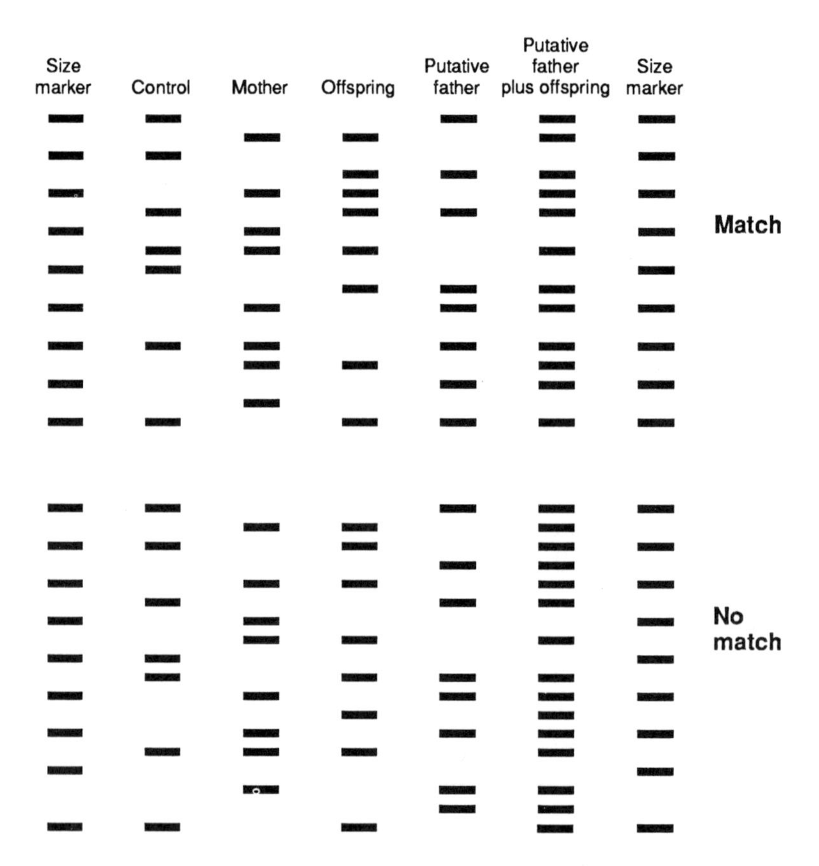

Figure 9-2. Diagrammatic composite DNA profiles for a paternity case.

true size of the allele. The actual sizes may be difficult to determine. Measurements must include well-established and accepted methodology.

Precision refers to reproducibility of the measurement. If random errors in a technique are large, this will be reflected in large standard deviations or coefficients of variation (*CV*); thus, smaller *SD* or *CV* values imply greater assay precision. If both accuracy and precision are high, the procedure is considered reliable.

Every assay is subject to error because it is impossible to duplicate every condition on a daily basis. (On occasion an uncontrollable variable may be encountered in the form of the analyst; this situation must be handled fairly but firmly.) The permissible range for control results is referred to as the acceptable error limit. Charts are usually plotted (Figure 9-3) with $\pm$ 1, 2, and 3 *SD* limits and from these, determinations are made, according to Shewhart rules or a modification thereof, as to when an assay is not under control. It should be remembered that if,

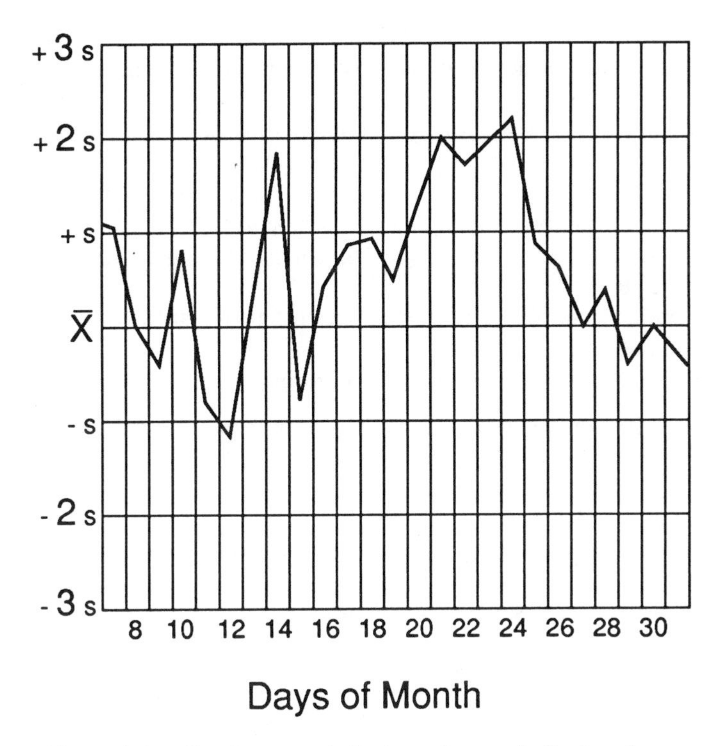

Figure 9-3. Quality control chart used as an indicator of assay result variation.

for example, ± 2 *SD* is chosen as a limit, 5% of the control results are expected to fall outside of this range simply due to chance. The assay in this case is not out of control, and there is little sense in pursuing a problem when none exists. QC data must be examined with an experienced eye. Internal and external QC sample patterns are also useful indicators of whether an incorrect or mutated probe is being used.

SHEWHART SYSTEM

Shewhart quality control is a multirule system based on relatively simple statistical mean and standard deviation calculations. Charts are plotted for ease of observation. There is a low level of rejection, improvement in error detection, and indication of the type of error present.

Multirule: When a control rule is violated, the run is usually rejected. A decision to accept a run requires that there be no violations of the rules except as outlined later.

Control materials: Controls must (1) behave as test samples, (2) be of sufficient quantity for at least a 12-month period, (3) be suitable over the period of use, and (4) have little variation between aliquots.

Control statistics: Data are usually accumulated over at least 20 days at one run per day with one of each control per run. The mean and standard deviation are calculated after any outliers have been rejected. The general rejection rule, after ranking the data from low to high, is to determine if the difference between the apparent outlier and the next value is greater than one-third of the data range; if so, then reject the outlier.

Control limits: Control limits are determined in terms of standard deviations and the rules outlined below.

Control rules (for two to four controls per run):

One 2_s (warning) one control exceeds; $\overline{X} \pm 2$ *SD*; inspect further.
One 3_s (reject) one control exceeds; $\overline{X} \pm 3$ *SD*, often a random error.
Two 2_s (reject) two controls exceed; $\overline{X} + 2$ *SD* or -2 *SD* in the same run or the same control in two runs.
Two 4_s (reject) two controls in the run are over 4 *SD* apart, often a random error.
Four 1_s (reject) four consecutive controls exceed the same limit of 1 *SD*.
Ten$_{\overline{X}}$ (reject) ten consecutive controls fall on the same side of the mean.

Procedure: If a one 2_s situation develops, observe for any of one 3_s through ten$_{\overline{X}}$. If none develops, accept the run; if any one occurs, reject the run and determine the problem. Note that one 3_s and two 4_s are often the result of random errors, whereas two 2_s, four 1_s, and ten$_{\overline{X}}$ are often due to systematic errors.

Decision to report without statistical control: Results may be reported if (1) the problem is due to the control materials, (2) the control problem resulted from an isolated event such as interchange of samples, (3) the analysis is under control in the range of the test samples, and (4) the error size is small relative to the usefulness of the results (e.g., the interpretation of allele size is not affected).

Problem resolution: If quality control problems exist with more than one control material, the problem likely involves the standards, instrument calibration, reagent blanks, or other similar factors. Random errors are more likely due to inconsistencies (e.g., with the analyst's techniques, such as pipetting). Control data from a rejected run should not be included in mean and standard deviation calculations.

Performance characteristics: Any test has inherent random error due to factors such as machine noise. QC procedures are established to test for errors superimposed on this source.

Experience: Experienced analysts are often able to make good judgements from observation of the pattern of points on control charts without use of rigid rules such as those outlined. Skill in interpretation acquired through experience should neither be underestimated nor overestimated. The rules presented provide a guide for interpretation and will require adaptation to the type of assay system and the reproducibility of the specific test.

RESULT INTERPRETATION

Perhaps the best overall indicator of correct result interpretation follows from participation in an external QC program. If errors are made in the QC specimens, then the likelihood of errors in service specimens is great and the DNA analysis program must be critically scrutinized. Once a laboratory gains a reputation for producing questionable results, it is extremely difficult, if not impossible, to shed this negative image; thus, preventive measures in the form of a solid quality assurance accredited program are essential.

REFERENCES

AABB Standards Committee. 1989. P7.000 DNA polymorphism testing. In *Standards for Parentage Testing Laboratories.* American Association of Blood Banks, Arlington, VA.

Mudd JL, Hartmann JM, Kuo MC, Nelson MS, Presley LA, and Stuver WC. 1989. Guidelines for a quality assurance program for DNA restriction fragment length polymorphism analysis. *Crime Laboratory Digest* 16:41–59.

Shewhart WA. 1931. *Economic control of quality of manufactured product.* Van Nostrand, New York.

Westgard JO, Barry PL, and Hunt MR. 1981. A multirule Shewhart chart for quality control in clinical chemistry. *Clin. Chem.* 27:493–501.

Wetli CV. 1989. On being an expert witness. *Lab. Med.* 20:545–550.

CHAPTER 10

Legal and Ethical Considerations

Kenneth E. Melson

IMPACT OF DNA TYPING

The legal system, in both the criminal and civil arenas, may well be revolutionized by the advent of forensic DNA typing. One state trial judge has written that DNA typing "can constitute the single greatest advance in the 'search for truth,' and the goal of convicting the guilty and acquitting the innocent, since the advent of cross-examination." *People v. Wesley*, 140 Misc.2d 306, 533 N.Y.S.2d 643 (Co. Ct. 1988). A prominent professor in the field of law and forensic sciences believes that "DNA analysis will be to the end of the 20th century what fingerprinting was to the 19th."[1] The Washington state legislature has found the accuracy of DNA identification to be superior to that of any presently existing technique,[2] and the Maryland General Assembly has proclaimed DNA identification to have been refined to a level of scientific accuracy that approaches an infinitesimal margin of error.[3]

Indeed, the forensic applications of DNA typing are limited only by the circumspection of the criminal mind. Regardless of the type of crime committed, whenever trace evidence appropriate for DNA analysis is left behind by the perpetrator and later recovered by the police, the test results can be an important investigative tool. Most frequently, such evidence will be found as a result of violent crimes. With 92,490 rapes and 20,680 homicides in the United States in 1988, the forensic application of DNA typing should significantly increase the arrest and conviction rates.[4]

The use of DNA typing is not confined to those cases in which body fluids, hairs, or tissue are left at the crime scene or on the victim by the perpetrator, thereby connecting a suspect to the scene or victim. Just as common is the situation where evidence is left by a victim on the suspect or the suspect's belongings, which will establish previous contact between the accused and the victim.

The Joseph Castro case is an example of this situation. He was accused in New York of stabbing to death a 20-year-old woman who was seven months pregnant and her 2-year-old daughter. At the time of his arrest, the detectives seized a wristwatch worn by Castro because of apparent bloodstains on the watch. The prosecution, hoping to prove that the origin of the bloodstains was the adult victim and not the defendant, sought the introduction of DNA identification tests. For reasons discussed later in this chapter, the New York court excluded the DNA identification evidence. *People v. Castro*, 545 N.Y.S.2d 985 (Sup. Ct. 1989).

In other criminal cases, DNA typing has been used to establish parentage. DNA analysis was used by police in Oklahoma to link a mother to her newborn baby, found in a garbage can.[5] A rape victim in New York aborted her fetus, which was conceived as a result of the rape. The prosecution successfully sought the admission of DNA test results proving that the defendant was the one who had impregnated the victim.[6]

DNA typing may also assist investigators in identifying the victims of homicides, when other means of identification are not possible or where body parts from multiple victims are found. In one instance, human tissue found on the grille of an automobile was determined to have come from a prospective victim by comparing the sample with the DNA of the victim's parents.[7]

In circumstances where several crimes exhibiting a common pattern are committed, DNA analysis may indicate whether the crimes are serial in nature, committed by a single individual, or whether different persons are responsible. If a string of serial crimes is identified, it may be possible to eliminate some of the crimes as the work, for example, of copycat criminals.[8] Other case applications are set forth in Chapter 11.

The collateral benefits of DNA analysis for the criminal justice system are also impressive. The likelihood that defendants will enter pleas of guilty when faced with DNA test results should increase, since in most cases there will be a corresponding increase in the weight of the government's case against the defendant. It has been reported that most rape defendants confronted with DNA test results implicating them have pled guilty.[9] This encouraging trend should also be seen in connection with other crimes, except when the stakes are a matter of life and death for the defendant. Timothy Spencer, indicted for capital murder and rape, went to trial despite evidence indicating that his DNA "characteristics would be present in one of every 135 million black individuals. There are approximately 10 million adult black males in the United States." *Spencer v. Commonwealth*, 238 Va. 275, 384 S.E.2d 775, 777 (1989), *cert. denied*, __U.S.__, 58 U.S.L.W. 3429 (Jan. 9, 1990). Spencer was sentenced to death.

As a corollary to the increase in guilty pleas because of overwhelming DNA evidence, victims may be more willing to report rapes to police. One deterrent to reporting is the additional trauma incurred by victims while testifying at a trial. The increased likelihood of guilty pleas generated by the DNA analysis may diminish that deterrent. Moreover, reports have indicated that rape victims are more likely to file a complaint if there is a high probability of conviction, and they are less likely to make a report when the probability is low.[10] DNA testing should increase dramatically the convictions in nonacquaintance rape cases.

Just as important as using DNA typing to inculpate a suspect is its use to exculpate a suspect. Although conclusive DNA test results will not always eliminate an individual as a suspect, particularly in cases with multiple perpetrators, it can exonerate the defendant and allow the police to focus their investigative efforts elsewhere. Such was the case in England in 1986, when a man charged with murder and rape was exonerated after DNA testing. Another suspect, linked to the crime by a DNA analysis of the sperm recovered from the victims, was later convicted.[11] The legal profession in Great Britain has widely accepted DNA printing.[12]

Exoneration of a defendant may also occur postconviction as a result of the use of DNA analysis. In 1979 Gary Dotson was sent to prison for allegedly raping Cathleen Crowell Webb. Although Webb recanted her story in 1985, saying she had wrongly accused Dotson, Illinois Governor Thompson refused to pardon him. His sentence was instead commuted to time served.

In 1988 a semen stain collected as evidence at the time of the alleged rape was sent to two laboratories. One laboratory reported the semen to be too degraded for analysis. The other laboratory's report allowed counsel for Dotson to argue that his client was completely eliminated as a source of the semen. A state court judge granted Dotson's motion for a new trial, and the state declined to retry him.[13]

Gary Yorke was not so fortunate. Four years after a rape was committed, and after being convicted of that crime and related offenses, Yorke filed for a new trial based on a claim of newly discovered evidence. The new evidence was a laboratory report concluding that Yorke's DNA pattern did not match the pattern in the vaginal washings obtained from the 15-year-old victim at the time of the crime. The origin of the questioned evidence could not be determined at all.

The appellate court, reviewing the trial court's denial of the motion for new trial, agreed that the evidence would not have produced a different verdict. Since the test results did not show the origin of the DNA, and since the victim did not know if the rapist ejaculated, the absence of Yorke's DNA did not exclude him as the criminal agent. *Yorke v. Maryland*, 315 Md. 578, 556 A.2d 230 (1989).

DNA analysis will also make a contribution in several areas of civil litigation. Tort actions in which the tortfeasor's identity is disputed may be settled earlier if DNA testing leaves only the issue of damages to be resolved. Plaintiffs in paternity cases will also benefit from DNA analysis. Rather than establishing paternity by the process of elimination, DNA testing positively identifies the father.[14] As a result, many alleged fathers are withdrawing their denials of paternity.[15]

That DNA typing has been favorably recognized in the scientific and legal literature is an understatement. All of the journal articles, save a couple, have proclaimed DNA's theory and technique as accepted by the scientific community.[16] However, if the theory, technique, procedure, and legal process are not also accepted by the courts, DNA will not be admissible as evidence, despite the acceptance of DNA testing among molecular biologists, geneticists, and other applicable scientific disciplines. When all is said and done, its admission into evidence rests solely upon principles of law.[17]

THE LEGAL STANDARDS FOR ADMISSIBILITY OF DNA TYPING

Although the theory underlying DNA typing is neither controversial nor of recent vintage, its application to forensic identification in legal proceedings, particularly in the criminal justice system, is new. To the extent that the theory and technique have not been judicially sanctioned in a forensic setting, DNA typing is novel scientific evidence. As such, the proponents of DNA typing, upon its debut in each jurisdiction, will, prior to its judicial acceptance, have to satisfy the test for admissibility prevailing in the particular jurisdiction in which the evidence is to be offered. As noted later in this chapter, once the test has been met, any proponent of similar evidence thereafter will be relieved of the burden of repeatedly showing that DNA typing passes the jurisdiction's test for admissibility. The foundational predicates, also discussed later, will still have to be properly presented.

Generally, it can be said that there are two primary tests used by the courts for determining the admissibility of novel scientific evidence: the *Frye* test and the relevancy test. A number of jurisdictions, however, apply a variation of the judicial themes represented by those tests.

The Frye Test

The majority of jurisdictions apply the test enunciated in *Frye v. United States*, 293 F. 1013 (D.C. Cir. 1923), or a variation thereof.[18] In assessing the evidence before it, the District of Columbia Court of Appeals held:

> Just when a scientific principle or discovery crosses the line between the experimental and demonstrable stages is difficult to define. Somewhere in this twilight zone the evidential force of the principle must be recognized, and while the courts will go a long way in admitting expert testimony deduced from a well-recognized scientific principle or discovery, *the thing from which the deduction is made must be sufficiently established to have gained general acceptance in the particular field to which it belongs.*

293 F. at 1014.

The ultimate objective of the *Frye* test is to ensure that only scientific evidence that is reliable will be admitted. It does so by requiring that the underlying theory and technique have been sufficiently used and tested within the scientific community to have gained general acceptance.[19]

There are numerous articles in legal journals debating the attributes of the *Frye* test.[20] Despite this debate among legal scholars, the fact is that DNA typing has already jumped into the "*Frye*[ing] pan"; in numerous jurisdictions throughout the country DNA typing is being subjected to a *Frye* analysis.

The admissibility of DNA typing pursuant to the *Frye* test is heavily dependent upon the manner in which the court applies *Frye*,[21] although no reported case to date has held DNA identification to be inadmissible using that legal standard. *Castro*, 545 N.Y.S.2d at 987. Application of *Frye* requires that the court focus on three issues.[22]

1. *In which field(s) within the scientific community does the novel scientific evidence belong?*

Selecting the appropriate field may in some cases create difficulties because multiple disciplines may be involved in application of the novel scientific evidence. Selection of a broad field may include scientists who have no experience in the forensic application of the theory and technique, whereas a narrow field might limit the consensus to those who are the proponents of a new theory and technique and who apply them commercially in forensic settings.[23]

With respect to DNA typing, designating the relevant field may depend on whether a court requires the general acceptance of just the underlying principle or, in addition, the technique by which the principle is applied. The relevant fields for DNA theory have been said by one commentator to be molecular biology and genetics.[24] If the reliability of population frequency statistics also goes to admissibility, then the specialties of population genetics and statistics also appear to be included in the relevant fields.[25]

The court in *People v. Wesley*, 140 Misc. 2d 306, 533 N.Y.S.2d 643 (Co. Ct. 1988), which considered the *Frye* test to include both theory and technique, found that "[t]he particular scientific fields that govern DNA Fingerprinting are molecular biology, genetics and a specialized branch of genetics known as population genetics" and later added chemistry, biology, and biochemistry to the list of relevant fields. 533 N.Y.S.2d at 645 and 659. Although not specifically delineating the relevant scientific fields, the court in *Castro* heard testimony from experts in molecular biology, genetics, population genetics, and DNA technology and procedures in assessing the general scientific acceptance of both the theory and technology of DNA identification.

2. *What must be generally accepted by the relevant scientific fields?*

As mentioned above, courts may differ on whether only the underlying scientific principle must be accepted by the scientific community or whether in addition the technique applying that principle must also be generally accepted.[26] To date, the reported cases applying the *Frye* test to DNA typing have required both theory and technique to be generally accepted.

In *Castro*, the first two prongs of a three-prong test used by the court dealt "strictly and exclusively with the *Frye* issue." 545 N.Y.S.2d at 988. These two prongs were

> *Prong I.* Is there a theory, which is generally accepted in the scientific community, which supports the conclusion that DNA forensic testing can produce reliable results?
> *Prong II.* Are there techniques or experiments that currently exist that are capable of producing reliable results in DNA identification and which are generally accepted in the scientific community?

545 N.Y.S.2d at 987. The same was true in *People v. Wesley*, also a New York case.

A Maryland appellate court has been less clear on what is required. Relying on a previous Maryland case that adopted *Frye* as its standard, the court in *Cobey v. Maryland*, 80 Md. App. 31, 559 A.2d 391 (1989) reiterated that "[t]he Frye standard requires that the *method* at issue be 'generally accepted as reliable' in the relevant scientific field or community before the test results derived therefrom may be admitted into evidence." 559 A.2d at 392 (emphasis added). The only findings of general acceptance by the court were with respect to the "procedures . . . constituting DNA fingerprinting" (559 A.2d at 396) and the "method by which Cellmark identified the unique links of Cobey's DNA." 559 A.2d at 398.

3. *What constitutes general acceptance?*

Once a court has determined the scientific fields from which the experts are to be drawn and what must be generally accepted, it must then decide what general acceptance means. Although most courts seem to have addressed this issue inadequately, the definition of "general acceptance" has a great impact on whether a novel scientific technique gains judicial acceptance.[27] A requirement of absolute unanimity among experts may preclude from a jury's consideration new scientific evidence that is nonetheless reliable. On the other hand, tolerating a high degree of divergence among experts may allow the admission of evidence that most scientists in the community find unacceptable.[28] The New York courts, without specifically defining "generally," qualified the term by stating that "the test is not whether a particular procedure is *unanimously* endorsed by the scientific community, but whether it is generally acceptable as reliable." *Castro*, 545 N.Y.S.2d at 986–987 (emphasis added). The Minnesota Supreme Court, considering the admissibility of DNA typing, remarked that it had rephrased the *Frye* standard to require that experts in the field generally agree that the evidence is trustworthy and reliable. *State v. Schwartz*, 447 N.W.2d 422, 424 (Minn. 1989).

Counting scientific heads to determine general scientific acceptance, whatever that standard means, is pragmatically unnecessary, it appears, in the area of DNA typing. There is such a unanimity of acceptance in the scientific and legal field about both the theory and technique of DNA typing that it can safely be predicted no court will be put in the position of counting heads. In *Castro*, the court found unanimity among all the scientists and lawyers as well. No witness challenged the validity of the scientific principle or technology in *Wesley; Cobey; Andrews v. State*, 533 So. 2d 841 (Fla. App. 5 Dist. 1988); *Spencer v. Commonwealth* (Spencer I), 238 Va. 275, 384 S.E.2d 775 (1989), *cert. denied*, __U.S.__, 107 L.Ed. 2d 1073 (1990) or *Spencer v. Commonwealth* (Spencer II), 238 Va. 295, 384 S.E.2d 785 (1989), *cert. denied*, __U.S.__, 110 S. Ct. 759 (1990).

Ascertaining General Acceptance

The proponent of novel scientific evidence has the burden of presenting to the court evidence from which general acceptance can be ascertained. The most common sources of information establishing general acceptance are expert witnesses, scientific and legal literature, and other judicial opinions.

Although the number of reported opinions is currently quite limited, each court publishing an opinion has found DNA typing reliable in both theory and technique.[29] In those cases where *Frye* has been applied, there has been no hesitancy in finding general acceptance of DNA identification.[30]

Scientific literature, and to a degree legal literature also, is heavily relied upon by the courts in assessing scientific acceptance. The *Castro* court cites 17 articles in scientific journals for the proposition that the procedures utilized in DNA typing have gained scientific acceptance. 545 N.Y.S.2d at 990, fn. 3. In *Andrews*, the court noted that one expert witness testified he was unaware of any scientific journal article that argued against the reliability of DNA typing. 533 So. 2d at 850.

No other form of evidence, however, is likely to be as persuasive as a live witness. But even with the expert, there are several issues to consider. The first is the qualifications of the witness. The technician who performs the laboratory procedures and who might normally testify as to the results probably is not qualified to testify to general acceptance, since only a scientist would be sufficiently knowledgeable about opinions held by others in the scientific field.[31]

Some courts also require impartiality on the part of expert witnesses who testify as to reliability and general acceptance.[32] When a witness who testifies is a leading proponent of a particular technique, some courts question whether he or she can be fair and impartial about the reliability or acceptance of that technique. The requirement of impartiality was added by the California Supreme Court in *People v. Kelly*, 17 Cal. 3d 24, 549 P.2d 1240, 130 Cal. Rptr 144 (1976), to establish the *Kelly-Frye* rule.[33] The trial court in *People v. Lopez*, N.Y. L.J., January 6, 1989, p.29, col. 1, commented, as it was issuing guidelines for future DNA cases, that "[g]iven the particular pecuniary interest a commercial company has in seeing that their tests are ratified by a court of law, their objectivity is certainly called into question." On the other hand, the Florida Court rejected such a requirement in *Andrews*, noting that "[n]either *Frye* nor our evidence code require impartiality." 533 So. 2d at 849, fn.9. In *Castro*, one of four experts upon whom the court relied was the Director of Forensic and Paternity Testing at Lifecodes Corp., the examining laboratory in that case. 545 N.Y.S.2d at 986.

Proper Application of the Technique

Several of the courts dealing with the admissibility of DNA evidence have quoted Professor Giannelli for the proposition that the *Frye* test for the admission of novel scientific evidence requires an analysis of three factors: (1) the validity of the underlying principle, (2) the validity of the technique applying that principle, and (3) the proper application of the technique on a particular occasion.[34] In actuality, however, only the first two factors relate strictly to the admission of novel scientific evidence.[35]

The third factor applies whenever scientific evidence is admitted in evidence, whether for the first time or any time thereafter. As discussed later in this chapter, the third factor often affects merely the weight of the evidence or, in other words, the degree of reliance put on the evidence by the fact-finder, rather than the admissibility of the evidence.

Particularly with respect to DNA typing, however, the courts are scrutinizing the laboratory procedures and protocols carefully and hinging admissibility on the proper application of the DNA technique to the evidence before the court. As Professor Giannelli points out, one important flaw in the *Frye* test is that it obscures critical problems in the use of a particular technique by focusing on the general acceptance issue.[36]

Thus, the courts have shifted the focus of their analysis to the appropriateness of the procedures employed by the examining laboratory. The Minnesota Supreme Court, while recognizing that DNA typing is generally accepted, nevertheless noted that specific DNA test results are only as reliable and accurate as the procedures used by the testing laboratory. *Schwartz*, 447 N.W.2d at 426. In part because the laboratory conducting the tests in *Schwartz* did not comply with appropriate standards and controls, the court found the test results to lack foundational adequacy and consequently held them to be inadmissible. 447 N.W.2d at 428.

In each reported decision, any existing controversy over the DNA evidence has centered around the application of the technique. Moreover, each time DNA evidence has been rejected, it has been because of the failure by the examining laboratory to follow accepted procedures. *See Castro* and *Schwartz*. More discussion on the subject follows later in this chapter.

The Relevancy Test of Admissibility

An alternate test for admissibility of novel scientific evidence, applied by a minority of courts, is the relevancy test, primarily espoused by the Federal Rules of Evidence and many state court evidence codes modeled after the Federal Rules.[37]

Under the Federal Rules, which favor the admission of relevant evidence, novel scientific evidence is treated like other expert testimony and is subjected to a balancing test that weighs the probative value of the evidence against the dangers of admitting it.[38] In *United States v. Baller*, 519 F.2d 463, 466 (4th Cir. 1975), *cert. denied*, 423 U.S. 1019 (1975), the court noted that "[u]nless an exaggerated popular opinion of the accuracy of a particular technique makes its use prejudicial or likely to mislead the jury, it is better to admit relevant scientific evidence in the same manner as other expert testimony and allow its weight to be attacked by cross-examination and refutation."

The Third Circuit Court of Appeals in *United States v. Downing*, 753 F.2d 1224 (3d Cir. 1985), became a leading advocate of the relevancy approach. While concluding that the Federal Rules of Evidence neither incorporate nor repudiate the *Frye* test, the court found it appropriate to adopt the more flexible approach to the admissibility of novel scientific evidence suggested by the Federal Rules. In doing so, the court recognized that "some scientific evidence can assist the trier of fact in reaching an accurate determination of facts in issue even though the principles underlying the evidence have not become 'generally accepted' in the field to which they belong." 753 F.2d at 1235. Noting, however, that the Federal Rules require a "quantum of reliability," the court fashioned a relevancy/reliability approach that focuses on reliability as a critical element of admissibility. 753 F.2d at 1235 and 1238.

In Florida, the District Court of Appeals applied this approach in *Andrews*. As *Downing* suggests, the Florida Court assessed the reliability of DNA typing by considering the "novelty" of the new technique, which the court explained as the technique's relationship to more established modes of scientific analysis. The long and extensive use of DNA testing was considered persuasive evidence of reliability. 533 So. 2d at 849.

As with establishing general acceptance, the existence of specialized literature concerning DNA testing also assisted the *Andrews* court in determining reliability. *Andrews*, 533 So. 2d at 850. This fact, as well as the nonjudicial use of the testing procedure, help assure the courts that DNA typing has been exposed to critical scientific scrutiny. *Downing*, 753 F.2d at 1239.

Another consideration promoting the reliability of DNA typing under the relevancy/reliability test is the frequency with which the process leads to erroneous results. *Andrews*, 533 S.2d at 850 and *Downing*, 753 F.2d at 1239. The experts appear to agree that flawed DNA testing will produce no result rather than an erroneous result. Bolstering this built-in safeguard is the utilization of control samples, which permits errors to be discovered. *Andrews*, 533 So. 2d at 850.

Two other indicia of reliability were considered by the *Andrews* court. One was the "qualifications and professional stature of the expert witnesses." *Andrews*, 533 So. 2d at 847. The trial court accepted the witnesses before it as "eminently qualified." *Andrews*, 533 So. 2d at 849. The other factor, implicitly considered by the court, was the acceptance of the technique by the scientific community. Even under the relevancy approach, the acceptance of a novel technique by its relevant scientific fields is one of the variety of factors that may be examined by the court. According to *Downing*, the scientific acceptance of the technique "may be decisive, or nearly so." 753 F.2d at 1238.

The trial court in *Andrews* heard testimony that the principles underlying DNA print identification are generally accepted in the scientific community. 533 So. 2d at 848. Moreover, the appellate court agreed with the prosecutors that, in the case before it, DNA print identification would have met the *Frye* standard as well as the relevancy test. 533 So. 2d at 847, fn. 6.

Because the judicial use of DNA typing was in its infancy at the time, the court in *Andrews* could not avail itself of another indication of reliability suggested in *Downing*, that is, the use of judicial notice of expert testimony in previous judicial proceedings to support or dispute the reliability of DNA typing. *Downing*, 753 F.2d at 1239. At the time of the *Andrews* hearing, there had been few, if any, DNA admissibility hearings. In the future, however, with the proliferation of admissibility hearings, and the resulting increase in appellate court decisions, the use of judicial notice may ease the time and burden of these hearings.

The central question for the *Andrews* court in its relevancy analysis of DNA typing was "whether the probative value of the testimony and test is substantially outweighed by its potential prejudicial effect." 533 So. 2d at 849. Arguably, in some cases, scientific evidence exhibiting substantial reliability may nevertheless confuse or mislead a jury or overwhelm it. *Downing*, 753 F.2d at 1239. But, with respect to DNA print identification, in light of the strong evidence of reliability and with no evidence to the contrary, the scales clearly weigh in favor of admissibility.

Another Approach

The Virginia Supreme Court in *Spencer I* approved the admission of DNA typing using its own more lenient reliability test.[39] Finding that the evidence before the trial court was undisputed, the court concluded that "DNA testing is a reliable scientific technique and that the tests performed in the present case were properly conducted." 384 S.E.2d at 783.

Thus, under Virginia's test, reliability, as opposed to the broader concept of relevancy, appears to be the linchpin of admissibility.[40] At least in *Spencer I*, where the evidence of reliability and scientific acceptance was so overwhelming that it met even the stricter *Frye* test,[41] the court apparently felt comfortable in reviewing the application of a two-prong test requiring only fact-finding by the trial court. Unlike the relevancy test, the Virginia approach does not require the court to employ policy considerations concerning the effect of the evidence on the fact finder.

LEGISLATED ADMISSIBILITY OF DNA TYPING

While the legal community has been grappling with the various issues pertaining to judicial acceptance of DNA typing, struggling along with the scientific community through extended admissibility hearings,[42] and awaiting the ultimate approval of appellate courts,[43] the legislatures of several states have delivered a preemptive strike to the admissibility question. To date, laws have been passed in Maryland, Minnesota, Nevada, and Louisiana authorizing the use of DNA typing evidence in criminal and civil cases.[44]

In *Cobey*, the Maryland appellate court, although approving the trial court's admission of DNA typing evidence, did not wholeheartedly embrace the concept. The court made crystal clear that DNA fingerprinting was not "admissible willy-nilly in all criminal trials." 599 A.2d at 398. Despite the court's reluctance to fully concede the admissibility of DNA, Maryland law, as of January 1, 1990, provides that "in any criminal proceeding, the evidence of a DNA profile is admissible to prove or disprove the identity of any person."[45] The Nevada and Minnesota statutes are broader in that they provide for admissibility in both civil and criminal cases. The Louisiana law, on the other hand, merely establishes DNA evidence as *relevant* proof of the identity of an offender of any crime.

The four statutory schemes also vary as to the purpose for which the DNA typing evidence may be used. Minnesota appears to have no limitations, whereas Maryland and Nevada allow its use to determine the identity of any person. Nevada is more expansive, however, by also including proof of parentage and identification of corpses as proper uses of DNA evidence.

On the other hand, Louisiana limits the scope of its statute by providing statutory admissibility only "to establish the identity of the *offender* of any crime" (emphasis added).[46] How narrowly this restrictive language will be interpreted remains to be seen. It would certainly seem proper to argue that if DNA typing evidence establishes that human remains found in a place under the offender's control were those of the victim, it establishes not only the identity of the remains but, circumstantially, the identity of the offender as well.

Minnesota and Nevada have also legislatively authorized the admissibility of statistical population frequency evidence. The objections to that type of evidence, as found in cases like *Schwartz, Castro* and *Andrews*, will now clearly go to the weight the fact-finder will give the evidence in those states, as opposed to its admissibility.

One other aspect of these statutes is the presence of procedural requirements. The Maryland statute requires notice to the adversary of intent to use DNA evidence at least 15 days prior to the criminal proceeding, discovery of the laboratory report, and production of chain of custody witnesses. Under the Nevada scheme, the court may order that the testimony of experts and of the persons examined be taken by deposition. This provision should presumably be used only in civil litigation, since traditionally in criminal cases depositions of witnesses are not normally allowed, particularly those of victims of sexual assaults. Provisions are also made in the Nevada statute for the issuance of a court order requiring a person to submit to the taking of known exemplars for testing.

In many states, legislation is also the basis for allowing DNA typing in paternity disputes. The Uniform Parentage Act, adopted in 17 states, is broad enough to encompass DNA testing.[47] In other states with parentage legislation requiring court-ordered blood or tissue testing, the language is also probably broad enough that no new legislation is required for DNA typing.[48]

In New York, the Family Court Act directs the court to order the parties to submit to one or more genetic marker tests and authorizes the results of the blood genetic marker tests to be received in evidence. In the case of *In re the Adoption of "Baby Girl S,"* 140 Misc.2d 299, 532 N.Y.S.2d 634 (Sur. Ct. 1988), the court noted that there is no requirement in paternity cases in New York to hold a *Frye* hearing on the admissibility of DNA profiling because the statute specifically provides for the admission of blood genetic marker tests. Consequently, since the DNA probe is undisputedly a blood genetic marker test, it was admitted as proof of parentage under the clear language of the Family Court Act. 532 N.Y.S.2d at 637.

DNA DATA BANKING

It did not take long for politicians in several states to recognize a benefit of DNA typing analogous to that of tactile fingerprinting. Statutes authorizing the creation of DNA data banks have been passed in nine states.[49] When forensic samples of blood, semen, tissue, or hair have been recovered and DNA-typed, the genetic data base can be searched for a match, thereby identifying a suspect. Although the ultimate potential for this investigative tool may not be reached until more states establish data banks that become uniform nationwide,[50] those states that have such a process provide their citizens a degree of enhanced security, if for no other reason than the deterrent effect it will have on potential recidivists.[51]

The FBI is in the process of developing uniform standards for DNA typing that will assist in establishing a nationwide DNA data bank.[52] (See Chapter 6 and

Appendix II). The Washington statute already requires that the system established by that state be compatible with the system used by the FBI.[53]

Even with nationwide data banks, at least one commentator feels that the evidence derived therefrom will not be particularly probative. If an individual is first identified because of a match obtained through the data bank, it is argued, the match will be much weaker evidence of guilt than if the initial identification were by other means not related to genetics. The reason for the lack of probative value of the evidence is purportedly clear when one applies Bayes's theorem.[54]

There have been concerns voiced by some that DNA data banking is a dangerous step threatening the privacy and security of the individual and leading to "genetic redlining."[55] And it has been surmised that these civil libertarian concerns have prevented more states from enacting responsible statutes providing for data banking.[56]

Each of the statutes now in existence is narrowly circumscribed for law enforcement purposes and is triggered only after conviction, in some instances only upon discharge, parole, probation, or other form of release. Moreover, the class of convicts subject to the law is limited in all but three states to those convicted of sex-related crimes. Washington's statute covers individuals convicted of violent crimes as well as sex offenses, and the California statute also covers those convicted of murder or felony assault or battery.[57] In Iowa, the court and parole board are left with the discretion to require a specimen for DNA profiling when placing a defendant who has been convicted of a public offense on probation, work release, or parole.[58]

Legislators have acted in a responsible manner by drafting laws responsive to the concerns over abuse of DNA data banking. Moreover, the current technique for DNA typing reveals virtually nothing about the genetic make-up of an individual.[59] The concern over DNA data banking promoted by some does a disservice to a society suffering from a constant rise in violent crime. John Hicks, Deputy Assistant Director, Laboratory Division, FBI, is reported to have stated that "[t]he use of DNA technology in law enforcement when applied in a responsible and well-coordinated manner with due consideration for the needs of our criminal justice system will enhance the effectiveness, efficiency and productivity at all levels" and "enhanc[e] the Nation's effort to combat violent crime."[60]

LEGAL FOUNDATION FOR GENERAL ADMISSIBILITY

Any party who seeks to introduce scientific evidence at a trial must establish a proper foundation. The nature of this foundation may vary from case to case, and, depending on the circumstances, the failure to lay it properly may affect the weight to which the trier of fact gives the evidence or may go to the very issue of its admissibility.

In general, issues concerning a proper foundation may arise even after a scientific theory and technique are no longer novel. One is whether the government in a criminal case properly obtained the questioned biological samples and the known exemplars for comparison with each other. A second is whether the proponent of the evidence can properly authenticate the substances that have been submitted for scientific analysis. Finally, the laboratory procedures, protocol, and quality control must be shown to have been appropriate and to have been followed with respect to the analysis of the evidence proffered for admission.

Obtaining Known and Unknown Biological Samples

As a foundation matter, the prosecution, if it is the proponent of DNA identification evidence, must satisfy the court, if challenged, that the physical evidence upon which the testing was conducted was obtained in accord with the Fourth Amendment to the Constitution, which proscribes unreasonable searches and seizures.[61] The constitutionality of the seizure of evidence is usually argued and decided at a pretrial suppression hearing, although sometimes the defense may not object to the evidence until it is offered at trial.

Collecting forensic evidence such as blood, hair, saliva, skin, or semen from a victim or a crime scene usually will not involve Fourth Amendment issues because the defendant will be unlikely to have a reasonable expectation of privacy in the object or place from which the specimens are seized.[62]

Collecting known samples from the defendant, however, may involve Fourth Amendment rights.[63] Prior to arrest, samples may be collected from a suspect pursuant to a search warrant, based upon probable cause and issued by a magistrate. Post arrest, the prosecution may obtain either a warrant or a court order directing that the defendant submit to the taking of biological samples. Refusal by the defendant to submit to a collection of specimens may be admissible at trial as circumstantial evidence of guilt.[64]

The more difficult situation arises when the police have no particular individual they suspect as the perpetrator of the crime they are investigating, or when they have a suspect but lack sufficient evidence to establish probable cause for a search warrant. The police must then resort to methods justifiable on a standard of proof less than that which has traditionally been required for a search warrant.

Nine states have enacted statutes or court rules that allow nontestimonial identification evidence to be obtained without a showing of probable cause.[65] These provisions can be used to gather the required samples of body fluid, skin, or hair without probable cause.

One commentator has argued in favor of a rule similar to a proposed Federal Rule of Criminal Procedure, which had previously been suggested by the Committee on Rules of Practice and Procedure of the Judicial Conference of the United States, but never adopted.[66] With several safeguards built into the Rule, it is argued that the concerns over privacy, harassment, and misuse of DNA information could be mollified. A commentator with an opposing view, however, suggests that the unique informational aspect of DNA profiling adds a new dimension to the Fourth Amendment analysis of what is reasonable in the context of compulsory identification procedures on less than probable cause, requiring heightened protection.[67] In that commentator's view, collecting evidence from a suspect for DNA profiling would be inappropriate on anything less than probable cause.

Chain of Custody

Before an expert will be allowed to testify at a trial as to the results of the DNA analysis, the proponent of the expert witness must establish a proper chain of custody for the substances subjected to the analysis. This requirement applies regardless of which party in a criminal or civil case is offering the expert.

The chain of custody requirement has two objectives. The first is to lay a proper foundation connecting the evidence to the defendant or to a place or object that is relevant to the case. This is a form of authentication that establishes the relevance of the evidence, one of the prerequisites to admissibility.

The second purpose of a chain of custody for physical evidence is to ensure that the object is what its proponent claims it to be. This is accomplished by ruling out any tampering with, and substantial alteration or substitution of, the evidence. If the substance analyzed for the presence of DNA has been tampered with or altered in a significant way, it becomes, in effect, a substance different from the one originally seized, and its relevance to the case disappears.

Alterations performed as a result of testing the substance, of course, do not affect the chain of custody. In most cases, the critical links in the chain of custody are those from the time the evidence was obtained to the time it was scientifically analyzed, since the latter is the time at which the integrity of the evidence is of paramount importance.

Similarly, if another substance is substituted for that which was originally obtained, the evidence is no longer authentic and cannot be admitted into evidence. By showing the proper custody of the evidence, any question of substitution, alteration, or tampering can be dispelled.

Particularly with DNA typing, a proper chain of custody is required because the substances analyzed are not identifiable in a singular sense. Consequently, simple recognition of the container or of the initials on the container holding the substance is insufficient. Mere identification in this sense cannot be submitted for proof that the substance has not been tampered with or that alterations or substitutions have not occurred.

Failure to adequately prove the chain of custody of the substance subjected to DNA testing will undoubtedly cause the exclusion of the evidence. Once this factual predicate cannot be met, the expert's testimony becomes irrelevant and consequently inadmissible.

The party offering the evidence shoulders the burden of proving the chain of custody. Once a showing has been made to a reasonable certainty, even though there may be some weak links, the burden shifts to the opponent to establish that the chain has been broken. If unable to successfully show a missing link, opposing counsel is left with arguing that the weight the judge or jury gives the evidence, including the expert's testimony, should be greatly diminished because of the possibility that the integrity of the evidence has been compromised.

Proper Laboratory Procedures

According to Professor Giannelli, evidence derived from a scientific principle obtains its reliability from the validity of the underlying theory, the validity of the technique applying that principle, and the proper application of that technique on a particular occasion.[68] The validity of the scientific theory and the technique are factors relevant to the admission of novel scientific evidence. Once the courts have accepted the theory and technique, they can be accorded judicial notice by the courts, and the proponent of the evidence is no longer required to establish them at each trial.[69]

However, the proper application of the technique by the laboratory with respect to the particular evidence being offered and the qualifications of the experts are issues that should be raised by the opposing counsel whenever scientific evidence is to be introduced. If the technique is not properly applied by qualified experts, or if laboratory procedures and protocols are not followed and quality control devices ignored or compromised, the reliability of the test results is diminished or destroyed, and relevance is lost.

The degree to which there is deviation from appropriate laboratory procedures, protocols, and quality controls normally goes to the weight given the expert's testimony by the trier of fact. For example, in *Andrews*, the defendant contended that the DNA testing was unreliable because new gel used in the electrophoresis process is tested only against the old gel and if the old gel worked improperly, that error would be carried over to the new gel. Although not accepting the defendant's argument, the court indicated that the "probative value of the evidence is for the jury." *Andrews*, 533 So. 2d at 849.

Similarly, in *Wesley*, the defendant, although acknowledging that the theory underlying DNA typing is valid and generally accepted in the scientific community, attacked the laboratory's procedures, methodology, and quality controls as inadequate and objected to the population studies used as a basis for a claimed power of identification for results under the laws of population genetics. Recognizing that these objections probably go "to the weight of the evidence, a matter of resolution by the trier of facts," as opposed to admissibility, the court found that it did not have to reach that question since neither objection was supported by the evidence presented at the pretrial hearing. *Wesley*, 533 N.Y.S.2d at 650–51.

The deviation from the acceptable procedures, protocols, and quality controls may be so gross, however, that a trial judge in his discretion may exclude the test results and testimony interpreting those results from evidence entirely. In such a case, then, the failure to properly apply the technique on that particular occasion would affect admissibility, rather than just the weight of the evidence. *See Castro,* 545 N.Y.S.2d at 999.

In *Castro*, the court viewed DNA identification as a complex area having a powerful impact on the jury, thus requiring an approach different from simply applying the standard admissibility tests. The focus of the judicial scrutiny was an examination of the actual testing procedures performed in that particular test. The court observed that "a scientist may have no trouble accepting the general proposition that DNA typing can become reliable, yet still have doubts about the reliability of the test performed by a particular laboratory." 545 N.Y.S.2d at 995.[70]

In the *Castro* case, the examining laboratory conducted DNA profiling on a bloodstain extracted from the defendant's watch. The accused had been charged with a double homicide of a mother and daughter. Representatives from the laboratory claimed the stain profile matched the adult victim's and the pattern frequency was estimated at 1 to 1 billion in the Hispanic population. The defendant claimed the stain was his own blood.

The trial court found that the defense was successful in demonstrating that "the testing laboratory failed in its responsibility to perform the accepted scientific techniques and experiments in several major respects." *Castro*, 545 N.Y.S.2d at 996. Because of the potential ramifications of this ruling to all cases involving DNA profiling, it is instructive to consider the specific areas of controversy, as reported by one of the defense witnesses in the *Castro* case.[71]

1. Two additional DNA profile bands were detected in the stain relative to the mother's profile produced from a whole blood sample. The prosecution claimed these bands were contaminants of nonhuman origin but that they were unidentifiable. Two extra bands were also observed in the daughter's DNA profile when compared with the stain profile. These were not reported.
2. Profile matches were declared by visual observation and were not confirmed by objective measurement criteria. Some bands did not truly match when later measured. The laboratory then attempted to modify the measurement system. This last-minute methodology change was not deemed acceptable by the court.
3. No positive control (male DNA) was assayed when a Y-chromosome-specific probe was hybridized with the stain DNA. As anticipated with a female sample, hybridization did not occur; however, there was no proof that the hybridization procedure was in fact successful.
4. A homozygous band of greater than 10 kb size was reported even though the stain DNA was degraded. The defense queried whether a larger band was perhaps not detected because of the poor quality of the specimen. The prosecution then suggested that an Alu repeat sequence at 23 kb could be observed and this was proof that larger fragments were detectable. The

court did not find this reasoning acceptable because only single-copy sequences were hybridized with the forensic probe, whereas the Alu probe detected a massive amount of repeat sequence.

5. At least one of the human probes was known to be contaminated with bacterial sequences, and bacterial and plasmid probes were contaminated with human sequences. Apparently no attempt had been made to eliminate these contaminants.
6. The procedure followed to calculate the probability that a match might have arisen by chance apparently included unwarranted assumptions and inconsistent procedures. Different match rules were used for the forensic and the population allele size database. This could account for at least an 8,000-fold error. Large deviations were discovered between the observed and expected population genotype frequencies for Hardy-Weinberg equilibrium. These results suggested the population used to determine allele frequencies was heterogeneous—consisting of two or more subpopulations with different allele and genotype frequencies. If Hardy-Weinberg and linkage equilibria do not exist, there is no reliable technique to convert allele frequencies into overall genotype frequencies. Disregard for this fact can result in spuriously low probabilities of a match.

Although the court ruled that the DNA identification evidence of inclusion was inadmissible, it did find that the evidence of exclusion—that is, that the bloodstain was not contributed by the defendant—was admissible as a question of fact for the jury. The defendant finally pled guilty to two counts of second degree murder and stated that the blood got on the watch while he was stabbing one of the victims.[72]

To reinforce its concern that analyzing laboratories in DNA cases perform the experiments and calculations in a manner as to yield "results sufficiently reliable to be presented to the jury," the *Castro* court suggested that pretrial hearings be conducted in each case. 545 N.Y.S.2d at 998–99.

The following pretrial procedures were recommended:

1. Notice of intent to offer DNA evidence should be served as soon as possible.
2. The proponent, whether defense or prosecution, must give discovery to the adversary, which must include: (1) copies of autorads, with the opportunity to examine the originals; (2) copies of laboratory books; (3) copies of quality control tests run on material utilized; (4) copies of reports by the testing laboratory issued to proponent; (5) a written report by the testing laboratory setting forth the method used to declare a match or nonmatch, with actual size measurements, and mean or average size measurements, if applicable, together with standard deviation used; (6) a statement by the testing lab, setting forth the method used to calculate the allele frequency in the relevant population; (7) a copy of the data pool for each of the loci examined; (8) a certification by the testing lab that the same rule used to declare a match was used to determine the allele frequency in the population;

(9) a statement setting forth observed contaminants, the reasons therefore, and tests performed to determine the origin and the results thereof; (10) if the sample is degraded, a statement setting forth the tests performed and the results thereof; (11) a statement setting forth any other observed defects or laboratory errors, the reasons therefore and the results thereof; and (12) chain of custody documents.

3. The proponent shall have the burden of going forward to establish that the tests and calculations were properly conducted. Once this burden is met, the ultimate burden of proof shifts to the *adversary* to prove, by a preponderance of the evidence, that the tests and calculations should be suppressed or modified (citation omitted). 545 N.Y.S.2d at 999.

With respect to whether the results of the pretrial hearing affect the weight of the evidence or the admissibility, the *Castro* court said, "It is noted that issues of fact which arise as a result of the hearing concerning the reliability of any particular test, or the size or ratio of the population frequency, relates to the weight of the evidence and not its admissibility. However, where the results are so unreliable, as was demonstrated in this case, the results are inadmissible as a matter of law." 545 N.Y.S.2d at 999.

PROCEDURAL AND ETHICAL CONCERNS

Procedural Safeguards

As a number of commentators have suggested, the implementation of several procedural safeguards could well inhibit unreliable evidence from reaching the legal threshold required for judicial acceptance. Professor Giannelli notes that an underlying concept inherent in the relevancy test is the ability of the adversary process to expose unreliable techniques.[73] The safeguards he recommends to ensure the effectiveness of the adversary process include (1) requiring the proponent of the new technique to give advance notice to the opponent of the intent to offer the evidence, (2) full pretrial disclosure of all information necessary to prepare effective cross-examination, (3) deposition of experts, (4) preservation of evidence for retesting by the adversary's experts, and (5) appointment of defense expert witnesses.[74]

The pretrial procedures outlined in *Castro* and set forth in the previous section were designed by the court to ensure the admission of reliable evidence. Comporting in part with the suggestions by Professor Giannelli, they overlap with procedural safeguards required by courts in other DNA cases.

The trial court in *Lopez*, for example, also established guidelines to facilitate the admission of DNA evidence in criminal cases. First, the state must shoulder the burden of introducing the examining laboratory's protocol, which must be an accepted scientific protocol. The scientist who performed the test should establish that the protocol was followed in the particular test, the results of which are to be offered in evidence. A complete chain of custody must also be established.

Joining with the *Castro* and *Lopez* courts, the Minnesota Supreme Court has also expressed concerns over unreliable test procedures in its decision in *Schwartz*. A representative from "Cellmark," the examining laboratory, admitted that the laboratory did not meet the minimum guidelines of the Technical Working Group on DNA Analysis Methods (TWGDAM) coordinated by the FBI or of the California Association of Crime Laboratory Directors (CACLD). Moreover, "Cellmark" falsely identified two samples as coming from the same subject during a proficiency test administered by CACLD. 447 N.W.2d at 426–27.

Particularly in cases where the laboratory procedures are questionable, but even where they are reliable enough to ensure accurate results, the court in *Schwartz* will not admit DNA test results when data relied upon by the laboratory in performing the tests are not available to the adversary for review and cross-examination. 447 N.W.2d at 427. Although the court acknowledged that DNA samples cannot practicably be saved for the defense to test, the next best safeguard is to provide liberal access to the data, methodology, and actual test results. Only then can the defense have an adequate opportunity for independent review. 447 N.W.2d at 427.

Yet another reason for full access exists, according to the *Schwartz* court: "The validity of testing procedures and principles is assessed in the scientific community by publishing data in peer review journals." 447 N.W.2d at 427. Because "Cellmark's protocols" have not been adequately made public, efforts to assess its methodology have been unsatisfactory. Thus, in Minnesota, not only is admissibility of DNA test results dependent on compliance with appropriate standards and controls, but also on the accessibility of the testing data and results. 447 N.W.2d at 428.

Because of the persuasiveness of frequencies attributable to particular probes, the *Lopez* court further suggested that any probe used in the test should be shown to have passed a peer review process. This process is to be a review of the scientific research by "an anonymous panel of at least two scientists eminently qualified in their designated field." N.Y. L. J., Jan. 6, 1989, at 29, Col. 1.

With respect to the persuasiveness of the frequency testimony that concerned the *Lopez* court, the state in *Schwartz* argued that adequate cross-examination and proper limiting instructions would prevent unfair prejudice. The court was not persuaded, however, that those safeguards would be adequate. On the contrary, it felt that juries in criminal cases would give undue weight and credence to statistical evidence. Consequently, the testimony as to population frequency statistics was held inadmissible. 447 N.W.2d at 428–29.[75]

Commentators in the legal literature have echoed the concerns over the establishment of standards and controls. One such author has recommended that the following controls and standards be developed prior to admitting DNA test results: (1) controls to ensure the accurate interpretation of results, (2) standards for declaring matches, (3) standards for the choice of and number of polymorphic sites studied, (4) standards for determining probability of a coincidental match and for determining the relevant population studies, (5) standards for record keeping, and (6) standards for proficiency testing and licensing.[76]

Reacting to concerns over the reliability of the procedures utilized in DNA typing, courts, legislatures, and organizations have begun to review and establish guidelines, procedures, and protocols. The Office of Technology Assessment is preparing a report on technical issues in DNA profiling, and both the National Academy of Sciences and the National Institute on Standards and Technology are exploring reliable standards and procedures for DNA testing.[77]

The Minnesota Supreme Court, as a result of the *Schwartz* case, has referred the task of recommending appropriate procedures and standards for DNA testing to the Supreme Court Advisory Committees on Rules of Evidence and Criminal Procedure. *Schwartz*, 447 N.W.2d at 428, fn. 4. And as noted in *Schwartz*, both the TWGDAM and the CACLD have developed standards for DNA typing.

Finally, another commentator argues that indigent defendants should be entitled to free DNA printing at a facility of their choice. He also encourages investigators to save enough of the forensic sample to allow the defense to perform its own DNA test.[78] The *Schwartz* court agrees that saving a portion of the specimen would be ideal. But, as stated before, the court recognizes that realistically it will infrequently happen; the samples, because of their size, are almost always entirely consumed in testing. 447 N.W.2d at 427.

Ethical Concerns

Several policy/ethical considerations have already been touched upon. They include broader discovery, greater disclosure of laboratory protocol and procedure, timely notice of intent to use DNA evidence, enhanced protection under the Fourth Amendment, and increased protection of privacy interests, all of which involve the tremendous potential DNA typing has for use and abuse in and out of the courtroom.

An example of what some consider overreaching on the part of law enforcement is related in *Florida v. Cayward*, Fla. Dist. Ct. App. 2d Dist., No. 89-00702, November 15, 1989, 58 U.S.L.W. 2347 (Dec. 19, 1989). There the police, with the knowledge of the prosecutor, fabricated two scientific reports and used them as ploys to interrogate a suspect in a sexual assault/murder case. One false report was prepared on official stationery of the Florida Department of Criminal Law Enforcement, the other on the stationery of the Lifecodes Corp., a DNA testing laboratory.

As a result of being confronted with these documents, which purportedly established that the semen stains on the victim's underwear came from the suspect, he confessed. The trial court suppressed that part of the statement given by the defendant subsequent to the use of the documents or the information contained in them by the police.

On appeal, the appellate court agreed that such conduct on the part of law enforcement cannot be sanctioned. Although a clear majority of federal and state courts agree that police deception does not render a confession involuntary per se, outrageous government misconduct violates due process and thereby bars the government from invoking judicial process to obtain a conviction.

A qualitative difference existed, in the court's mind, between verbal artifices and falsely contrived scientific documents, the latter impressing one as being inherently more permanent and outwardly reliable. Several practical concerns also made a difference to the court. It feared that the false documents might be retained and filed in police paperwork, eventually making their way to the courtroom or into the press. Finally, the use of false documents erodes the public's confidence in law enforcement generally.

CONCLUSION

DNA typing has gained tremendous popularity in the United States since its introduction in 1987.[79] Its possibilities are limitless. And with its claims of mystic infallibility and virtual certainty of identification, it is not surprising that the law enforcement community has fully embraced the technique. But some caution is advised. In most of the initial cases involving DNA printing, there were no challenges to the reliability of the theory and technique, and only very weak opposition to the specific laboratory procedures.

Now, however, concerns over the reliability of laboratory procedures are being expressed in the legal and scientific literature, and media accounts of DNA typing are no longer entirely favorable.[80] Legal attacks in court have become more intense, with judges rejecting DNA evidence in some cases because of procedural irregularities.

The dialogue concerning the reliability of forensic applications of DNA typing is healthy for all involved. Introspection by the scientific community and critical examination by the legal community will help refine DNA typing to a point where no one will lack confidence in its use. Then its true potential can be fulfilled. In the meantime, DNA laboratories should be encouraged to follow the lead of the FBI, which through a spokesman has indicated it will treat its DNA data "in a way that's very, very conservative."[81]

NOTES

1. Comment, *DNA Fingerprinting and Paternity Testing*, 22 U.C.D. L. REV. 609, 635 n. 98 (1989), quoting Professor James E. Starrs, Professor of Law and Forensic Science, The George Washington University, The National Law Center.
2. *See* legislative finding (1989 C.350) accompanying WASH. REV. CODE 43.43.752 (1989).

3. *See* Preamble to MD. ANN. CODE § 10-915 (Supp. 1989) found in 1989 LAWS of MD., ch. 430.
4. Dept. of Justice & FBI, *Uniform Crime Rep.: Crime in the U.S.* 47 (1989); Comment, *DNA Identification Tests and the Courts*, 63 WASH. L. REV. 903, 905 n.2 (1988).
5. Comment, *supra* note 1, at 640.
6. Pearsall, *DNA Printing: The Unexamined "Witness" in Criminal Trials*, 77 CALIF. L. REV. 665, 693 (1989); *People v. Bailey*, 140 Misc.2d 306, 533 N.Y.S.2d 643 (Co. Ct. 1988).
7. Comment, *supra* note 4, at 905 n2.
8. *Id.* at 905 n2.
9. Comment, *Admit It! DNA Fingerprinting Is Reliable*, 26 HOUS. L. REV. 677, 704 (1989); Comment, *supra* note 1, at 630 n75.
10. Comment, *supra* note 9, at 705.
11. Comment, *supra* note 4, at 905 n2.
12. Comment, *supra* note 9, at 703.
13. 75 A.B.A. J. 19 (Oct. 1989); Pearsall, *supra* note 6, at 694–95.
14. Comment, *supra* note 1, at 613.
15. *Id.* at 629.
16. *See generally* Comment, *DNA Identification Tests and the Courts*, 63 WASH. L. REV. 903 (1988); Brown, *DNA and Kelly-Frye: Who Will Survive in California?*, 11 CRIM. JUST. J. 1 (1988); Burk, *DNA Fingerprinting: Possibilities and Pitfalls of a New Technique*, 28 JURIMETRICS J. 455 (1988); Note, *The Dark Side of DNA Profiling: Unreliable Scientific Evidence Meets the Defendant*, 42 STAN. L. REV. 465 (1990); Pearsall, *supra* note 6 at 665; Comment, *supra* note 1 at 609; Comment, *supra* note 9 at 677; Note, *DNA Typing: A New Investigatory Tool*, 1989 DUKE L.J. 474 (1989); Thompson and Ford, *DNA Typing: Acceptance and Weight of the New Genetic Identification Tests*, 75 VA. L. REV. 45 (1989); Williams, *DNA Fingerprinting: A Revolutionary Technique in Forensic Science and Its Probable Effects on Criminal Evidentiary Law*, 37 DRAKE L. REV. 1 (1987–88). The scientific literature is referred to liberally in these articles and in *People v. Castro*, 545 N.Y.S.2d 985, 990 n.3 (Sup. Ct. 1989).
17. Burk, *supra*, note 16, at 467 n. 54.
18. Thompson and Ford, *supra*, note 16 at 53.
19. Id.
20. *See generally* Borders, Jr., *Fit to Be* Fryed: Frye v. United States *and the Admissibility of Novel Scientific Evidence*, 77 KY. L. J. 849 (1988–89); Giannelli, *The Admissibility of Novel Scientific Evidence:* Frye v. United States, *a Half-Century Later*, 80 COLUM. L. REV. 1197 (1980); Moenssens, *Admissibility of Scientific Evidence—Should the Frye Rule Be Maintained?*, 25 WM. & MARY L. REV. 545 (1984); MOENSSENS, INBAU, and

STARRS, SCIENTIFIC EVIDENCE IN CRIMINAL CASES (3rd ed. 1986); Rules for Admissibility of Scientific Evidence, 115 FRD 79 (1987); Starrs, *"A Still-Life Watercolor"*: Frye v. United States, 27 J. FOR. SCI. 684 (July 1982).

21. Thompson and Ford, *supra* note 16, at 101.
22. Giannelli, *supra* note 20, at 1197–1208.
23. *Id.* at 1208–10.
24. Thompson and Ford, *supra* note 16, at 56.
25. *Id.* at 57.
26. *See generally* Giannelli, *supra* note 20.
27. Giannelli, *supra* note 20, at 1210–11.
28. *Id.* at 1211.
29. Andrews v. State, 533 So. 2d 841 (Fla. App. 5 Dist. 1988); Cobey v. State, 80 Md. App. 31, 559 A.2d 391 (1989); Spencer v. Commonwealth (Spencer I), 238 Va. 275, 384 S.E.2d 775 (1989), Spencer v. Commonwealth (Spencer II), 238 Va. 295, 384 S.E.2d 785 (1989), *cert. denied*, State v. Schwartz, 447 N.W.2d 422 (Minn. 1989); West Virginia v. Woodall, People v. Castro, 545 N.Y.S.2d 985 (Sup. Ct. 1989); People v. Wesley and People v. Bailey, 140 Misc. 2d 306, 533 N.Y.S.2d 643 (Co. Ct. 1988); In re the Adoption of "Baby Girl S," 140 Misc.2d 299, 532 N.Y.S.2d 634 (Sur. Ct. 1988); People v. Lopez, N.Y. L. J. p. 29, Col. 1 (Jan. 6, 1989).
30. *Cobey, supra* note 29; *Castro, supra* note 29; *Wesley* and *Bailey, supra* note 29; *Lopez, supra* note 29; Although the courts in *Andrews, supra* note 29, and *Spencer, supra* note 29, did not apply the *Frye* test, both courts indicated the evidence before them would satisfy the *Frye* test. Cellmark Diagnostics publishes a quarterly newsletter entitled BENCHMARK, in which the company lists recent DNA court appearances and whether the *Frye* test was applied.
31. Giannelli, *supra* note 20, at 1214–15; Comment, *supra* note 4, at 944; Brown, *DNA and* Kelly-Frye: *Who Will Survive in California?*, 11 CRIM. JUST. J. 1, 54 (1988).
32. *Andrews v. State, supra* note 29, at 849 n 9.
33. Brown, *supra* note 31, at 3.
34. *See, e.g., Andrews, supra* note 29, at 843 and *Lopez, supra* note 29, Col. 1..
35. Giannelli, *supra* note 20, at 1202.
36. *Id.* at 1226.
37. *Id.* at 1228–29, 1235; Rules 401, 402, and 403, Federal Rules of Evidence. There has been some disagreement as to whether the Federal Rules of Evidence supersede the *Frye* test. A number of federal courts and state courts with evidence codes similar to the Federal Rules nevertheless continue to apply the *Frye* test. Comment, *supra* note 4, at 934 2nd n156; Giannelli, *supra* note 20, at 1229; United States v. Downing, 753 F.2d 1224, 1234 (3rd. Cir. 1985).

38. Comment, *supra* note 4, at 934.
39. The Virginia Supreme Court had previously rejected the *Frye* test in O'Dell v. Commonwealth, 234 Va. 672, 695–96, 364 S.E.2d 491, 504 (1988), *cert. denied*, __U.S.__, 109 S.Ct. 186 1988.
40. The relevancy approach recognizes relevancy as the linchpin of admissibility, while also ensuring that only reliable scientific evidence is admitted. *Andrews, supra* note 29, at 841–846.
41. *Spencer I, supra* note 29, at 783 n10.
42. In *Castro, supra* note 29, the admissibility hearing extended over a 12 week period, resulting in a 5,000 page transcript.
43. Even after hearings as extensive as in *Castro, supra*, note 29, the appellate court may never have an opportunity to review the trial court's opinion. In *Castro*, for example, the defendant pled guilty after the admissibility hearing. The entry of a plea of guilty ends the matter in most jurisdictions and neither side is able to obtain a review of the trial court's decision.
44. LA. REV. STAT. ANN. tit. 15 § 441.1 (Supp. 1990); MD. ANN. CODE § 10-915 (Supp. 1989); MINN. STAT. ANN. §§ 634.25 and 634.26 (Supp. 1989), and NEV. REV. STAT. § 56.020 (Supp. 1989). For a comprehensive overview of DNA legislation, *see* Starrs, *Twists and Turns in DNA's Legislative Ladder*, 2 BENCHMARK 1 (Fall 1989). BENCHMARK is published quarterly by Cellmark Diagnostics.
45. *See* note 44, *supra*.
46. *See* note 44, *supra*.
47. *See* Starrs, *supra* note 44.
48. *Id.*
49. ARIZ. REV. STAT. ANN. § 31-281 (Supp. 1989); CAL. PENAL CODE § 290.2 (Deering 1989), amended ch. 1304 (1989); COLO. REV. STAT. ANN. § 17-2-201(5)(g)(I)(Supp. 1989); FLA. STAT. § 943.325 (1989); IOWA CODE ANN. §§ 13.10, 901.2, and 906.4 (Supp. 1989); MINN. STAT. ANN. §§ 299C.155 and 609.3461 (Supp. 1990); NEV. REV. STAT. ANN. §§ 176.111 and 179A.075 (Supp. 1989); VA. CODE § 53.1-23.1 (1950 as amended); WASH. REV. CODE 43.43.752, 43.43.754, 43.43.756, and 43.43.758 (1989).
50. Brown, *supra* note 31, at 21.
51. A legislative recognition of the deterrent effects of DNA data banking can be found in the Iowa statute, *supra* note 49.
52. Brown, *supra* note 31, at 21.
53. *See* note 49, *supra*.
54. Thompson and Ford, *supra* note 16, at 100–101. "According to Bayes's theorem, evidence is probative to the extent it is more likely to exist if the defendant is guilty than if he is innocent. In this case a match between the suspect and perpetrator is nearly certain if the suspect is in fact guilty. But a match between the suspect and perpetrator is also certain if he is not guilty,

because he was chosen based on the fact he matches. Hence, Bayesian analysis suggests that this evidence has *no* probative value" (emphasis in original).

55. Note, *The Dark Side of DNA Profiling: Unreliable Scientific Evidence Meets the Criminal Defendant*, 42 STAN. L. REV. 465, 533 (1990). "Genetic redlining" is the differentiated treatment of individuals based upon apparent or perceived human variation. Also *see* Pearsall, *supra* note 6; Comment, *supra* note 9; Barinaga, *DNA Fingerprinting Database to Finger Criminals*, 331 *NATURE* 203 (1988).
56. Starrs, *supra* note 44, and *see* note 49, *supra.*
57. *See* note 49, *supra.*
58. *See* note 49, *supra.*
59. Burk, *supra* note 17, at 471; Note, *DNA Typing: A New Investigatory Tool,* 1989 DUKE L. J. 474, 483 (1989).
60. Note, *supra* note 55, at 537.
61. The Fifth Amendment privilege against self-incrimination would not be involved. *See* Williams, *DNA Fingerprinting: A Revolutionary Technique in Forensic Science and Its Probable Effects on Criminal Evidentiary Law*, 37 DRAKE L. REV. 1, 23, and 27 (1988). But *see* Note, *supra* note 55, at 533 and n392.
62. However, if the crime occurs in a place where the defendant has a reasonable expectation of privacy, such as his home, then the Fourth Amendment may be implicated. *See* Mincey v. Arizona, 437 U.S. 385, 98 S.Ct.2408, 57 L.Ed.2d 290 (1978). Also, if unknown biological specimens are removed from the defendant at the time of his arrest, the lawfulness of the arrest and warrantless seizure of the specimens will have to be established.
63. For a more detailed discussion of this topic, *see* Williams, *supra* note 61, at 20–31.
64. Williams, *supra* note 61, at 29.
65. ALASKA R. CT. 16(c)(1)–(2) (1988); ARIZ. REV. STAT. ANN. 13-3905 (1978); COLO. R. CRIM. P. 41.1 (1984); IDAHO CODE ANN. 19-625 (1987); IOWA CODE ANN. 810.1–.2 (West 1978 & Supp. 1988); NEB. REV. STAT. 29-3301 to -3307 (1985); N.C. GEN. STAT. §§ 15A-271 to -282 (1983); UTAH CODE ANN. 77-8-1 to -4 (1982); VT. R. CRIM. P. 41.1 (1983). Note, *supra* note 59, at 489. The Nebraska Supreme Court has interpreted its statute, however, to require probable cause. State v. Evans, 215 Neb. 433, 338 N.W.2d 788 (1983).
66. Proposed Rule 41.1 of the Federal Rules of Criminal Procedure reads in part as follows:

(c) BASIS FOR ORDER. An order shall issue only on an affidavit or affidavits sworn to before the federal magistrate and establishing the following grounds for the order:

(1) that there is probable cause to believe that an offense has been committed;

(2) that there are reasonable grounds, not amounting to probable cause to arrest, to suspect that the person named or described in the affidavit committed the offense; and

(3) that the results of specific nontestimonial identification procedures will be of material aid in determining whether the person named in the affidavit committed the offense.

(l) DEFINITION OF TERMS. As used in this rule, the following terms have the designated meanings: . . .

(3) "Nontestimonial identification" includes identification by fingerprints, palm prints, foot prints, measurements, blood specimens, urine specimens, saliva samples, hair samples, or other reasonable physical or medical examination, handwriting exemplars, voice samples, photographs, and lineups.

52 F.R.D. 409, 462 (1971); Note, *supra* note 59, at 476.

67. Note, *supra* note 55, at 531.
68. Giannelli, *supra* note 20, at 1200–01.
69. *Id.* at 1202–03.
70. Citing Thompson and Ford, *supra* note 16, at 57–58.
71. Lander, *DNA Fingerprinting on Trial*, 339. NATURE 501–505 (1989). Dr. Lorne Kirby contributed the summary of the six areas of controversy.
72. Verified with the Bronx County prosecutor.
73. Giannelli, *supra* note 20, at 1239–40.
74. *Id.* at 1240–1245.
75. However, as of August 1, 1989, a legislative enactment provided for the admissibility of statistical probability evidence. MINN. STAT. ANN. § 634.26 (Supp. 1989).
76. Note, *supra* note 55, at 479.
77. Note, *supra* note 55, at 494.
78. Pearsall, *supra* note 6, at 677–78.
79. Note, *supra* note 55, at 477.
80. Kolata, *Some Scientists Doubt the Value of "Genetic Fingerprint" Evidence*, N. Y. Times, Jan. 29, 1990, at 1.
81. *Id.*

CHAPTER 11

Case Applications

A number of forensic and family relationship cases, as well as medical, animal science, wildlife poaching investigation, and plant science applications are presented in this chapter. As suggested by the titles and headlines from various journal and newspaper articles (Figure 11-1), the process of identification using recombinant DNA technology has proven to be very practical. Semen from a rape case, a hair follicle from a homicide, blood stains at a break-in, chorionic villi from a prenatal diagnosis, blood cells from a transplant patient, tumorous tissue, a big game animal gut pile, a freezer steak, rare condor blood, a whale skin biopsy, plant tissue, and ancient human and other animal remains are some of the sources of DNA used for typing.

Perhaps the most apparent indicator of application potential can be deduced from the number of recent patent applications covering recombinant DNA processes and products. In addition, many new government and commercial ventures have been established to accommodate the anticipated service load.

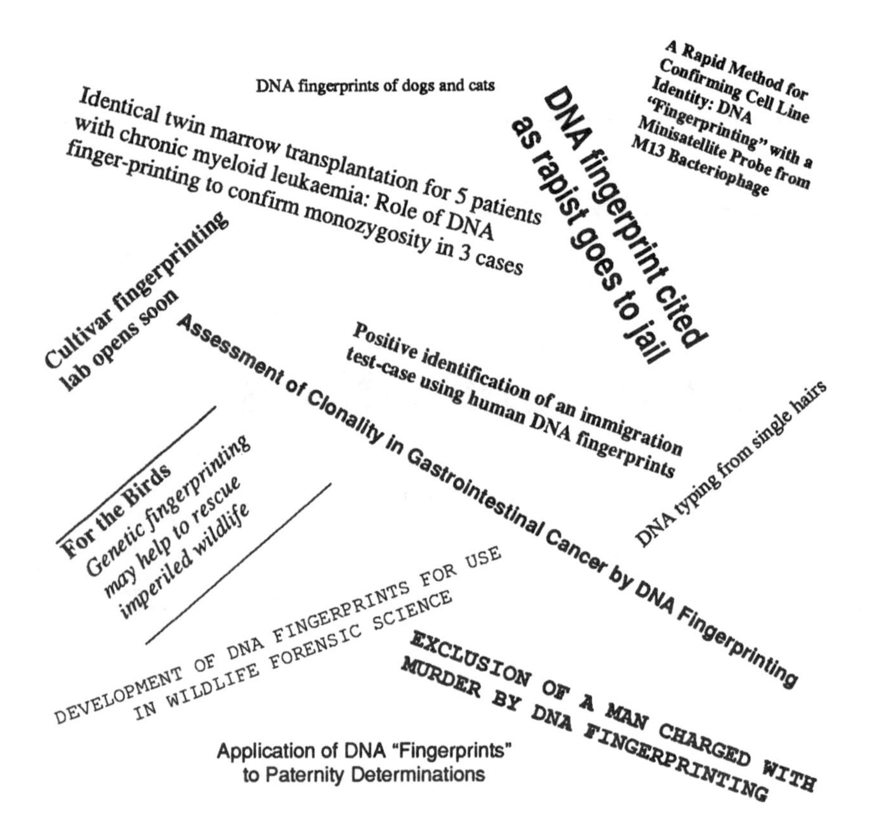

Figure 11-1. Titles and headlines from journal and newspaper articles illustrating the applications of DNA fingerprinting.

FORENSIC

The analysis of DNA is providing hard evidence for the resolution of serious criminal acts and other difficult identification problems in homicide, rape, accident, missing persons, break-ins, and hit-and-run cases (Anderson 1989; Barinaga 1989; Conner 1988; Dodd 1985; Fowler 1988; Fox 1989; Fukushima 1988; Gill 1987; Giusti 1986; Hicks 1989; Higuchi 1988; Howlett 1989; Jeffreys 1988; King 1989; Kobayashi 1988; Lander 1989*a*; Lewin 1989; McElfresh 1989; Marx 1988; Merz 1988; Newmark 1987, 1987*a*; Norman 1989; Ross 1989; Taylor 1989; Yokoi 1989). The determination of whether a series of crimes is serial or copycat, that is, committed by one or more than one perpetrator, is critical to the investigation of many cases. If DNA profiles match for specimens from different crime sites, this suggests that the same individual was involved and investigators can then concentrate their efforts on the hunt for one person.

The forensic scientist first prepares a DNA identity profile of the crime (evidence), suspect, and victim specimens (Figure 11-2). Specimens might include a semen stain from the site of a rape and a whole blood sample from a suspect and victim, or in a poaching case, a piece of animal gut from a kill site and a section of steak from a freezer or tissue from a trophy head. Two alternative questions can be advanced regarding the evidence: *(a)* What is the probability of obtaining the DNA fragment patterns observed if the samples are from the same source? *(b)* What is the probability if they are from different sources? Also, a calculation can be made to determine the percentage of the population that could contribute a specific profile for which the suspect is included. (This approach is followed by the FBI.) It should be noted that the forensic DNA expert is not attempting to answer the question: "What is the probability that this suspect committed a particular crime?" This broader question usually involves many other pieces of evidence that the case investigator must pursue and the judge or jury consider. The probability of obtaining the same DNA patterns for tissues from the same source is one. If the sources are different, but the patterns identical, the probability that the tissue derived from a different source equals the population frequency for that specific profile pattern.

The statistical data used to calculate the probabilities must be determined from specific population locus genotype frequency studies. If the frequencies of the locus genotypes that make up an identity pattern are 1/50, 1/75, 1/100, and 1/200, the frequency of that composite genotype in the population is the product of this series or 1/75,000,000. The incriminating value *(IV)* of this evidence is very high and could be an important factor in the determination of guilt provided other evidence points to the suspect as the perpetrator. (*IV* is the ratio of the probability of a pattern match if the crime and suspect specimens are from the same source to the probability of a match if they derive from different sources.) Samples consisting of DNA from more than one source, such as sperm from two or more males in a rape case, can usually be detected. A maximum of two bands are possible with any single-locus probe, provided the endonuclease used to digest the DNA does not recognize a cleavage site within the allele. If more bands are observed, the specimen is undoubtedly a mixture.

A number of techniques are available for sex determination. One is the hybridization of genomic DNA with Y chromosome- (male-) specific probes. Simple dot-blot procedures (Tyler 1986) may be applicable if the sequence to be detected is highly repetitive (Figure 11-3), or if only a single copy sequence is present, PCR amplification is first carried out. A 49-bp portion from a 3.4-kb Y chromosome-specific tandem repeat has been amplified for human sex determination from minute samples, such as single hairs (Higuchi 1988). Because this

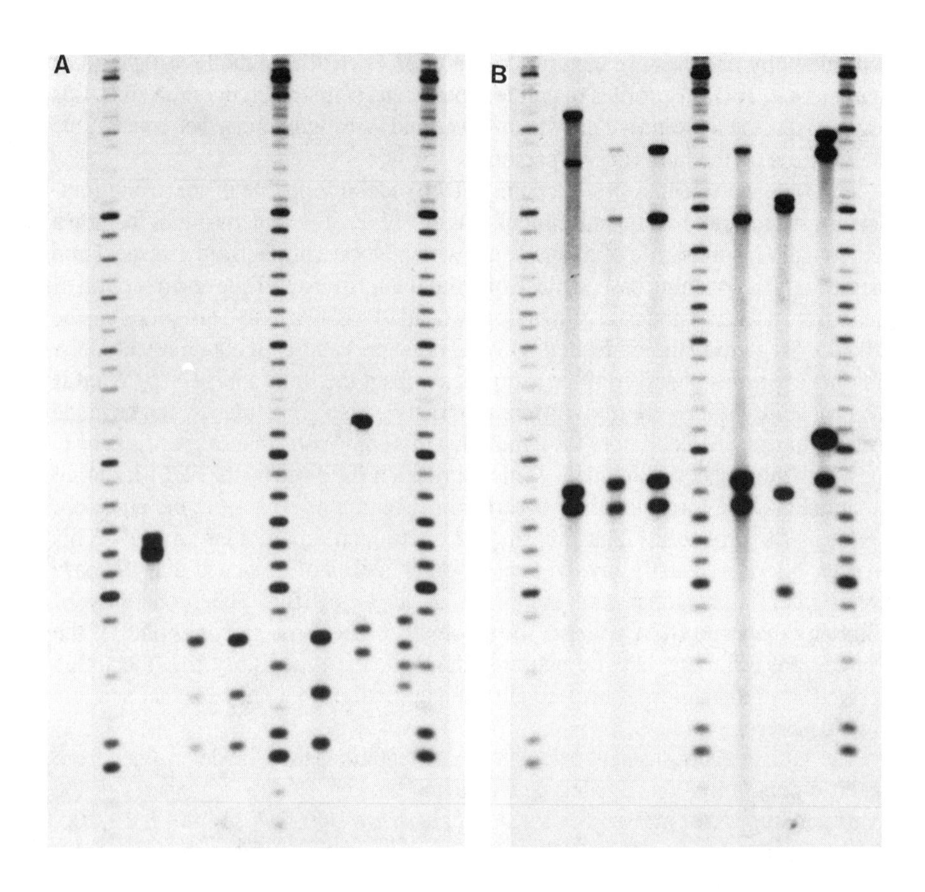

Figure 11-2. Autoradiograms of DNA-PRINT test results from a rape case. The lanes from left to right in each autoradiogram are (1) marker, (2) suspect 1, (3) evidence, (4) suspect 2, (5) marker, (6) evidence, (7) victim, (8) control, and (9) marker. The evidence lanes represent DNA from vaginal swabs (3) and semen stains (6). The suspect and victim DNA specimens were isolated from blood samples. The DNA on the left blot (A) was hybridized with the probes for loci D2S44 and D17S79. The DNA on the left blot (B) was hybridized with a probe for locus DXYS14. Suspect 1 can be excluded in this case because a number of bands in his blood sample do not match those in the evidence specimen. Suspect 2 cannot be excluded because his DNA profile matches the profile in the evidence specimens. (Autoradiograms courtesy of Lifecodes Corporation, Valhalla, New York.)

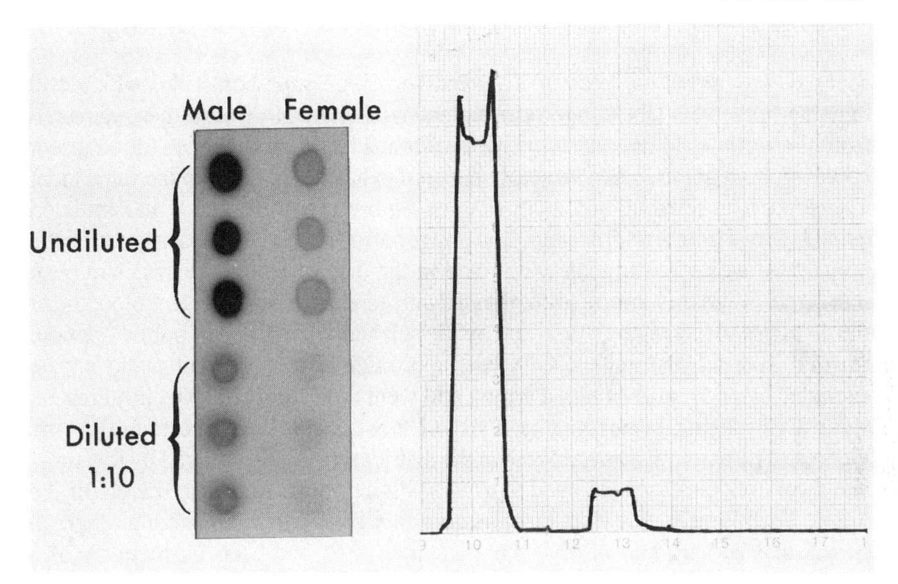

Figure 11-3. Dot-blot and densitometer tracings of human male and female DNA hybridized with a Y chromosome-specific repetitive sequence probe. Equal concentrations of DNA from three males and three females were applied to the membrane.

specific Y material is not present in females, no amplification will occur. Simple dot-blot procedures can then be used to determine the presence or absence of the sequence. The danger exists that a male specimen might be mistakenly identified as female if the amplification process fails for other reasons. This potential problem can be solved by using a co-amplification protocol where a sequence such as that found at the HLA DQα (also designated DQA1) locus, found in both males and females, is included.

Considerable care in the areas of chain of custody (continuity), quality control, and the population data base used in probability calculations must be exercised in case work; otherwise, unreliable conclusions may be drawn and the expert witness accused of presenting worthless evidence. These considerations are discussed in Chapter 10 and in the following cases.

Cases

Switched specimens. The first reported use (1987, The Narborough Village murders) of genetic fingerprinting in a criminal case in the United Kingdom involved the exclusion of a seventeen-year-old male charged with the murder of a schoolgirl. The victim had been sexually assaulted and killed in July, 1986. The accused was also a suspect in a November, 1983 rape-homicide, which had remained unsolved. DNA fingerprint patterns of sperm DNA from semen recovered from the victims did not match the pattern from a blood sample from the accused; however, both of the semen samples appeared to have derived from the same man. The youth was released and a considerably more involved search was initiated using Jeffreys' fingerprint system. The police had other evidence that suggested the perpetrator was a young male who lived in the district where the girls had been murdered. With this information, they advanced a proposal to collect blood from the district's suspect population, and by the substantial pressure of public opinion, secured over 5,500 samples. Of these, approximately 40 percent could not be excluded by conventional blood typing and were DNA profiled. The print results were discouraging because none matched those from the stains found on the victims. The emotional atmosphere in the district was highly charged by this time, and most conversations included at least a passing reference to the search for the killer. Fortuitously, one such conversation in a local pub led to a comment by a patron that he had donated two blood samples; one had been in the name of a coworker (Colin Pitchfork) who had been "unable" to donate. The coworker was readily identified and arrested, his DNA analyzed, and the pattern found identical to that of the sperm specimens. The suspect confessed to both crimes and was sentenced to life in prison. (Cellmark Diagnostics 1989, 1989*a*; Merz 1988.)

A polio victim. In another British case, Robert Melias was sentenced (November, 1987) to eight years in jail for the January, 1987 rape of a 43-year-old woman crippled by polio. The perpetrator had broken into the victim's apartment, raped her, and fled. DNA fingerprints from sperm present on the woman's clothing gave a positive match to the suspect's DNA (Associated Press, November 14, 1987).

No mistaken identity. The first conviction in the United States based on DNA profiling evidence occurred in Florida in a November, 1987 rape trial. (State v. Andrews, Case N. 87-1565, Orange County, Florida.) Tommie Lee Andrews was sentenced to 22 years in prison for this crime. (*The New York Times,* November 21, 1988). Since that time, DNA evidence has figured in hundreds of additional cases.

In October 1988, a light-skinned Hispanic, Victor Lopez, the so-called "Forest Hills rapist," was on trial in New York accused of sexually assaulting three women. The women had told the police their assailant was black. Was the accused a victim of mistaken identity? DNA profiling suggested otherwise. The accused's blood pattern matched that of sperm left at the scenes. Lopez was found guilty of the attacks (*The New York Times,* November 21, 1988; *Time,* October 31, 1988).

The widowed pensioner. The first case in Canada (Regina vs. McNally) in which DNA identity data produced in this country were accepted into evidence (April, 1989), and this evidence led to a conviction, involved the repeated rape of a widowed pensioner. The suspect initially attempted to challenge the evidence by pleading not guilty; however, RCMP laboratory analysis of semen stains from the crime scene together with a blood sample from the suspect gave a positive DNA profile match. There was a chance of only 1 in 70 billion that someone else's DNA would match that from the suspect—he pleaded guilty and received a seven-year jail term.

A burglar's fall. A case involving a homicide that occurred approximately a year before DNA testing was carried out illustrates the use of this new technology for analyses of old dried specimens in unsolved murders. A social worker in New York had been murdered, and although bloodstains were found in her apartment, there was insufficient evidence to link these with any suspect. A year later, a burglar in the same area died from a fall and, based on a suspicion that the burglar could also have been the murderer, the investigators had DNA prints prepared from his blood. The profile matched those from the apartment stains. These results provided considerable evidence that the dead burglar was at the murder scene a year earlier (Lifecodes Corporation 1986).

The bus driver. In 1987 in the State of Washington, a 57-year-old woman, incapable of lucid communication because of Alzheimer's disease, was sexually assaulted. She had been transported to and from an adult day-care facility by bus on the day of the attack, but the driver maintained adamantly that he knew nothing about the rape. The police investigation, however, pointed to him as the only suspect. The only piece of evidence was a semen stain found on the victim. Conventional antigen typing analysis indicated the grouping, although the same as that of the bus driver, was also present in 20% of the Washington male population. A portion of the sample was then sent to Lifecodes Corporation for DNA profiling. Only 1 in 3.5 million carried the DNA pattern found in the sperm, and the bus driver had that profile (Beeler, 1988; *The New York Times,* November 21, 1988). The driver pleaded guilty and was given a 10-year prison term (State v. Haynes, No. 87-1-02309-7, Pierce County, Washington).

Crematorium bloodstain. DNA analysis results were used as part of the trial evidence in a rather bizarre homicide in Wichita, Kansas (1988). A mortuary worker, Randy D. Pioletti, was accused of the murder and incineration of his estranged wife at the mortuary crematorium. The worker and his wife had been seen together the day before her death. The day following the murder, he had deposited a closed pail of charred bones (claiming it was rags) at the home of a friend. The bones were later identified as almost certainly those of the mortuary attendant's wife. Although the evidence to date was highly incriminating, the man continued to deny that he had killed his wife and even maintained that he had not taken her to the mortuary on the day of her disappearance. Investigators later discovered a bloodstain on the side of the crematorium oven. The specimen was DNA profiled and found to match tissue from the dead woman. The mortuary worker was convicted of first-degree homicide and aggravated kidnap (*The New York Times*, November 21, 1988).

Amplified evidence. The DQα and DPß loci are components of the class II HLA gene region. Considerable sequence variation exists in alleles at these loci and, based on this fact, oligonucleotide probes have been designed for dot-blot and reverse dot-blot hybridization with specific PCR-amplified DQα and DPß DNA sequences (Erlich 1989, von Beroldingen 1989). Because amplified alleles are scored on a +/- (present/absent) basis on the blot, probability calculations are simplified considerably (Figure 11-4). The complications associated with measurement of DNA fragment length do not arise as with the RFLP analysis; however, the statistical power does not yet approach that of the VNTR system.

A case example is presented by von Beroldingen (1989) where in vitro amplification products from the DQα region of DNA extracted from evidence specimens were analyzed to determine whether a suspect could be excluded in a sexual assault case. The procedures followed in this case are quite instructive because the victim had sexual intercourse with her husband only nine hours prior to the alleged assault. This resulted in a mixed vaginal swab consisting of female epithelial cells plus semen from two males. The suspect DNA typed identically with sperm isolated from the vaginal swab; however, the same pattern was found in 2.6% of the male population. The husband's sperm was also isolated from the swab specimen and typed—the pattern differed from that of the suspect. The suspect could not be excluded.

Finally exonerated. A Bronx, New York rape case (New York v. Neysmith) illustrates how errors can creep into the analysis or chain of custody procedure and result in unreliable conclusions. A suspect was first charged, based on the victim's identification, and later released because his DNA profile did not match that of a semen specimen collected at the crime site. The exclusion was challenged by the prosecution on the grounds that the suspect had perhaps submitted someone else's blood for analysis. The analyzing laboratory initially determined that the original and the resubmitted samples were from different sources. The defendant was, however, finally excluded and exonerated after a third specimen was analyzed by a different laboratory and found not to match the semen stain. The first laboratory later determined that an error had been made (Lander 1989).

67,500,000,000:1. A pending state of Georgia rape-murder trial (George v. Caldwell) provides an example of a case in which proper DNA quality control procedures apparently were not followed. A father was accused of raping and killing his daughter. The analysis laboratory determined that DNA profiles from the father's blood specimen and a semen stain from the victim matched with 67,500,000,000:1 odds against the match arising at random in the population. Only

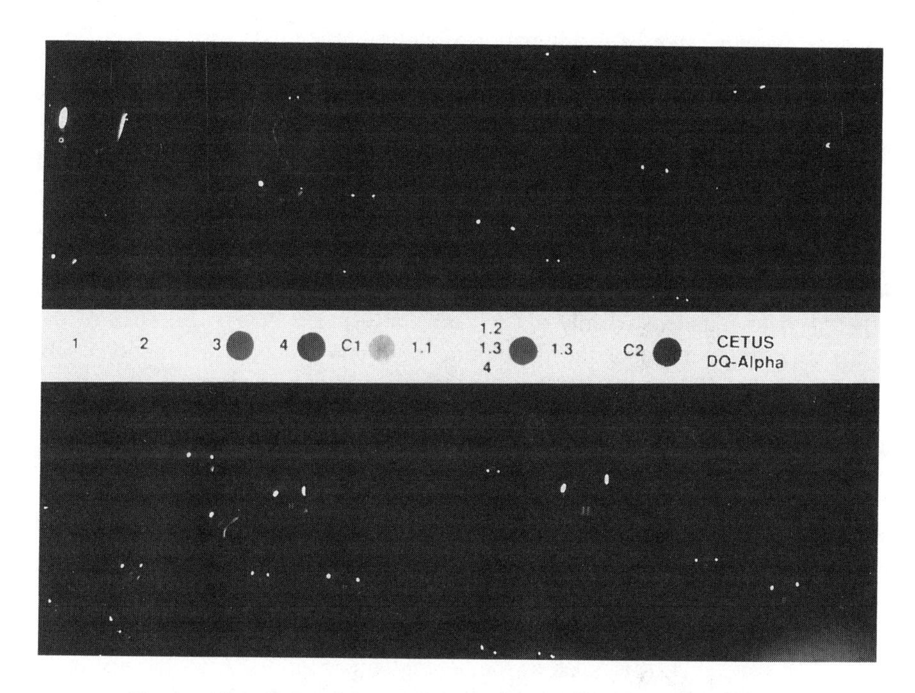

Figure 11-4. Reverse dot-blot format for DQα allele typing. Oligonucleotide probes (ranging in size from 17 bp to 33 bp) designed to recognize sequence variations in the DQα alleles are immobilized on the nylon membrane strip. Primers have been designed that flank the variant region in the DQα gene. Target DNA, 242 bp in length, is amplified using these primers and this DNA is hybridized to the probe strip. 1, 2, 3, and 4 represent immobilized probes to alleles A1, A2, A3, and A4. 1.1 and 1.3 represent probes to sub-type allele A1 into A1.1, and A1.3. 1.2/1.3/4 represents a probe that recognizes A1 sub-types A1.2 and A1.3; this probe also recognizes allele A4. C1 represents a probe that hybridizes to all of the alleles and C2 a probe that hybridizes to all alleles except the A1.3 sub-type. C1 and C2 act as positive controls in the analysis system.
In the sample blot illustrated, the target DNA hybridized to 3, 4, C1, 1.2/1.3/4, and C2. The individual must, therefore, be heterozygous for alleles A2 and A4. (Blot courtesy of Cetus Corporation, Emeryville, California.)

visual comparisons were used for the crime and suspect profiles, and one probe result suggested there may in fact not have been a match. The laboratory claimed that the semen stain DNA electrophoresis gel lane ran faster than that of the blood sample. No internal DNA controls were included to support this conclusion (Lander 1989).

A small town murder. The population allele frequency data base used in DNA profile match probability calculations must be carefully chosen. The following case summary (Texas v. Hicks) is an example of the use of questionable population allele frequencies.

A rape-murder occurred in a small Texas town. Part of the evidence submitted to the court consisted of DNA profiles from the suspect, victim, and crime specimens. The odds of the declared suspect-crime match occurring at random were reported to be 1 in 96,000,000. The probability calculations, however, apparently took no account of the fact that the crime was committed in a small inbred town founded by only a few families (Lander 1989). Because of the relatively small gene pool, the allele frequencies and genotype distribution could be considerably different from those in the broader population.

The defendant in this case was convicted.

PARENTAGE

The need to determine true biological relationships in disputed parentage, missing persons, and immigration cases is common. The value of individual DNA characteristics in identity determination arose during the analysis of families for inherited genetic defects, such as the hemoglobinopathies. After the results of recombinant DNA tests on a number of family members had been assimilated, it became obvious in certain cases that nonpaternity was the only reasonable conclusion consistent with the inheritance patterns. Needless to say, these determinations resulted in no small degree of marital discord. Perhaps such unexpected observations during analysis for metabolic defects should have been anticipated; estimates of nonpaternity apparently range as high as 25% in certain seaport cities and in the United States general population, an average of 10% has been quoted.

Highly informative tandem repeat sequence probes have been developed specifically for paternity and other identity analysis (Figure 11-5). Not only can putative fathers be readily excluded, but the true biological parents can also be determined with practical certainty provided all of the appropriate specimens are available (Baird 1986, 1987; Helminen 1988; Jeffreys 1986; Odelberg 1988; Roberts 1987; Wells 1988).

Thousands of paternity cases have been resolved using DNA analysis techniques and of these, on average, less than 0.1 percent continue to court.

Cases

Arranged killing. A pregnant teenager was accused of arranging the killing of her father. She did not deny this fact, but she claimed that the father had sexually abused her and had made her pregnant. The girl later aborted and tissues from the fetus were typed by DNA analysis. Specimens from the dead father were also typed. The

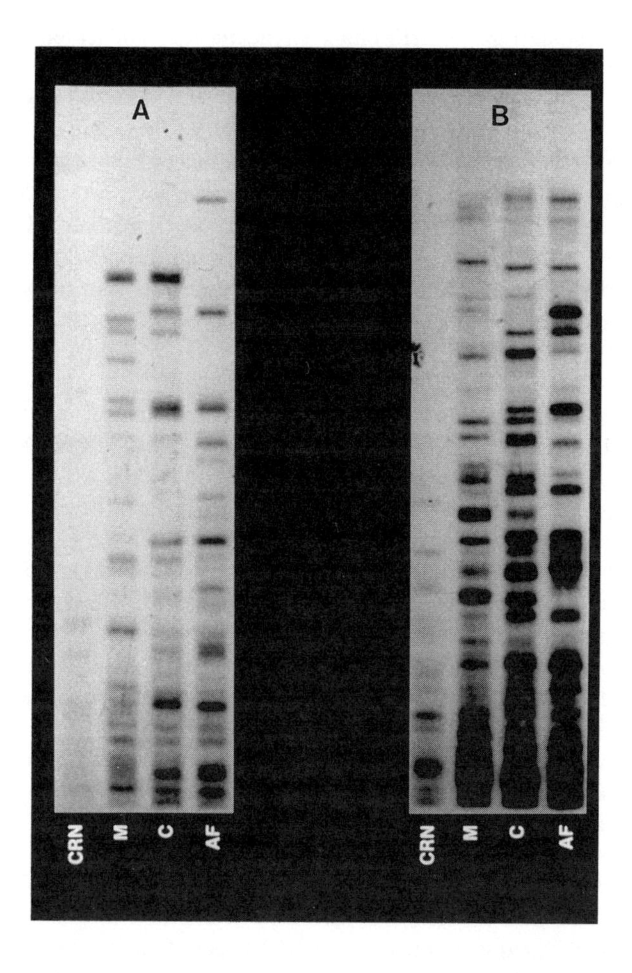

Figure 11-5. Autoradiograms of DNA Fingerprint results from a criminal paternity case. The lanes form left to right on each autoradiogram are human control (CRN), mother (M) child (C), and alleged father (AF). DNA from blood samples was digested with the restriction enzyme Hinf I and hybridized with the multilocus probes 33.15 (left autoradiogram, A) and 33.6 (right autoradiogram, B).
A stepfather was charged with sexual abuse of a stepdaughter. Evidence included a demonstration of criminal paternity by the defendant (AF) with the stepdaughter's older sister (M) who had also been sexually abused by the defendant. For each band detected in the child (C), a band of the same size was detected in (M) or (AF). (Autoradiograms courtesy of Cellmark Diagnostics, Germantown, Maryland.)

resultant prints proved that the girl's father was not the father of her fetus (Lifecodes Corporation 1986).

A victim's estate. The former girlfriend of an automobile accident victim claimed she was pregnant with the dead man's child. According to Kentucky state law, the child was entitled to the victim's estate. After the woman gave birth, a blood sample from the infant was analyzed and DNA prints prepared. Prints were also prepared from an approximately ten-month-old stored specimen from the putative father. Comparisons of prints of the mother, child, and deceased man showed that the victim was not the father and no claim was allowed on his estate (Lifecodes Corporation 1986).

An abandoned baby. In February, 1988 the president of an insurance company in Maryland had her automobile towed to a repair shop. The mechanic who worked on the vehicle claimed he discovered a dead infant on the rear seat. The woman stated emphatically to police investigators that she had not been pregnant and supported her claim with medical examination evidence. Tissue samples from the woman and the infant were analyzed with both single-locus and multilocus DNA probes, with the result that maternity was highly suggested. Murder charges were not filed because of evidence that the baby was stillborn on or about February 4, 1988. The woman gave birth later that year (October 24) to a full-term infant. The date of conception was calculated to be on or about January 29, according to sonograms taken by her obstetrician.

Was the maternity test result true or false? The analysis results were clear with all the probes used; however, the evidence in cases such as this should be reanalyzed by an independent laboratory (Lander 1989).

The real parents. DNA analysis techniques have been used in a number of cases involving missing and unidentified persons. In addition to genome analysis, mitochondrial DNA sequences also can be used in identification. Mitochondria are transmitted only by the mother and are shared identically by all siblings. This phenomenon is of considerable value when the missing person's mother is dead but maternal aunts or uncles are available for testing.

A tragedy of appalling proportions involved the disappearance of over 9,000 Argentinians between 1975 and 1983 (Anon 1989*b*; Diamond 1987; King 1989, 1989*a*; Orrego 1988, 1990). These individuals were abducted by the military and police in the name of "the defense of national security and honor." Approximately 200 children, most of whom were born to abducted pregnant women in captivity, are currently being sought for reunification with their biological relatives. The parents, for the most part, are known or presumed dead and many of the children are thought to be living with military couples who claim to be the biological parents. Some of these individuals may even have been involved in the murders of the real parents. When a suspect family comes to light, attempts are made to obtain blood samples from the putative biological relatives, especially the grandparents, the disputed child, and if possible, the military couple. The inheritance of different alleles is determined and the probability of relationship calculated. These cases can present major puzzles for the investigators; however, the power of techniques such as DNA analysis have made the project feasible. To date, almost 50 cases have been resolved.

An immigration test. The use of genotyping in disputed immigration cases soon should be routine in a number of countries (Johnston 1987, Newmark 1988). United Kingdom law states that resident immigrants are entitled to request resident status for specific close relatives. In 1986 over 12,000 immigration applications were received from the wives and children of Pakistani and Bangladeshi men resident in Britain. Of these more than one-quarter were rejected because of insufficient proof of relationship. It is forecast that DNA profiling will expedite the decision process and prevent wrongful dismissal of applicants.

One immigration case involved a boy born in the U.K. who later emigrated to Ghana to be with his father and other relatives. When he returned to Britain, the immigration authorities had reason to believe he was not the boy who had emigrated but was either totally unrelated or was perhaps the son of a sister of the putative mother. Initial non-DNA blood typing analysis indicated the boy was, in fact, the son or perhaps a nephew. To complicate matters, the father was not available for testing, and even if he had been, the mother was not sure he was the biological father. DNA fingerprinting was finally performed on the boy, his siblings, and the mother. The results provided unequivocal evidence that maternity as well as paternity were correct as originally stated. Family reunion in Britain was granted (Hill 1986; Jeffreys 1985, 1986*a*).

MEDICAL

The applications of DNA profiling in the medical setting include posttransplant cell population identification, twin zygosity determination, tumor analysis, identification of microorganisms, tissue culture cell line identification, paternity determination, and the detection of maternal and fetal tissue mixtures in chorionic villi specimens (CVS).

Applications

Posttransplant cell population identification. Identifying the origin of a posttransplant cell population after a bone marrow transplant is important in monitoring engraftment (Keable 1988, Thein 1986). In addition to the conventional methods of karyotyping and antigen analysis, DNA profiling may provide a useful approach. A specific cell population of as little as one percent of the total number of cells can be detected in a mixture provided the host and donor are typed before the transplant is performed. Only donor cells were initially detected in one patient; however, by 12 months posttransplant, the host-donor populations were approximately 50:50. In another case, the highly informative 33.15 minisatellite probe was used to profile DNA from a young boy with acute lymphoblastic leukemia who had received a bone marrow transplant from his sibling (Thein 1986). Pretransplant DNA sample profiles indicated seven fragments in the donor not present in his recipient brother. These fragments were then used as markers in posttransplant samples from the recipient to determine engraftment. At three weeks posttransplant, there was no evidence of engraftment as measured using conventional techniques. DNA testing confirmed this observation.

Twin zygosity determination. This is important for epidemiological and genetic studies in addition to possible transplant matching. Also, identical twins have on average more medical complications than do fraternal twins. Approximately 30% of Caucasian twins are of different sex and are, therefore, dizygotic. Twenty percent are monochorionic and must be monozygotic. The remaining 50% usually require genetic analysis to determine zygosity (Figure 11-6). A false positive identity rate of less than 1 in 10^8 can be achieved for siblings (who have half their genomes in common) when two multilocus probes such as those of Jeffreys et al. (1986) are used. This suggests that zygosity can be determined with near certainty using this new analysis approach (Figure 11-7).

A number of studies have been reported concerning determination of twin zygosity using Jeffreys' 33.15 and 33.6 minisatellite probes. Hill et al. (1985) analyzed DNA from 12 sets of twins; the zygosity was already known in 7 sets by sex or placental examination. DNA results agreed with all of the previous calculations. Azuma et al. (1989) analyzed DNA from 2 sets of triplets. All three males in one set had different profiles, whereas in the second set, consisting of one female and two males, the males had identical profiles. Jones et al. (1987) confirmed twin monozygosity for the purpose of syngeneic bone marrow transplant in three patients with chronic myeloid leukemia. In two of the cases, multiple blood transfusions had been administered prior to referral for transplant, therefore invalidating conventional RBC analysis for zygosity determination.

Tumor analysis. DNA profiling provides a new strategy for detecting somatic changes in human cancer DNA in terms of genomic rearrangement, clonality, and tumor development (deJong 1988, Layton 1987, Thein 1987). Fey et al. (1988) prepared DNA profiles, using 33.15, 33.6, and 3' HVR probes, for patients with gastrointestinal cancer and were able to demonstrate clonal somatic mutations in many of the tumors. The changes consisted of both a loss and a gain of fragment bands relative to the patient's peripheral blood and normal mucosa samples. Smit et al. (1988) analyzed tumor heterogenicity in a patient with synchronously occurring female genital tract malignancies. All tumor locations were found to

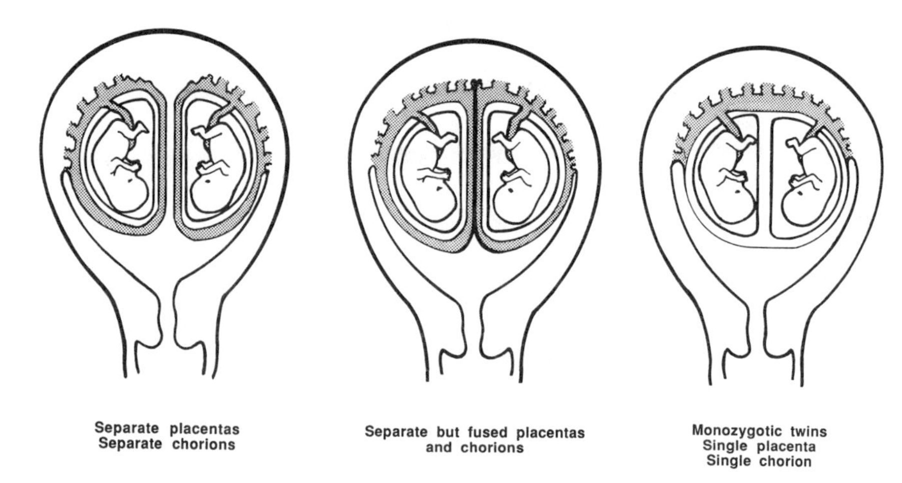

Figure 11-6. Placenta and chorion arrangements in twins. A single placenta and a single chorion but separate amniotic sacs is diagnostic of monozygotic twins. The other arrangements as diagrammed can occur in either monozygotic or dizygotic twins. DNA fingerprinting may be useful to confirm zygosity in these cases.

have a common change in the DNA profile pattern compared with the patient's constitutional DNA.

Identification of microorganisms. A technique for rapidly identifying disease-causing microorganisms including many mycobacteria, viruses, and parasites is still lacking. The length of time required to detect a pathogen can be critical in terms of fatality because of a severe infection. Recently, DNA probe kits have become available for faster detection of legionella, herpes simplex virus, and other microorganisms; however, many problems remain to be solved, including simplifying the assay procedures and increasing the quantity of DNA sample available for analysis (Anon 1989*a*). The use of chemiluminescent tags in place of radioisotopes (Bronstein 1989), in situ hybridization, and PCR amplification of the DNA (Impraim 1987) are potential solutions to these problems.

Analysis of profile patterns from 13 natural *E. coli* strains, using bacteriophage M13 and minisatellite-derived oligonucleotide probes, have revealed differences within strain pairs ranging from 0 to 31 fragments with an average of 19 differences (Huey 1989). This finding offers considerable potential for characterizing and rapidly identifying bacterial strains as well as for studying population dynamics and evolution.

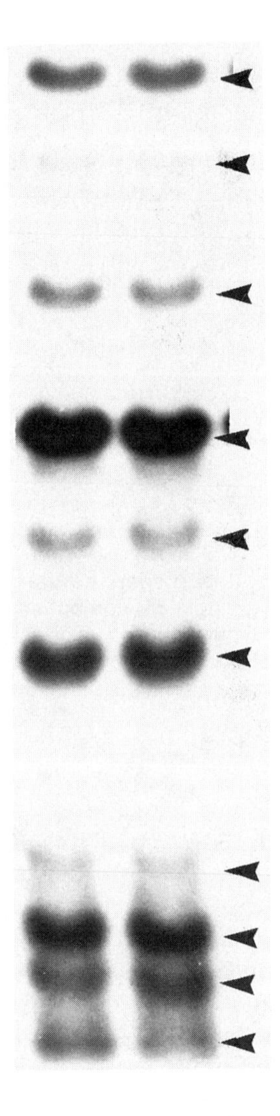

Figure 11-7. Autoradiogram of DNA profiling results from a set of twins. Blood specimen DNA was digested with the restriction enzyme Hinf I and hybridized with the multilocus probe 3'HVR. The profiles in both lanes are identical as anticipated for monozygotic twins.

Tissue culture cell line identification. Tissues can be cultured and the cultures stored to provide materials for analysis of genetic defects or other experimental work such as assessing the divergence of animal or plant sublines. One group (Devor 1988) has used a minisatellite DNA fingerprinting probe derived from M13 bacteriophage to identify a mislabeled cell line. Another group (Thacker 1988) used the 33.15 and 33.6 minisatellite probes to fingerprint a number of stocks of Chinese hamster cells to avoid cell line cross-contamination, to check cell hybrids, and to assess subline divergence from one another.

Paternity determination. Pedigree analysis is usually required to determine the probability of recurrence of a genetic defect. Antenatal diagnosis may be performed, at approximately 16 weeks gestation, on an amniotic fluid specimen or at approximately 11 weeks on chorionic villi aspirated from the placenta. DNA from these tissues can be analyzed directly or the specimens cultured and the resultant fibroblast DNA processed.

Inherited defects can be transmitted to one or more siblings from either or both parents depending on the type of mutation. If a trait is autosomal recessive, each parent is a carrier and has a 0.5 chance of passing the mutant gene to an offspring. One-quarter of the children on average will be affected, one-half will be carriers like the parents, and one-quarter will receive all normal alleles (Figure 11-8). Cystic fibrosis, phenylketonuria, and the hemoglobinopathies are examples of autosomal recessive diseases. An autosomal dominant defect such as Huntington chorea is transmitted from only one parent with, on average, one-half of the offspring affected (Figure 11-9). This disease can be especially devastating as it often does not manifest itself until mid-life. Parents have by then completed their families and possibly transmitted the mutant gene to one or more offspring. In X-linked recessive defects, the mutant genes are usually passed only from female carriers; one-half of their daughters will be carriers and one-half of the sons affected (Figure 11-10). Hemophilia is perhaps the best known example of an X- or sex-linked disease. Knowledge of paternity is extremely important for families with a child having an autosomal recessive defect. If nonpaternity existed, the probability of having a second child with the same disease is negligible unless the child was fathered by the same man. If paternity is evident, the chance is 0.25.

CVS analysis. Profiling of CVS and maternal tissue, such as blood, may be warranted on occasion since maternal-fetal tissue mixtures may persist in chorionic villi samples after in vitro separation. If all maternal fragments are observed in the CVS extract, then contamination is confirmed because only one-half should be present.

ANIMAL SCIENCES

The applications of DNA profiling in the field of animal sciences are many and varied, as illustrated by recent analyses with monkeys (Dixson 1988, Weiss 1988); birds (Burke 1987, 1989, 1989*a*; Gyllensten 1989, 1990; Hill 1987; Howlett 1989; Lewin 1989; Longmire 1988; Quinn 1987; Wetton 1987); big game species

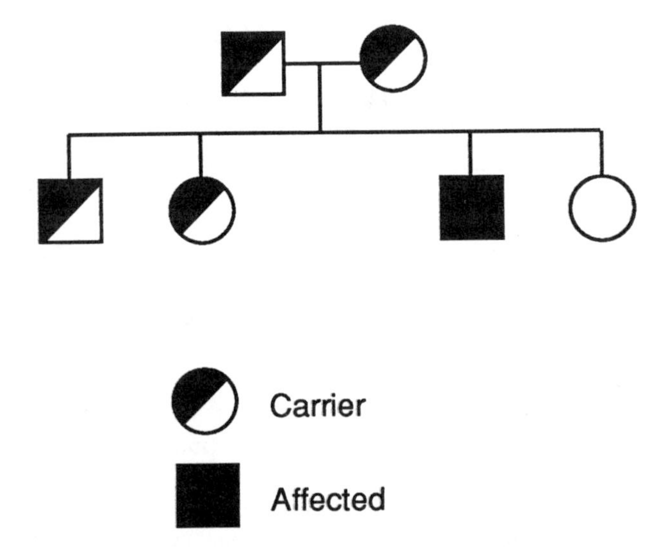

Figure 11-8. Pedigree illustrating autosomal recessive inheritance. Each parent is a carrier and each offspring has a 0.5 probability of inheriting the defective gene from each parent.

(Thommasen 1989); whales (Amos 1989, Hoelzel 1988); fish (Fields 1989; Ryskov 1988, 1988*a*; Turner 1989); mice (Jeffreys 1987*a*); rats (Ivanov 1989), and domesticated species of cattle, horses, cats, dogs, and fowl (Georges 1988; Jeffreys 1987; Kuhnlein 1989; Ryskov 1988, 1988*a*; Vassart 1987). Areas of usefulness include determination of lineage to avoid excessive inbreeding and the appearance of detrimental recessive traits; paternity exclusion or inclusion; studies of evolution; classification into genus, species, or subspecies; individual identity testing; breeding markers for valuable traits such as meat and milk production; sex determination; pathogen detection; the verification of semen specimens for artificial insemination; and identification of the origin of fish populations such as salmon for the purpose of meeting obligations outlined in the Pacific Salmon Treaty between Canada and the United States.

mtDNA has proven to be a rich source of sequence variation for differentiating animal species, examining patterns of base substitution during evolution, and following gene frequency changes through time. Three primers directed toward conserved regions of the DNA have been used to amplify homologous segments from more than 100 species including mammals, birds, amphibians, fishes, and invertebrates (Kocher 1989, 1989*a*). Only nanogram or microgram quantities of unpurified mtDNA from fresh or preserved and dried specimens were PCR-amplified—using a one hundredfold reduction of one primer—to provide sufficient single-stranded DNA for direct sequencing.

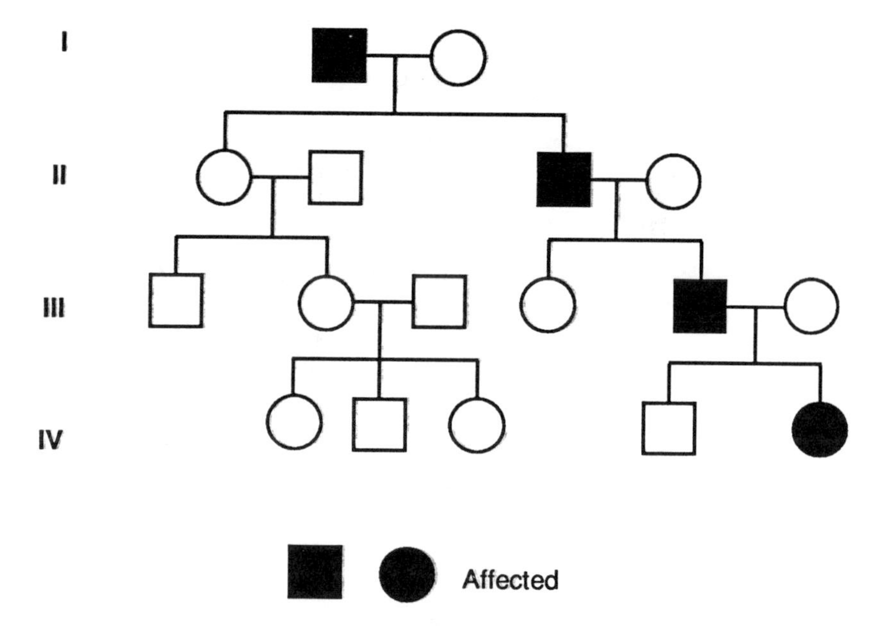

Figure 11-9. Pedigree illustrating autosomal dominant inheritance. One parent is a "carrier" and each offspring has a 0.5 probability of inheriting the defective gene from that parent.

Cases

Canine and feline ID. Pedigree analysis or proof of paternity may be necessary before animals can be officially registered. Individual identification and verification of semen used in artificial insemination may also be required where expensive breeds are involved. The minisatellite probes 33.15 and 33.6 cross-hybridize to multiple polymorphic DNA fragments in dog and cat genomes (Jeffreys 1987; Figure 11-11). Statistical calculations useful in determining the probability of a match between DNA profiles can be carried out as described in Chapter 8.

The stud. A dog used for stud purposes was discovered to be infertile; his semen revealed the absence of spermatozoa. During the previous 15 months, he had mated 27 bitches with two apparent successes. In light of the semen results, DNA fingerprinting was carried out on the putative sire, one dam, and her puppies. A number of profile bands were observed in the puppies but these bands were not present in the dam or the putative sire. Apparently, the bitch had been courted by at least one other male and the stud dog in question was truly infertile (Morton 1987).

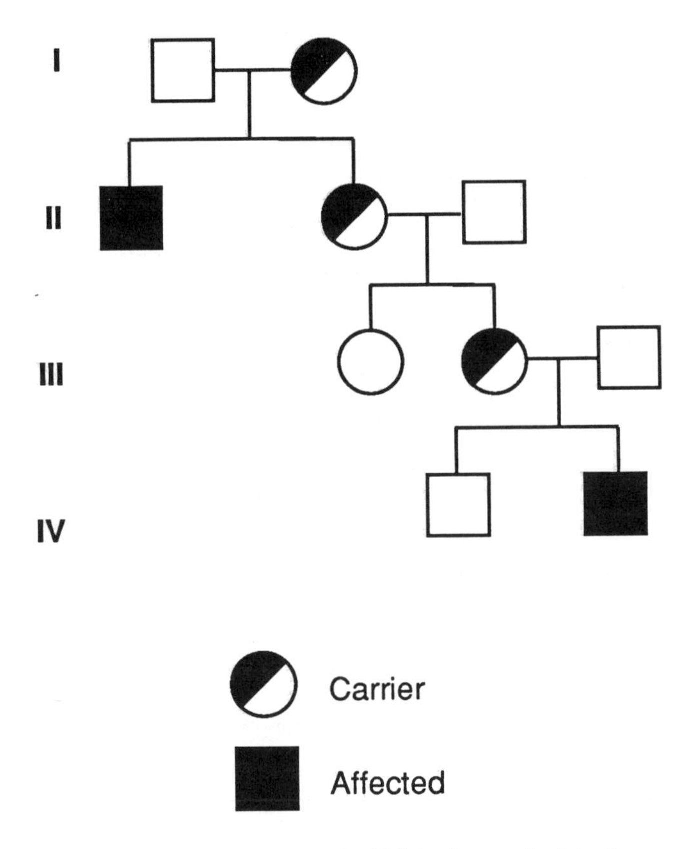

Figure 11-10. Pedigree illustrating X-linked recessive inheritance. Carrier females have a 0.5 probability of transmitting the mutant gene to an offspring. The males are affected and the females are carriers like their mother.

Domesticated fowl. DNA fingerprinting has been used to determine genetic distances between strains of poultry (Kuhnlein 1989). Genomic samples from White Leghorn and New Hampshire strains as well as French Broiler Breeder lines were hybridized with a bacteriophage M13 repetitive sequence probe. Up to 35 DNA bands were observed on each bird profile. The variability of the fingerprint patterns within the poultry strains decreased, as expected, with increased inbreeding. The authors suggest that measurements of genetic distances, using the fingerprinting technique, may prove useful for optimizing hybrid vigor, which is associated with egg production traits.

Falcons and cranes. DNA profiling of endangered species, such as the peregrine falcon, Mauritius Kestrel, and the whooping crane, is proving to be invaluable.

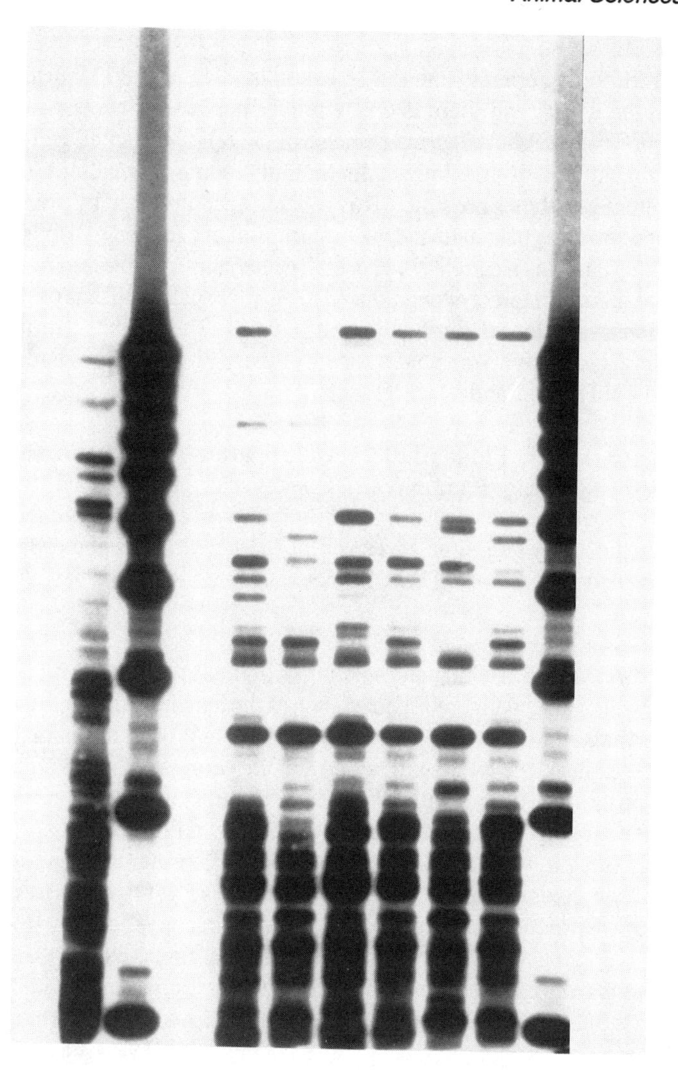

Figure 11-11. Autoradiogram of DNA Fingerprint results from a canine paternity case. The lanes left to right are (1) λDNA plus a characterized human control, (2) 1-kb size markers, (3) blank, (4) bitch, (5) pup 1, (6) pup 2, (7) pup 3, (8) alleged sire I, (9) alleged sire II, and (10) 1-kb size markers. DNA from blood specimens was digested with the restriction enzyme Hinf I and hybridized with the multilocus probe 33.15. Alleged sire I was excluded and II was included as a possible sire. Pups #1 and #2 have bands not accounted for in the bitch or putative sire. These bands could be a result of new mutations or a result of involvement of a third male. (Autoradiogram courtesy of Cellmark Diagnostics, Germantown, Maryland.)

Only 146 whooping cranes from the 16 extant in 1941, and 80 Mauritius Kestrels remain in the world, and there is concern over the lack of genetic diversity and consequent fitness of the offspring because of the relatively small gene pool. One program objective is to maximize genetic outbreeding by mating birds with the most different genotypes possible. This is being accomplished by preparing DNA profiles and selecting pairs with the maximum number of different bands (Longmire 1989). The Mauritius Kestrels are especially interesting in this regard because the numbers of this small falcon were reduced to only two pairs and thus passed through an evolutionary bottleneck during the past 30 years. The banding profiles together with coding and comparison charts for 19 of these birds are presented in Figure 11-12 and Tables 11-1 and 11-2.

Old Orange Band. The Florida dusky seaside sparrow became extinct in 1987 with the passing of old Orange Band at Walt Disney World. Those who had toiled to save these birds were surprised when it was subsequently discovered by genetic analysis, that in fact, the dusky individuals were simply darker versions of the very common standard seaside sparrows. They did not constitute a subspecies as many had argued.

Laundered macaws. Poaching and stealing of animals presents an increased threat to the survival of an endangered species such as the hyacinth macaw. It is estimated that fewer than 3,000 of these birds remain in the wild and although capture for private sale is illegal, nests are still raided and the young sold to the highest bidder. Breeding programs, using legally obtained birds, are designed to provide macaws for sale on the open market; however, as with many wild species, reproduction in captivity is fraught with difficulties and the few offspring produced meet only a fraction of the demand. Some traders will purchase wild young to sell later as legitimate birds produced in captivity. Wildlife officials have considerable difficulty distinguishing a bird hatched in the wild from a captive one; thus, although suspected, the culprit traders continue their illegal activities unhindered. Jeffreys' multilocus minisatellite probes have been used recently to determine whether particular birds were the products of legitimate matings. In one case, a number of macaws were sold as offspring from a specific group of breeding birds. When DNA fingerprints of the "family" members were analyzed, both nonpaternity and nonmaternity were discovered, confirming the suspicion that the young were of direct wild origin. The guilty were convicted and appropriately sentenced (Anon 1989, Forman 1989).

Sexually liberated. Wetton et al. (1987) and Burke (1989) have demonstrated mixed parentage among nestlings of house sparrows by analyzing DNA with Jeffreys' 33.15 and 33.6 probes, and Quinn et al. (1987) have isolated a series of probes from a lesser snow goose genomic library and determined that multiple maternity and paternity exist in single broods of this socially monogamous species.

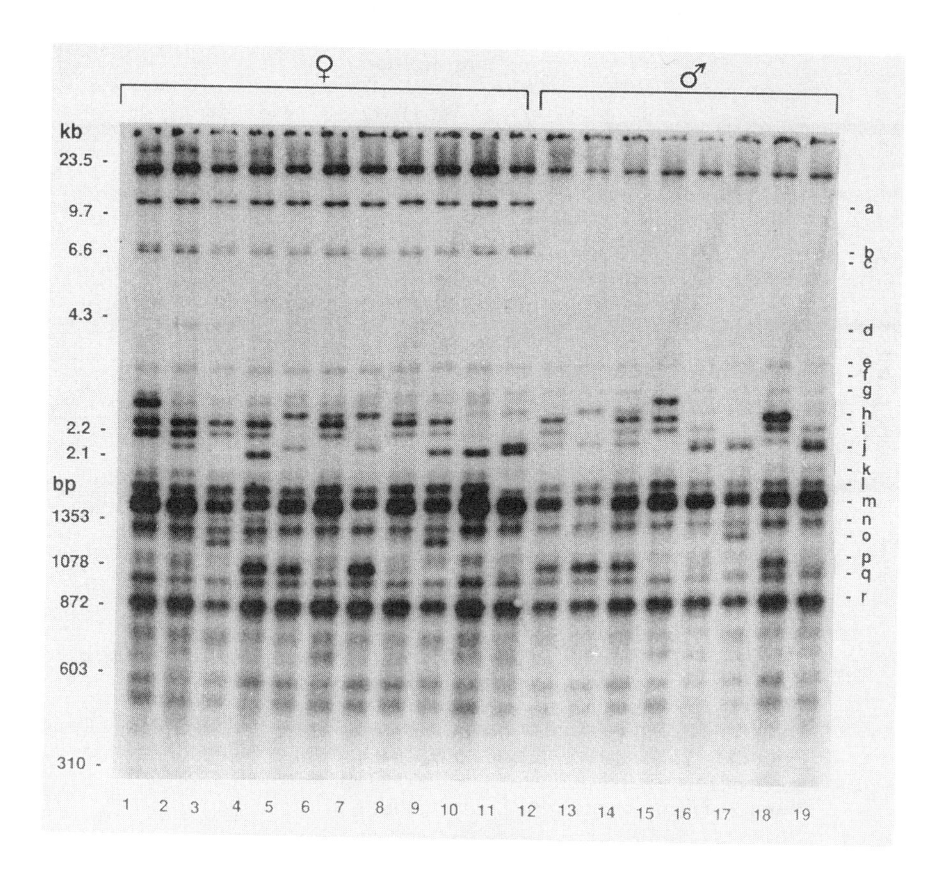

Figure 11-12. Autoradiogram of DNA profiles for 19 Mauritius Kestrels. DNA from blood was digested with the restriction enzyme Hae III and hybridized with a ^{32}P-labeled multilocus probe, M13. Hybridization was performed under the following conditions: 42°C in a solution consisting of 6 x SSC, 40% formamide, 5mM EDTA (pH 8.0) and 0.25% powdered milk. Wash conditions were carried out at 50°C in 1 x SSC, and 0.1% SDS. The females (which are the heterogametic sex in birds) display invariant as well as polymorphic bands. The relative positions of the invariant bands provide an indicator of band shifting. Each letter d to r corresponds to a polymorphic fragment that was scored. (Autoradiogram courtesy of J.L Longmire Genetics Group, Los Alamos National Laboratory, Los Alamos, New Mexico.)

Behavioral studies recently completed with the aid of minisatellite probes involved confirmation of maternity and paternity in nests of dunnocks, also known as hedge sparrows (Burke 1989*a*). These tiny birds apparently are quite liberated

Band number	Bird number 1	2	3	4	5	6	7	8	9	10	11	12	13	14	15	16	17	18	19
d	0	1	1	0	0	0	0	0	0	0	0	0	0	0	0	0	0	0	0
e	1	1	1	1	1	1	1	1	1	1	1	1	1	1	1	1	1	1	1
f	0	0	0	0	0	0	0	1	0	0	0	0	0	1	0	0	0	0	0
g	1	1	1	1	1	1	1	1	1	1	1	1	1	1	0	1	1	1	1
h	1	1	1	1	0	1	0	1	1	0	0	1	0	1	1	0	0	1	0
i	1	1	1	1	0	1	0	1	1	0	0	1	0	1	1	1	0	1	1
j	0	0	0	1	0	0	0	0	1	1	1	0	0	0	0	1	1	0	1
k	1	1	1	1	1	1	1	1	1	1	1	0	0	0	0	1	1	1	1
l	1	1	1	1	1	1	1	1	1	1	0	1	1	1	1	1	1	1	1
m	1	1	1	1	1	1	1	1	1	1	1	1	1	1	1	1	1	1	1
n	1	1	1	1	1	1	1	1	1	1	1	1	1	1	1	1	1	1	1
o	0	0	1	0	0	0	0	0	1	0	0	0	0	0	0	0	1	0	0
p	0	0	0	1	1	0	1	0	0	0	0	1	1	1	0	0	0	1	0
q	1	1	1	1	1	1	1	1	1	1	1	1	1	1	1	1	1	1	1
r	1	1	1	1	1	1	1	1	1	1	1	1	1	1	1	1	1	1	1

Table 11-1. Binary code of the 15 DNA bands (d to r) scored in Figure 11-12. "1" represents presence of a band and "0" represents absence of the band. For example, band d is present only in birds 2 and 3. (Table courtesy of J.L. Longmire.)

sexually in that relationships of monogamy, polygamy, cooperative polyandry (two males, one female), and polygynandry (two males, two or more females) can exist in the population. Males do not discriminate in multi-sired broods between their respective young except to feed the offspring to the degree that they had access to the female during the mating season. This behavior has been used as an indicator of paternity and has been confirmed with DNA fingerprint analysis.

A whale of a tale. Population data are often difficult to assimilate for conservation and management of endangered species such as the whale. This information is fundamental to ensure that sufficiently high numbers are maintained to provide a diverse genetic pool. Fortunately, a minimally intrusive means of identifying stocks and carrying out population censuses has been developed (Hoelzel 1988). This technique involves collecting 200 to 300 mg skin biopsies from free-ranging whales, then isolating DNA (0.5 to 1.0 mg) for fingerprinting. The samples are collected using dart-tipped arrows and the DNA later isolated and hybridized with nuclear (33.15 and 33.6) and dolphin mtDNA probes (Figure 11-13).

	1	2	3	4	5	6	7	8	9	10	11	12	13	14	15	16	17	18	19
1	0																		
2	1	0																	
3	2	1	0																
4	2	3	4	0															
5	3	4	5	3	0														
6	0	1	2	2	3	0													
7	3	4	5	3	0	3	0												
8	1	2	3	3	4	1	4	0											
9	2	3	2	2	5	2	5	3	0										
10	3	4	5	3	2	3	2	4	3	0									
11	4	5	6	4	3	4	3	5	4	1	0								
12	2	3	4	2	3	2	3	3	4	5	6	0							
13	4	5	6	4	1	4	1	5	6	3	4	2	0						
14	3	4	5	3	4	3	4	2	5	6	7	1	3	0					
15	2	3	4	4	5	2	5	3	4	5	6	2	4	3	0				
16	2	3	4	2	3	2	3	3	2	1	2	4	4	5	4	0			
17	4	5	4	4	3	4	3	5	2	1	2	6	4	7	6	2	0		
18	1	2	3	1	2	1	2	2	3	4	5	1	3	2	3	3	5	0	
19	2	3	4	2	3	2	3	3	2	1	2	4	4	5	4	0	2	3	0

Bird number

Table 11-2. The number of DNA band differences between all pairs of individuals in Figure 11-12 (sex disregarded). The number ranges from 1 to 7. Kestrels 11 and 14 exhibit the greatest diversity in their genomes in terms of the bands scored. If these markers are representative of the whole genome these two birds, fortunately a female and a male, should be mated because they are the most distantly related genetically. Their offspring have the greatest chance of not receiving the same deleterious mutant (recessive) genes from each parent. (Table courtesy of J.L. Longmire.)

DNA typing has also been used to resolve a case of "disputed" paternity among whales in captivity (Hewlett 1989). The Vancouver Public Aquarium has three killer whales *(Orcinus orca)*: a cow named Bjossa, age 14 yr, from Iceland; a bull named Finna, age 14 yr, from Iceland; and a bull named Hyak, age 22 yr, from Pender Harbour, British Columbia. On November 13, 1988 Bjossa (ironically a name meaning bull) gave birth to a calf and, along with the considerable excitement surrounding this event, a question arose as to which bull was the sire. Both bulls

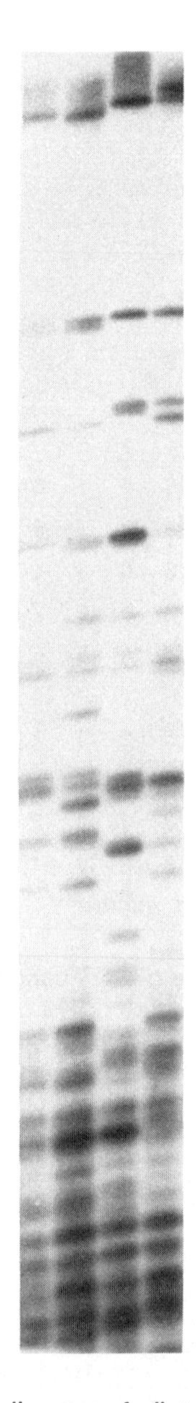

Figure 11-13. Autoradiogram of pilot whale DNA fingerprints. DNA was digested with the restriction enzyme Dde I and hybridized with the multilocus probe 33.15. The lanes from left to right are (1) mother #1, (2) fetus #1; and (3) mother #2, (4) fetus #2. (Autoradiogram courtesy of Bill Amos, Department of Genetics, University of Cambridge, Cambridge, U.K.)

had access to Bjossa when she had conceived 17 months earlier so there was no obvious way of knowing the paternity; however, the odds were with Hyak since Finna was considered too young to reproduce. To resolve this conundrum, DNA from the four animals was isolated and fingerprints prepared (Figure 11-14). As observed in the autoradiogram, all of the obligate paternal fragments in the calf's fingerprint are present in Hyak but not Finna, identifying Hyak as the sire.

WILDLIFE POACHING

The Smoking Gun

Thousands of big game animals are killed illegally each year in North America. A major problem in the battle against poaching is identification of seized parts and products of poached animals after the species-specific characteristics, such as antlers or hides, have been removed. At present, identification of wildlife tissues involves use of electrophoretic and immunological techniques. Species-specific markers can be determined using electrophoretic systems, and immunological tests are generally accurate to the family level. Such methods, however, are frequently nonspecific and positive identification has often been impossible even with the best of techniques. These methods allow only for a statistical association or probability exclusion. As a result, unless a hunter is present and in possession of a smoking gun, allegations of wildlife poaching frequently cannot be proven because vital evidence linking killed animals and poachers is inadequate.

Tissue samples have been analyzed from deer *(Odocoileus spp.)*, moose *(Alces alces)*, elk *(Cervus spp.)*, Holstein-Freisian cattle, mongrel dogs, and thoroughbred horses. For big game animals, DNA can be isolated from tissue recovered from a roadside gut pile and matched to an identical DNA pattern from tissue recovered from a poacher's freezer or from a pelt or mounted trophy head (Figure 11-15). Also, it should be possible to distinguish among wildlife, domesticated animal, and human patterns.

Profile autoradiograms (Figures 11-16 through 11-20) obtained with Hinf I and Hae III genomic digests, using ^{32}P random-labeled M13 and 3'HVR probes, show resolvable fragment (allele) sizes ranging from approximately 15 kb (top of figures) to 2 kb (bottom of figures). Band intensity varies directly with the number of tandem repeats and inversely to the degree of mismatch with the hybridized probe.

Pattern differences obtained with DNA from fresh white blood cells from humans and domesticated animals (related and unrelated individuals) are illustrated in Figures 11-16 and 11-17. Cows C1 and C2 are mother and daughter; C4 is unrelated. Horses H1 and H2 are the parents of H3; H4 is unrelated. Dogs D2 and D3 are offspring of D1. Humans Hu1 and Hu2 are unrelated; Hu3 and Hu4 are identical twins.

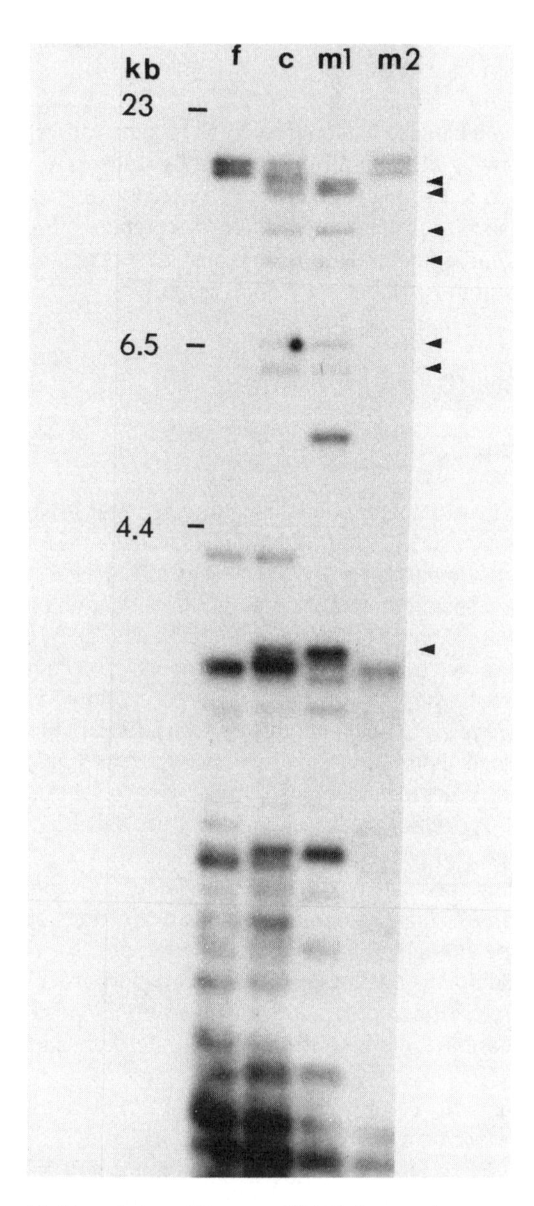

Figure 11-14. Autoradiogram of DNA fingerprint results from a Vancouver Public Aquarium killer whale paternity case. The lanes from left to right are Bjossa (f), the cow; the calf (c); Hyak (m1), a bull; and , Finna (m2) a bull. DNA samples were digested with the restriction enzyme Alu I and hybridized with the multilocus probe M13. Arrows on the autoradiogram indicate obligate paternal fragments in the calf's fingerprint. All of these fragments are present in m1 but not in m2, identifying Hyak as the sire of the calf. (Autoradiogram courtesy of D.A. Duffield, Department of Biology, Portland State University, Portland, Oregon.)

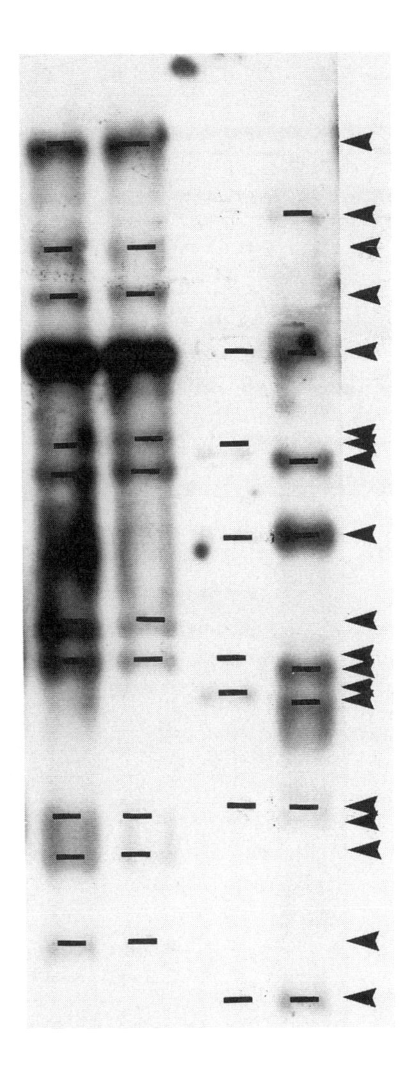

Figure 11-15. Autoradiogram of DNA fingerprints from four samples of moose tissue. The lanes from left to right are (1) liver from animal #1, (2) muscle from the same animal #1, (3) muscle from animal #2, and (4) muscle from animal #3. The DNA was digested with the restriction enzyme Hae III and hybridized with the multilocus probe M13. Animals #1, #2, and #3 can be readily distinguished by their profiles. As anticipated, the profiles in lanes (1) and (2) are identical. (Autoradiogram courtesy of Helix Biotech Corporation, Richmond, British Columbia, Canada.)

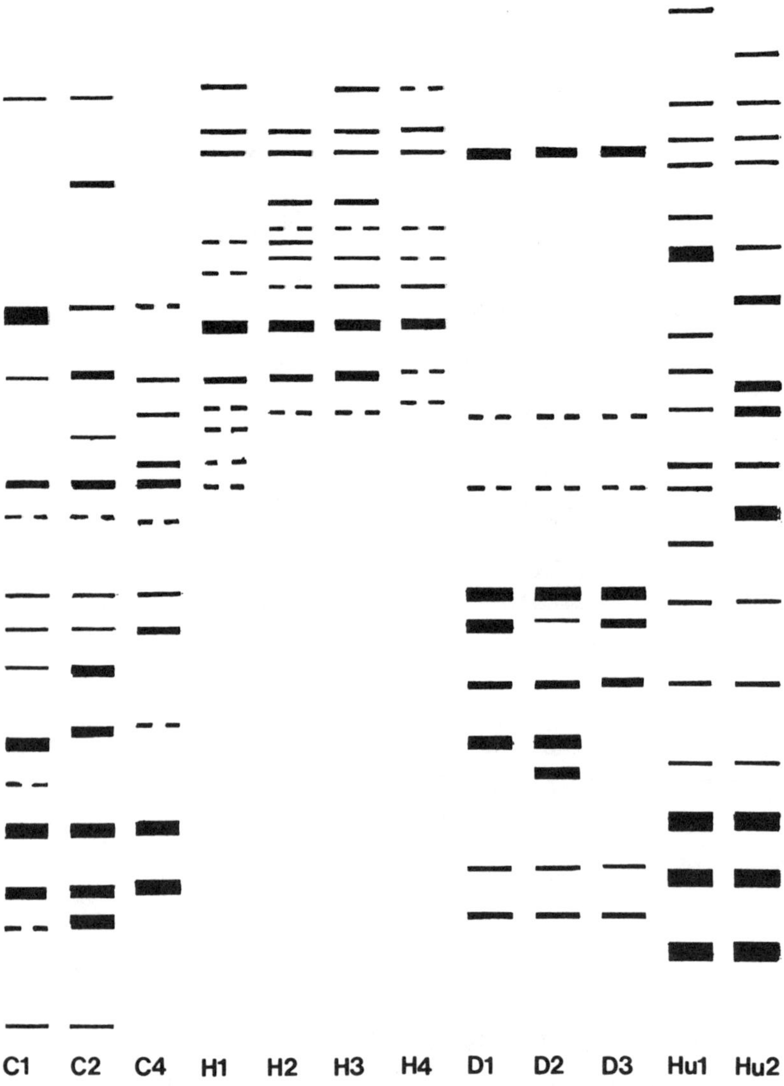

C1 C2 C4 H1 H2 H3 H4 D1 D2 D3 Hu1 Hu2

Figure 11-16. Autoradiogram of white blood cell genomic DNA samples digested with Hae III and hybridized with the M13 probe; C = cow, H = horse, D = dog, Hu = human. Solid lines represent dark bands and broken lines represent faint bands.

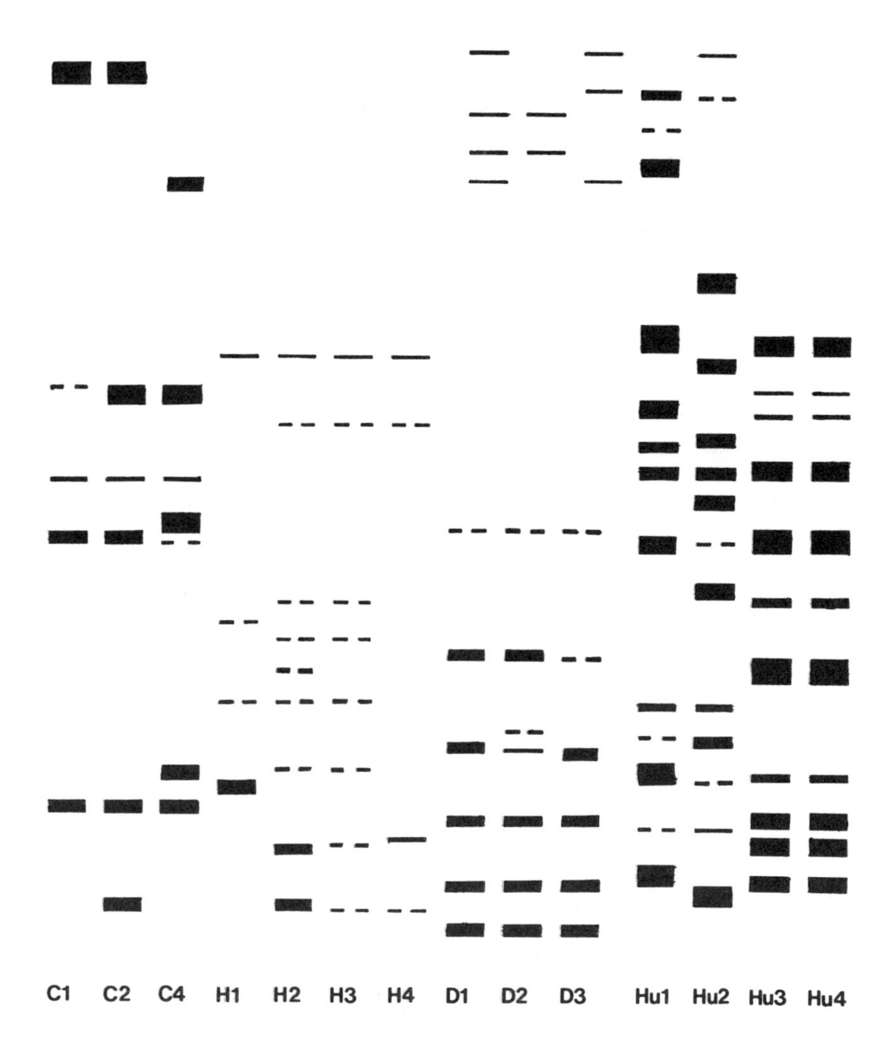

Figure 11-17. Autoradiograms of white blood cell genomic DNA samples digested with Hae III (Hu3 and Hu4 were digested with Hinf I) and hybridized with the 3'HVR probe. C = cow, H = horse, D = dog, Hu = human.

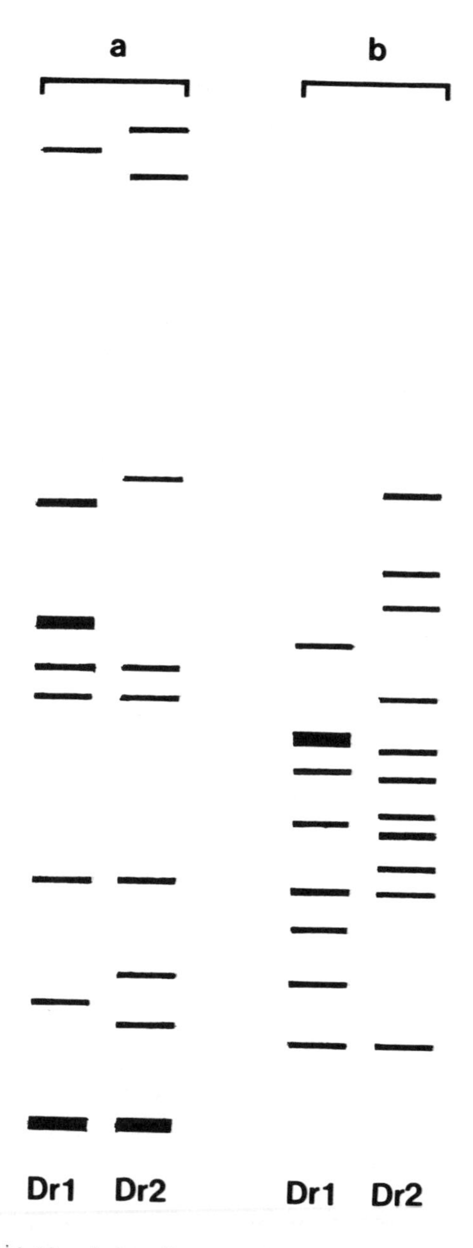

Figure 11-18. Autoradiograms of deer (Dr1, Dr2) muscle genomic DNA digested with Hinf I (a) and Hae III (b) and hybridized with 3'HVR (a) and M13 (b) probes.

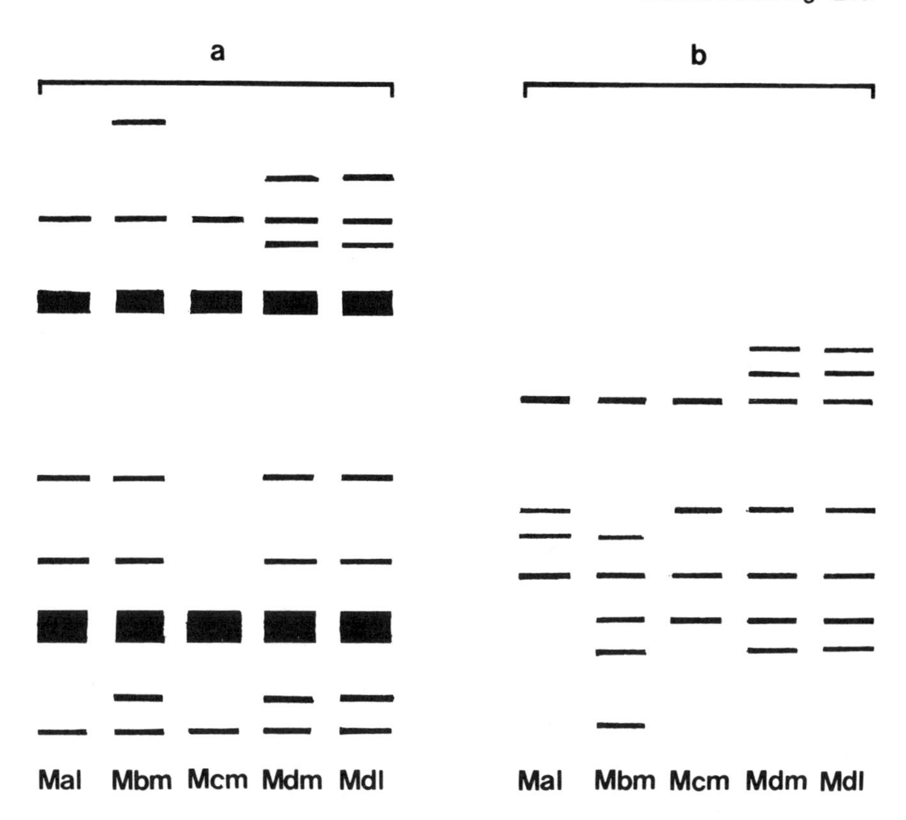

Figure 11-19. Autoradiograms of moose genomic DNA digested with Hae III (a) and Hinf I (b) and hybridized with M13 (a) and 3' HVR (b) probes. Ma, b, c, d are different animals; l = liver; m = muscle.

Figure 11-20. Autoradiogram of elk (E) and moose (M) muscle and cow (C) white blood cell genomic DNA digested with Hinf I and hybridized with the 3'HVR probe. E2= DNA extracted from liver (l), muscle (m), and heart (h) of the same animal; E11 and E12 = different DNA extracts from the same animal.

Fragment patterns obtained from muscle of unrelated deer, Dr1 and Dr2, show clear differences (Figure 11-18). Liver and muscle patterns from moose killed in the same vicinity during the 1983 to 1986 hunting seasons are shown in Figure 11-19. The degree of relationship between these animals was not known.

Autopsy liver, muscle, and heart tissue from the same elk (E2) showed identical patterns (Figure 11-20). Muscle samples from a different animal (E1) processed on two separate occasions illustrate a single band difference from E2 (Figure 11-20). For purposes of species comparison, moose and cow DNA were analyzed concurrently.

Print patterns differ considerably for specimens from the same animal when DNA is hybridized with different probes. This was expected because separate sets of alleles are hybridized with 3'HVR and M13. DNA extracted from different tissues of a specific animal, then digested with the same enzyme and hybridized with the same probe, results in identical patterns.

Individuals of known relationship have, on average, allele transmission patterns expected in a Mendelian system; for example, H3 is the foal of mare H1 and stallion H2 (Figures 11-16, 11-17). Each allele (band) observed in H3 is present in H1 or H2; however, the parents have alleles not present in the foal. This is logical because only one-half of each parental genotype is transmitted to an offspring. A mutation could result in a spurious fragment band in an offspring. However, when the overall number of bands is considered, especially if more than one probe is used, it is doubtful that a single mutation would significantly influence the conclusion concerning the degree of biological relationship.

Unrelated animals may have a number of alleles in common, as illustrated by the matching bands for deer Dr1 and Dr2 (Figure 11-18). The degree of this occurrence is a function of the allele frequency in the population. If an allele frequency is 0.01, then according to the Hardy-Weinberg law, approximately 2% of the animals will carry that allele.

These preliminary data support the hypothesis that recombinant DNA technology can be applied to wildlife forensic science. The overall patterns for various species appear to differ, thus, providing another possible tool for species identification.

Almost all incidences of poaching result in a large quantity of tissue left at the kill site. Matching DNA fingerprints from this tissue with that recovered from a poacher's possession should be feasible in most cases.

As single- and multilocus probes become available for each animal species, techniques such as the polymerase chain reaction should be useful for the amplification of specific DNA fragments in specimens where the DNA is minimal or the material is partially degraded. For all blood, muscle, and liver specimens analyzed to date, DNA quality has been good.

Considerable research is yet to be undertaken to improve autoradiogram band resolution, to identify other probes applicable to different poached species, and to assimilate sufficient statistical data to withstand standards of evidence. However, even at this stage it is possible to detect individual specific DNA fragments from several big game wildlife species using 3'HVR and M13 probes. Results to date suggest that DNA tissue identification may be used to confirm poaching occurrences. DNA typing of game farm animals could also provide a strong deterrent to the laundering of poached meat through these commercial operations (Thommasen 1989).

PLANT SCIENCES

Preliminary observations with multilocus probes, including M13 and the minisatellite series 33.15 and 33.6, suggest that individual plants within many species such as poplar and spruce trees can be distinguished (Rogstad 1988, 1988*a*; Wolfe 1989). Different cultivars of rice plants are distinguishable although the patterns between members of the same cultivar are not (Dallas 1988). Somatic stability (identical profile patterns) has also been noted between different tissues from the same plant.

Short-day and long-day season varieties of soya beans have been separated using genotyping techniques. This differentiation could be critical in the prevention of crop failures if the varieties were inadvertently mixed.

Profiling could also be important in policing plant patent infringements and in identifying trait markers for use in breeding programs. DNA markers are, for example, used by the brewing industry to identify yeast for patent purposes.

FUTURE

Automation and simplification of the DNA assay methodology is eagerly awaited. The present techniques of extraction, amplification, digestion, separation, transfer, and hybridization will, no doubt, seem archaic in a decade. Perhaps within this period the total human genome will have been sequenced, and simple base differences in highly variable regions directly analyzed in identity determinations.

Applications in animal and plant sciences will in all likelihood lag behind those for humans; however, a foundation will be established for more detailed analysis of these complex genomes for species preservation and improvements in disease resistance and food production.

A number of ethical issues remain to be resolved. In a utopian world, ethical concerns regarding DNA profiling for forensic analysis would not arise. Cases involving paternity, rape, homicide, poaching, and other crimes would simply not exist. Unfortunately, society is confronted constantly by such problems. Serious consideration must be given to the development of DNA profile data bases of, for example, convicted felons. Issues concerning informed consent versus compulsion, the use of results, confidentiality, and even the possibility of universal DNA typing will have to be resolved (Ad Hoc Committee 1990, Coghlan 1989, Garfield 1989, Knoppers 1989, Roberts 1989, Suzuki 1988). Misinformation and miscommunication lead inevitably to fear and mistrust. The difficult ethical questions that accompany the development of this new technology must be thoroughly explored in order to achieve a rational consensus on them.

The benefits of this new DNA technology, which had its beginnings in 1985, have barely been tapped. Positive individual identification from a single hair, minute semen stain, or a few particles of tooth pulp, mark only the beginning of this exciting era of DNA fingerprinting.

REFERENCES

Ad Hoc Committee on Individual Identification by DNA Analysis, The American Society of Human Genetics. 1990. Individual identification by DNA analysis: Points to consider. *Am. J. Hum. Genet.* 46:631–634.

Amos B. 1989. Conservation of fingerprint pattern between species — artefact or indication of underlying evolutionary mechanisms. *Fingerprint News* 3:14–15.

Anderson A. 1989. DNA fingerprinting, new technique on trial. *Nature* 339:408.

Anderson A. 1989*a*. Judge backs technique. *Nature* 340:582.

Anon. 1989. Bird dealers caught by genes. *IWCH Wildlife Trade Monitoring Traffic Bulletin* 10:2.

Anon. 1989*a*. DNA technology and rapid diagnosis of infection. *Lancet* 2:897–898.

Anon. 1989*b*. DNA points the finger at Argentina's past. *New Scientist* 121:29.

Azuma C, Kamiura S, Nobunaga T, Negoro T, Saji F, and Tanizawa O. 1989. Zygosity determination of multiple pregnancy by deoxyribonucleic acid fingerprints. *Am. J. Obstet. Gynecol.* 160:734–736.

Baird M, Balazs I, Giusti A, Miyazaki L, Nicholas L, Wexler K, Kanter E, Glassberg J, Allen F, Rubinstein P, and Sussman L. 1986. Allele frequency distribution of two highly polymorphic DNA sequences in three ethnic groups and its application to the determination of paternity. *Am. J. Hum. Genet.* 39:489–501.

Baird M, Wexler K, Clyne M, Meade E, Ratzlaff L, Smalls G, Benn P, Glassberg J, and Balazs I. 1987. The application of DNA-Print for the estimation of paternity. In *Advances in Forensic Haemogenetics 2*. Springer-Verlag, New York.

Barinaga M. 1989. DNA fingerprinting pitfalls come to light. *Nature* 339:89.

Beeler L and Wiebe WR. 1988. DNA identification tests and the courts. *Washington Law Review* 63:903–955.

Bronstein I and Voyta JC. 1989. Chemiluminescent detection of Herpes Simplex Virus I DNA in blot and in-situ hybridization assays. *Clin. Chem.* 35:1856–1857.

Burke T and Bruford MW. 1987. DNA fingerprinting in birds. *Nature* 327:149–152.

Burke T. 1989. DNA fingerprinting and other methods for the study of mating success. *Trends Ecol. Evol.* 4:139–144.

Burke T, Davies NB, Bruford MW, and Hatchwell BJ. 1989*a*. Parental care and mating behaviour of polyandrous dunnocks *Prunella modularis* related to paternity by DNA fingerprinting. *Nature* 338:249–251.

Cellmark Daignostics. 1989. The Narborough Village murders. *Benchmark* 2(2):1–5.

Cellmark Diagnostics 1989*a*. United Kingdom court cases.

Cherfas J. 1985. Geneticists develop DNA fingerprinting. *New Scientist* 105:21.

Coghlan A and Joyce C. 1989. Public debate grows over ethics of DNA fingerprinting. *New Scientist* 124:26.

Conner S. 1988. Genetic fingers in the forensic pie. *New Scientist* 117:31–32.

Dallas JF. 1988. Detection of DNA "fingerprints" of cultivated rice by hybridization with a human minisatellite DNA probe. *Proc. Natl. Acad. Sci. USA* 85: 6831–6835.

deJong D, Voetdijk BMH, Kluin-Nelemans JC, van Ommen GJB, and Kluin PM. 1988. Somatic changes in B-lymphoproliferative disorders (B-LPD) detected by DNA-fingerprinting. *Br. J. Cancer* 58:773–775.

Devor EJ, Ivanovich AK, Hickok JM, and Todd RD. 1988. A rapid method for confirming cell line identity: DNA "fingerprinting" with a minisatellite probe from M13 bacteriophage. *BioTechniques* 6:200–201.

Diamond JM. 1987. Abducted orphans identified by grandpaternity testing. *Nature* 327:552–553.

Dixson AF, Hastie N, Patel I, and Jeffreys AJ. 1988. DNA 'fingerprinting' of captive family groups of common marmosets *(Callithrix jacchus)*. *Folia. Primatol.* 51:52–55.

Dodd BE. 1985. DNA fingerprinting in matters of family and crime. *Nature* 318:506–507.

Erlich HA and Bugawan TL. 1989. HLA class II gene polymorphism: DNA typing, evolution, and relationship to disease susceptibility. In *PCR Technology Principles and Applications for DNA Amplification*, 193–208. Erlich HA, ed. Stockton Press. New York.

Evett IW, Werrett DJ, and Gill P. 1989. DNA fingerprinting on trial. *Nature* 340:435.

Fey MF, Wells RA, Wainscoat JS, and Thein SL. 1988. Assessment of clonality in gastrointestinal cancer by DNA fingerprinting. *J. Clin. Invest.* 82:1532–1537.

Fields RD, Johnson KR, and Thorgaard GH. 1989. DNA fingerprints in rainbow trout detected by hydridization with DNA of bacteriophage M13. *Trans. Am. Fish Soc.* 118:78–101.

Forman L, Kriss J, Sheridan K, Fristoe V, and Cotton R. 1989. Forensic applications of DNA fingerprinting to endangered species: Hyacinth macaw. *Northwest Association of Forensic Scientists Spring 1989 Meeting Abstracts.* National Fish and Wildlife Forensics Laboratory, Ashland, OR.

Fox JL. 1989. FBI embracing genetic fingerprints. *BioTechnology* 7:548.

Fowler JCS, Burgoyne LA, Scott AC, and Harding HWJ. 1988. Repetitive deoxyribonucleic acid (DNA) and human genome variation—A concise review relevant to forensic biology. *J. Forensic Sci.* 33:1111–1126.

Fukushima H, Hasekura H, and Nagai K. 1988. Identification of male bloodstains by dot hybridization of human Y chromosome-specific deoxyribonucleic acid (DNA) probe. *J. Forensic Sci.* 33:621–627.

Garfield E. 1989. DNA fingerprinting: A powerful law-enforcement tool with serious social implications. *The Scientist* 3:10–11.

Georges M, Lequarre AS, Castelli M, Hanset R, and Vassart G. 1988. DNA fingerprinting in domestic animals using four different minisatellite probes. *Cytogent. Cell Genet.* 47:127–131.

Gill P and Werrett DJ. 1987. Exclusion of a man charged with murder by DNA fingerprinting. *Forensic Sci. Int.* 35:145–148.

Giusti A, Baird M, Pasquale S, Balazs I, and Glassberg J. 1986. Application of deoxyribonucleic acid (DNA) polymorphisms to the analysis of DNA recovered from sperm. *J. Forensic Sci.* 31:409–417.

Gyllensten UB, Jakobsson S, Temrin H, and Wilson A. 1989. Nucleotide sequence and genomic organization of bird minisatellites. *Nucleic Acids Res.* 17:2203–2214.

Gyllensten UB, Jakobsson S, and Temrin H. 1990. No evidence for illegitimate young in monoganous of polygynous warblers. *Nature* 343:168–178.

Helminen P, Ehnholm C, Lokki ML, Jeffreys A, and Peltonen L. 1988. Application of DNA "fingerprints" to paternity determinations. *Lancet* 1:574–576.

Hewlett KG, Chamberlin-Lea J, Goldie PL, Duffield DA, and Stevens TA. 1989. Determination of paternity in killer whales *(Orcinus orca)* by chromosomes and DNA fingerprinting. *Eighth Biennial Conference on the Biology of Marine Mammals,* December 1989. The Society for Marine Mammology, Pacific Grove, CA.

Hicks JW. 1989. DNA typing: A unique weapon against crime. *The Scientist* 3:12–13.

Higuchi R, von Beroldingen CH, Sensabaugh GF, and Erlich HA. 1988. DNA typing from single hairs. *Nature* 332:543–546.

Hill AVS and Jeffreys AJ. 1985. Use of minisatellite DNA probes for determination of twin zygosity at birth. *Lancet* 2:1394–1395.

Hill WG. 1986. DNA fingerprint analysis in immigration test-cases. *Nature* 322:290–291.

Hill WG. 1987. DNA fingerprints applied to animal and bird populations. *Nature* 327:98–99.

Hoelzel AR and Amos W. 1988. DNA fingerprinting and "scientific" whaling. *Nature* 333:305.

Howlett R. 1989. Paternity of dunnocks by DNA fingerprinting. *Nature* 338:203.

Howlett R. 1989*a*. DNA forensics and the FBI. *Nature* 341:182.

Huey B and Hall J. 1989. Hypervariable DNA fingerprinting in *Escherichia coli:* Minisatellite probe from bacteriophage M13. *J. Bacteriol.* 171:2528–2532.

Impraim CC, Saiki RK, Erlich HA, and Teplitz RL. 1987. Analysis of DNA extracted form formalin-fixed, paraffin-embedded tissues by enzymatic amplification and hybridization with sequence-specific oligonucleotides. *Biochem. Biophys. Res. Commun.* 142:710–716.

Ivanov PL, Semyokhina AF, and Ryskov AP. 1989. Rat DNA fingerprinting for *Rattus norvegicus:* New approach in genetic analysis. *Genetika* 25:238–249.

Jeffreys AJ, Brookfield JFY, and Semeonoff R. 1985. Positive identification of an immigration test-case using human DNA fingerprints. *Nature* 317:818–819.

Jeffreys AJ, Wilson V, Thein SL, Weatherall DJ, and Ponder BAJ. 1986. DNA "fingerprints" and segregation analysis of multiple markers in human pedigrees. *Am. J. Hum. Genet.* 39:11–24.

Jeffreys AJ, Brookfield JFY, and Semeonoff R. 1986*a*. DNA fingerprint analysis in immigration test cases. *Nature* 322:290–291.

Jeffreys AJ, Hillel J, Hartley N, Bulfield G, Morton D, Wilson V, Wong Z, and Harris S. 1987. The implications of hypervariable DNA-regions for animal identification. *Anim. Genet.* 18 [Suppl 1]:141–142.

Jeffreys AJ, Wilson V, Kelly R, Taylor BA, and Bulfield G. 1987*a*. Mouse DNA "fingerprints": Analysis of chromosome localization and germ-line stability of hypervariable loci in recombinant inbred strains. *Nucleic Acids Res.* 15:2823–2836.

Jeffreys AJ, Wilson V, Neumann R, and Keyte J. 1988. Amplification of human minisatellites by the polymerase chain reaction: Towards DNA fingerprinting of single cells. *Nucleic Acids Res.* 16:10953–10971.

Johnston K. 1987. UK immigration authorities may use DNA fingerprinting. *Nature* 329:5.

Jones L, Thein SL, Jeffreys AJ, Apperley JF, Catovsky D, and Goldman JM. 1987. Identical twin marrow transplantation for 5 patients with chronic myeloid leukaemia: Role of DNA fingerprinting to confirm monozygosity in 3 cases. *Eur. J. Haematol.* 37:144–147.

Joyce C. 1989. The finger of doubt points at "foolproof" fingerprinting. *New Scientist* 122:24.

Keable H, Pico JL, Venvat AM, Charpentier F, Baume D, Hayat M, and Brison O. 1988. DNA fingerprint study of engraftment in bone marrow transplantation recipients for leukemia. *Bone Marrow Transplant* 3:232.

King M-C. 1989. Genetic testing of identity and relationship. *Am. J. Hum. Genet.* 44:179–181.

King M-C. 1989*a*. Genetics and the disappeared: The search for two generations. Presented plenary lecture, AAAS Annual Meeting, San Francisco, CA, January 1989, page 28.

Kinoshita J. 1989. Misprints, seeking new standards for forensic DNA typing. *Sci. Am.* 261:12–12B.

Knoppers BM and Laberge C. 1989. DNA sampling and informed consent. *Can. Med. Assoc. J.* 140:1023–1028.

Kobayashi R, Nakauchi H, Nakahori Y, Nakagome Y, and Matsuzawa S. 1988. Sex identification in fresh blood and dried bloodstains by a nonisotopic deoxyribonucleic acid (DNA) analyzing technique. *J. Forensic Sci.* 33:613–620.

Kocher TD, Thomas WK, Meyer A, Edwards SV, Paabo S, Villablanca FX, and Wilson AC. 1989. Dynamics of mitochondrial DNA evolution in animals: Amplification and sequencing with conserved primers. *Proc. Natl. Acad. Sci. USA* 86:6196–6200.

Kocher TD and White TJ. 1989*a*. Evolutionary analysis via PCR. In *PCR Technology Principles and Applications for DNA Amplification*, 137–147. Erlich AH, ed. Stockton Press, New York.

Kuhnlein U, Dawe Y, Zadworny D, and Gavora JS. 1989. DNA fingerprinting: A tool for determining genetic distances between strains of poultry. *Theor. Appl. Genet.* 77:669–672.

Lander ES. 1989. DNA fingerprinting on trial. *Nature* 339:501–505.

Lander ES. 1989*a*. Expert's report in *People v. Castro*. 41 pages + figures; Whitehead Institute for Biomedical Research and Harvard University.

Layton DM and Mufti GJ. 1987. Human cancer DNA fingerprint analysis. *Br. J. Cancer* 56:381.

Lewin R. 1989. Judging paternity in the hedge sparrow's world. *Science* 243:1663–1664.

Lewin R. 1989*a*. DNA typing is called flawed. *Science* 245:355.

Lifecodes Corporation. 1986. New genetic I.D. test aids forensic investigators. Case summaries.

Logtenberg H and Bakker E. 1988. The DNA fingerprint. *Endeavour New Series* 12:28–33.

Longmire JL, Lewis AK, Brown NC, Buckingham JM, Clark LM, Jones MD, Meincke LJ, Meyne J, Ratliff RL, Ray FA, Wagner RP, and Moyzis RK. 1988. Isolation and molecular characterization of a highly polymorphic centromeric tandem repeat in the family *Falconidae*. *Genomics* 2:14–24.

Longmire J. 1989. Molecular genetic studies of endangered species. *Northwest Association of Forensic Scientists Spring 1989 Meeting Abstracts*. National Fish and Wildlife Forensics Laboratory, Ashland, OR.

Lynch M. 1988. Estimation of relatedness by DNA fingerprinting. *Mol. Biol. Evol.* 5:584–599.

Marx JL. 1988. DNA fingerprinting takes the witness stand. *Science* 240:1616–1618.

McElfresh KC. 1989. DNA fingerprinting. *Science* 246:192.

McGourty C. 1989. New York State leads on genetic fingerprinting. *Nature* 341:90.

Merz B. 1988. DNA fingerprints come to court. *J. Am. Med. Assoc.* 259:2193–2194.

Morton OB, Yaxley RE, Patel I, Jeffreys AJ, Howes SJ, and Debenham PG. 1987. Use of DNA fingerprint analysis in identification of the sire. *Vet. Record* 121:592–594.

Newmark P. 1987. Dispute over who should do DNA fingerprinting in murder hunt. *Nature* 325:97.

Newmark P. 1987*a*. DNA fingerprinting at a price at ICI's UK laboratory. *Nature* 327:548.

Newmark P. 1988. DNA fingerprinting to be used for British immigrants? *Nature* 331:556.

Norman C. 1989. Caution urged on DNA fingerprinting. *Science* 245:669.

Odelberg SJ, Demers DB, Westin EH, Hossaini AA. 1988. Establishing paternity using minisatellite DNA probes when the putative father is unavailable for testing. *J. Forensic Sci.* 33:921–928.

Orrego C, Wilson AC, and King M-C. 1988. Identification of maternally-related individuals by amplification and direct sequencing of a highly polymorphic, noncoding region of mitochondrial DNA. *Am. J. Hum. Genet.* 43:A219.

Orrego C and King M -C. 1990. Determination of familial relationships. In *PCR Protocols: A Guide to Methods and Applications,* 416–426. Innis M, Gelfand D, Sninsky J, and White T, eds. Academic Press, New York.

Quinn TW, Quinn JS, Cooke F, and White BN. 1987. DNA marker analysis detects multiple maternity and paternity in single broods of the lesser snow goose. *Nature* 326:392–394.

Roberts DF, Papiha SS, and Bhattacharya SS. 1987. A case of disputed maternity. *Lancet* 2:478–480.

Roberts L. 1989. Ethical questions haunt new genetic technologies. *Science* 243:1134–1136.

Rogstad SH, Patton II JC, and Schaal BA. 1988. M13 repeat probe detects DNA minisatellite -like sequences in gymnosperms and angiosperms. *Proc. Natl. Acad. Sci. USA.* 85:9176–9178.

Rogstad SH, Patton JC, and Schaal BA. 1988*a*. A human minisatellite probe reveals RFLPs among individuals of two angiosperms. *Nucleic Acids Res.* 16:11378.

Ross AM and Harding HWJ. 1989. DNA typing and forensic science. *Forensic Sci. Int.* 41:197–203.

Ryskov AP, Jincharadze AG, Prosnyak MI, Ivanov PL, and Limborskaya SA. 1988. DNA fingerprints of different organisms using labelled M13 phage. *Genetika* 24:227–238.

Ryskov AP, Jincharadze AJ, Prosnyak MI, Ivanov PL, and Limborskaya SA. 1988*a*. M13 phage DNA as a universal marker for DNA fingerprinting of animals, plants, and microorgansims. *FEBS Lett.* 233:388–392.

Smit VTHBM, Cornelisse CJ, DeJong D, Dijkshoorn NJ, Peters AAW, and Fleuren, GJ. 1988. Analysis of tumor heterogeneity in a patient with synchronously occurring female genital tract malignancies by DNA flow cytometry, DNA fingerprinting, and immunohistochemistry.*Cancer* 62:1146–1152.

Suzuki D and Knudtson P. 1988. *Genethics: The Ethics of Engineering Life.* Stoddart Publishing, Toronto.

Taylor G. 1989. DNA fingerprinting. *Nature* 340:672.

Thacker J, Webb MBT, and Debenham PG. 1988. Fingerprinting cell lines: Use of human hypervariable DNA probes to characterize mammalian cell cultures. *Somatic Cell Mol. Genet.* 14:519–525.

Thein SL, Jeffreys AJ, and Blacklock HA. 1986. Identification of post -transplant cell population by DNA fingerprint analysis. *Lancet* 2:37.

Thein SL, Jeffreys AJ, Gooi H, Cotter F, Flint J, O'Connor NTJ, Weatherall DJ, and Wainscoat JS. 1987. Detection of somatic changes in human cancer DNA by DNA fingerprint analysis. *Br. J. Cancer* 55:353–356.

Thommasen HV, Thomson MJ, Shutler GG, and Kirby LT. 1989. Development of DNA fingerprints for use in wildlife forensic science. *Wildl. Soc. Bull.* 17:321–326.

Turner BJ, Elder JF, Laughlin TF. 1989. DNA fingerprinting of fishes: A general method using oligonucleotide probes. *Fingerprint News* 4:15–16.

Tyler MG, Kirby LT, Wood S, Vernon S, and Ferris JAJ. 1986. Human blood stain identification and sex determination in dried blood stains using recombinant DNA techniques. *Forensic Sci. Int.* 31:267–272.

Vassart G, Georges M, Monsieur R, Brocas H, Lequarre AS, and Christophe D. 1987. A sequence in M13 phage detects hypervariable minisatellites in human and animal DNA. *Science* 235:683–684.

Von Beroldingen CH, Blake ET, Higuchi R, Sensabaugh GF, and Erlich H. 1989. Applications of PCR to the analysis of biological evidence. In *PCR Technology Principles and Applications for DNA Amplification*, 209–223. Erlich HA, ed. Stockton Press, New York.

Weiss ML, Wilson V, Chan C, Turner T, and Jeffreys AJ. 1988. Applications of DNA fingerprinting probes to Old World monkeys. *Amer. J. Primatol.* 16:73–79.

Wells RA, Wonke B, and Thein SL. 1988. Prediction of consanguinity using human DNA fingerprinting. *J. Med. Genet.* 25:660–662.

Wetton JH, Carter RE, Parkin DT, and Walters D. 1987. Demographic study of a wild house sparrow population by DNA fingerprinting. *Nature* 327:147–149.

Wolfe KH, Gouy M, Yang Y-W, Sharp PM., and Li WH. 1989. Date of the monocot-dicot divergence estimated from chloroplast DNA sequence data. *Proc. Natl. Acad. Sci. USA* 86:6201–6205.

Yokoi T and Sagisaka K. 1989. Sex determination of blood stains with a recombinant DNA probe: Comparison with radioactive and non-radioactive labeling methods. *Forensic Sci. Int.* 41:117–124.

APPENDIX I

Guidelines for a Quality Assurance Program for DNA Restriction Fragment Length Polymorphism Analysis

Technical Working Group on DNA Analysis Methods (TWGDAM)
James J. Kearney, TWGDAM Chairman
FBI Laboratory
Quantico, Virginia
Reprinted by permission of the author

Quality Assurance Subcommittee

James L. Mudd, Subcommittee Chairman
FBI Laboratory
Quantico, Virginia

John M. Hartmann
Orange County Sheriff-Coroner Department
Santa Ana, California

Margaret C. Kuo
Orange County Sheriff-Coroner Department
Santa Ana, California

Mark S. Nelson
North Carolina State Bureau of Investigation
Raleigh, North Carolina

Lawrence A. Presley
FBI Laboratory
Washington, D.C.

Willard C. Stuver
Metro-Dade Police Department
Miami, Florida

CRIME LABORATORY DIGEST Vol. 16, No. 2, April-July 1989

INTRODUCTION

With the advent of DNA typing technology in the forensic laboratory, the forensic examiner now has the potential to individualize various body fluids and tissues. In addition, since the tests performed by crime laboratories can have a significant impact on the outcome of a trial, it is important that any test procedure used by the laboratory possess a high degree of accuracy and reproducibility. Consequently, the use of appropriate standards and controls is essential in order to ensure reliable results.

As any technology becomes more discriminating and precise, it is essential that the quality of the analytical data be more closely monitored. A detailed and flexible quality assurance program can assist in establishing a basis for scientifically sound and reliable forensic analysis.

Although often used interchangeably, quality assurance (QA) and quality control (QC) refer to different, specific quality functions (American National Standard ANSI/ASQC A3-1978; Kilshaw 1986, 1987a,b). The function of the QA program is to provide to all concerned, the evidence needed to establish with confidence that the QC function is being performed adequately. This is accomplished in part through the use of proficiency tests and audits. The QC measures are employed by the DNA analysis laboratory to ensure that the quality of the product (DNA typing) will meet and satisfy specified criteria.

Although the application of formal QA programs in forensic laboratories is currently not widespread and little information has appeared in the forensic science literature (Bradford 1980; Brunelle *et al.* 1982; Pereira 1985), a great deal has been written on the application of QA programs to clinical and federally operated laboratories (Alwan and Bissell 1988; Box and Bisaard 1987; Bussolini *et al.* 1988; Ford 1988; Gautier and Gladney 1987; Hay 1988; Kenney 1987; Kidd 1987; Simpson 1983; Taylor 1985, 1987; Whitehead and Woodford 1981).

In November, 1988, the first meeting of the Technical Working Group on DNA Analysis Methods (TWGDAM) was hosted by the FBI Laboratory at the FBI Academy. This group consisted of 31 scientists representing 16 forensic laboratories in the United States and Canada and 2 research institutions (see Appendix A for the names of the TWGDAM participants). The purpose of this group is: (1) to pull together a select number of individuals from the forensic science community who are actively pursuing the various DNA analysis methods; (2) to discuss the methods now being used; (3) to compare the work that has been done; (4) to share protocols; and (5) to establish guidelines where appropriate. During the first meeting, a subcommittee was established to formulate suggested guidelines for a QA program in crime laboratories conducting restriction fragment length polymorphism (RFLP) DNA analysis.

These guidelines represent the minimum QA requirements for DNA RFLP analysis and are intended to serve only as a guide to laboratory managers in establishing their own QA program for DNA RFLP analysis.

These QA guidelines were designed, using established quality functions (American National Standard ANSI/ASQC C1-1968; ANSI/ASQC Z-1.15-1979, ANSI/ASQC Q90-1987a,b; Juran 1979; Ruzicka 1979) to follow systematically the DNA RFLP typing procedure and cover all significant aspects of the laboratory process. In addition, they provide the necessary documentation to ensure that the DNA analysis process is operating within the established performance criteria, and they provide a measure of the overall quality of the results.

These guidelines form the basis of a quality assurance program for RFLP analysis and are subject to future revisions as the state of the art and experience dictate.

1. Planning and Organization

1.1 Goals: It is the goal of the (organization's name) to:

1.1.1 Provide the users of laboratory services access to DNA typing of selected biological materials associated with official investigations using RFLP DNA testing.

1.1.2 Ensure the quality, integrity and accuracy of the DNA typing data through the implementation of a detailed QA program.

1.2 Objectives: It is the objective of the QA Program to:

1.2.1 Monitor on a routine basis the analytical testing procedures for DNA typing by means of QC standards, proficiency tests and audits.

1.2.2 Verify that the entire DNA typing procedure is operating within the established performance criteria and that the quality and validity of the analytical data is maintained.

1.2.3 Ensure that problems are noted and that corrective action is taken and documented.

1.3 Authority and Accountability

1.3.1 Organization Structure: Defines the relationships within the laboratory between individuals, job responsibilities and operational units. It defines the relationship of the QA program to DNA analysis and related laboratory operations as well as to the laboratory management.

1.3.2 Functional Responsibilities: The job function and responsibility for each position within the laboratory should be clearly established. It should specify and describe the lines of responsibility for developing, implementing, recording and updating the QA system.

1.3.3 Levels of Authority: Clear lines of authority and accountability should be established between personnel responsible for the QA program and those assigned to manage and perform the DNA analysis. It should be established as to who may take what action, whether approval is required, and from whom approvals are needed.

2. Personnel

2.1 Job descriptions

The job descriptions for all DNA personnel should include responsibilities, duties and skills.

2.2 Qualifications

The criteria for qualifications and training of technical personnel within the DNA testing laboratory will be established by each laboratory. Each individual engaged in the typing of DNA should have, at a minimum, the education, training and experience as specified in Sections 2.2.1 through 2.2.3.

2.2.1 Supervisor

2.2.1.1 Education—*must* have a minimum of a BA/BS degree in a biological, chemical or forensic science.

2.2.1.2 Training—It is highly desirable that this individual have the following:

(a) Training in DNA analysis technique with individuals, agencies or other laboratories having an established training program that meets American Society of Crime Laboratory Directors (ASCLD) accreditation standards (ASCLD 1985), *and*

(b) Undergraduate or graduate courses in genetics, biochemistry and molecular biology (molecular genetics or recombinant DNA technology).

2.2.1.3 Experience—Technical supervisors should have a minimum of 2 years of experience as a forensic science analyst/examiner and 6 months of DNA laboratory experience.

2.2.1.4 Qualification—It is highly desirable that a supervisor meets the same requirements as those specified for an examiner/analyst in 2.2.2.4.

2.2.1.5 Continuing Education—Supervisors should stay abreast of developments within the field of DNA typing by reading current scientific literature and attending seminars, courses or professional meetings. Management should provide supervisors with an opportunity to comply with the above.

2.2.2 Examiner/Analyst

2.2.2.1 Education—*must* have a minimum of a BA/BS degree in a biological, chemical or forensic science.

2.2.2.2 Training—*must* include as minimum the following:

(a) Training in DNA analysis technique with individuals, agencies or other laboratories having an established training program that meets ASCLD accreditation standards (ASCLD 1985), *and*

(b) Undergraduate or graduate courses in genetics, biochemistry and molecular biology (molecular genetics or recombinant DNA technology).

2.2.2.3 Experience—Prior to any DNA typing and reporting on case work samples, the examiner/analyst *must* have a minimum of 6 months of forensic DNA laboratory experience. It is highly desirable that the examiner/analyst have minimum of 1 year forensic laboratory experience.

2.2.2.4 Qualification—The minimum requirements for qualification of examiner/analysts within a laboratory must be based on the experience, training and job knowledge of the individual and on the successful completion of a formal system of training which requires the passing of specified tests and a demonstration of a thorough knowledge of the theory and practice of DNA typing.

These minimum requirements also *must* include the successful completion of a series of proficiency samples.

2.2.2.5 In-House Qualification Standards for Examiner/Analyst—It is highly recommended that the examiner/analyst complete the requirements for any in-house certification which should include the following:

(a) Knowledge of the scientific literature and procedures with reference to DNA typing. This will be evaluated by written testing after training or from grades received in courses taken.

(b) Skills and mechanical abilities to perform the test can be evaluated by the observation of qualified personnel and by seeing if the proper test results are obtained.

(c) The ability to correctly interpret the test results is of paramount importance. All examiners should go through a well documented and rigorous set of proficiency test samples while in training. After completion of the training program, examiners should undergo periodic in-house and blind proficiency testing administered by an outside agency.

2.2.2.6 Continuing Education—Examiners/analysts *must* stay abreast of developments within the field of DNA typing by reading current scientific literature and through attendance at seminars, courses or professional meetings. Management *must* provide examiners/analysts with an opportunity to comply with the above activities.

2.2.3 Technicians

2.2.3.1 Technicians not performing analytical techniques should have the experience and education commensurate with the job description.

2.2.3.2 Technicians involved in performing analytical techniques related to DNA analysis should have a minimum of a BS/BA degree and receive on-the-job training by a qualified analyst. It is understood that technicians will not have the responsibility for the interpretation of results, preparation of reports, or providing testimony concerning such.

3. Documentation

The DNA laboratory *must* maintain documentation on all significant aspects of the DNA analysis procedure, as well as any related documents or laboratory records that are pertinent to the analysis or interpretation of results, so as to create a traceable audit trail. This documentation will serve as an archive for retrospective scientific inspection, reevaluation of the data, and reconstruction of the DNA procedure. Documentation will exist for the following topic areas:

3.1 Test Methods and Procedures for DNA Typing

3.1.1 This document *must* describe in explicit detail the protocol currently used for the analytical testing of DNA. It *must* include the date the procedure was adopted and the authorization for its use. Revisions *must* be clearly documented and appropriately authorized.

3.2 Population data base

3.3 Quality control of critical reagent and materials to include lot and batch numbers, manufacturer's specifications and in-house evaluations.

- 3.4 Case files/case notes—*must* provide foundation for results and conclusions contained in formal report.
- 3.5 Data analysis and reporting
- 3.6 Evidence handling protocols
- 3.7 Equipment calibration and maintenance logs
- 3.8 Proficiency testing
- 3.9 Personnel training and qualification records
- 3.10 Method validation records
- 3.11 Quality assurance and audit records
- 3.12 Quality assurance manual
- 3.13 Equipment inventory
- 3.14 Safety manuals
- 3.15 Material safety data sheets
- 3.16 Historical or archival records
- 3.17 Licenses and certificates

4. Materials and Equipment

Only suitable and properly operating equipment should be employed. Chemicals and reagents should be of suitable quality, correctly prepared, and demonstrated to be compatible with the methods employed.

- 4.1 Instruments and Equipment
- 4.1.1 Inventory—A list of equipment essential for DNA analysis which includes the manufacturer, model, serial number, agency inventory number, and purchase and replacement dates.
- 4.1.2 Operation Manual—The manufacturer's operation manual should be readily available.
- 4.1.3 Calibration and Maintenance Procedures—There should be written calibration and maintenance procedures and schedules.
- 4.1.4 Calibration and Maintenance Logs—There should be a permanent log of calibration and maintenance of equipment essential for DNA typing.
- 4.2 Chemicals and Reagents
- 4.2.1 Logs *must* be maintained of all commercial supplies and kits (such as probes and restriction enzymes) as indicated in Section 3.3.
- 4.2.2 Formulation—There *must* be a written procedure for the formulation of reagents, standards and controls.
- 4.2.3 Labelling Requirements—To include identity, concentration, date of preparation, identity of analyst preparing reagents, (storage requirements, and expiration date where appropriate).
- 4.2.4 Storage and Disposal—All chemicals and radioactive materials *must* be stored, used and disposed of in a manner conforming to established safety requirements.

4.2.5 Material Safety Data Sheets (MSDS)—There should be a file of MSDS received from the manufacturer for all chemicals used in the laboratory. These data sheets should be readily available to all laboratory personnel.

4.2.6 A current inventory should be maintained to include information on supplier, catalog number, lot number, date received and storage location.

4.3 Glassware and Plasticware

4.3.1 Preparation—There should be specific procedures for cleaning, preparation, and sterilization.

4.3.2 Disposal—There should be specific procedures for the safe disposal of contaminated or broken glassware or plasticware.

5. Validation of Analytical Procedures

5.1 DNA Probes

5.1.1 The DNA probe(s) selected for use in forensic analysis should be readily available to the forensic science community.

5.2 DNA Loci for RFLP Analysis (Baird 1989; AABB Standards Committee 1989)

5.2.1 Inheritance—DNA loci used in forensic testing shall have been validated by family studies to demonstrate that the loci exhibit Mendelian inheritance as reported in scientific communications. For those DNA loci used in parentage testing, the frequency of mutation and/or recombination should not be greater than 0.2 percent.

5.2.2 Gene Mapping—The chromosomal location of the polymorphic loci (single locus polymorphisms) used for forensic testing shall be submitted to or recorded in the Yale Gene Library or by the International Human Gene Mapping Workshop.

5.2.3 Documentation—The polymorphic loci shall be documented in the literature stating the restriction endonuclease and the probes used to detect the polymorphism.

5.2.4 Polymorphism—The type of polymorphism detected shall be known (that is, single locus or multiloci).

5.2.5 Population Studies—Population distributions for at least the commonly recognized racial groups should be determined for the restriction enzyme and locus combination.

5.3 Developmental Validation of the DNA Analysis Procedure

During the development of a RFLP procedure and prior to the adoption of the procedure by a DNA laboratory, validation studies *must* have been conducted by the scientific community. These validation studies form the basis for evaluating the validity, accuracy and reproducibility of a particular DNA analysis procedure. This validation should include the following (Budowle *et al.* 1988):

5.3.1 Standard Specimens—The typing procedure should be evaluated using fresh body tissues, and fluids that have been obtained and stored in a controlled manner. Determine if DNA isolated from different tissues from the same individual yields the same typing profiles.

5.3.2 Consistency—Using specimens obtained from donors of known phenotypes/genotype, evaluate the reproducibility of the technique both within the laboratory and among different laboratories.

5.3.3 Population Studies—Establish population distribution data in different racial groups for restriction fragment bands detected by a given restriction enzyme-DNA probe pair.

5.3.4 Reproducibility—Prepare dried stains using body fluids from donors of known phenotypes and analyze to ensure that the stain specimens exhibit accurate, interpretable and reproducible DNA typing profiles that match the profiles obtained on liquid specimens.

5.3.5 Time/Temperature Studies—Determine if the polymorphic patterns in dried stains change as a function of time and temperature.

5.3.6 Degradation Studies—Expose laboratory-prepared body fluid stains to a variety of commonly encountered substances to assess the impact of these substances on DNA profiles.

5.3.7 Nonprobative Evidence—Examine DNA profiles in nonprobative evidentiary stain materials. Compare the DNA profiles obtained for the known liquid blood versus questioned blood deposited on typical crime scene evidence.

5.3.8 Nonhuman Studies—Determine if DNA typing methods designed for use with human specimens detect DNA profiles in nonhuman source stains.

5.3.9 On-site Evaluation—Set up newly developed typing methods in the case working laboratory for on-site evaluation of the procedure.

5.3.10 It is essential that the results of experimental studies be shared as soon as possible with the scientific community through presentations at scientific/professional meetings. It is imperative that the complete details of the experimental study be afforded the opportunity for peer review through timely publications in scientific journals.

5.4 In-House Validation of Established Procedures (ASCLD 1986)
Prior to implementing a new RFLP procedure, or an existing RFLP procedure developed by another laboratory that meets the developmental criteria described under Section 5.3, the forensic laboratory *must* first validate this procedure in its own laboratory. This same prerequisite would apply to any existing procedure to which significant modifications have been made. This validation study forms the basis for assessing the specificity, reproducibility, and limitations of the particular RFLP procedure. This validation *must* include the following:

5.4.1 The method *must* be tested using known samples.

5.4.2 The method *must* be tested using proficiency test samples. The proficiency test may be administered internally, externally or collaboratively.

5.4.3 If a significant modification has been made to an analytical procedure, the modified procedure *must* be compared to the original using identical samples.

5.4.4 Measurement imprecision *must* be determined by repetitive analyses for establishing matching criteria.

6. Evidence Handling Procedures

Evidence and samples from evidence *must* be collected, received, handled, sampled and stored so as to preserve the identity, integrity, condition and security of the item. Destructive testing of evidence and evidence samples should be performed so as to provide the maximum utility from the most economical consumption.

6.1 Sample labeling—Each sample *must* be labeled with a unique identifier in accordance with agency policy.

6.2 Sample handling—Each agency will prepare a written policy to ensure that evidence samples will be handled so as to prevent loss, alteration or contamination.

6.3 Chain of custody—A clear, well documented chain of custody must be maintained from the time the evidence is first received until it is released from the laboratory (ASCLD 1986).

7. Internal Controls and Standard

7.1 Procedures for estimating DNA recovery
A procedure *must* be used for estimating the quality (extent of DNA degradation) and quantity of DNA recovered from the specimens. One or more of the following procedures may be employed to evaluate the effectiveness of the DNA recovery:

7.1.1 Yield Gel—Yield gels must include a set of high molecular weight DNA calibration standards for quantitative estimate of yield. This procedure provides an estimate of the quantity of DNA recovered and the degree of DNA degradation—*or*

7.1.2 UV Absorbance—260/280 mm absorbance to provide measures of the quantity of DNA extracted from liquid blood. This procedure provides an estimate of the quantity of DNA recovered and assesses organic solvent and protein contamination in the recovered DNA—*or*

7.1.3 Fluorescence—Approximate quantitation of extracted DNA by comparison with known concentration of high molecular weight DNA. This procedure provides only an estimate of the quantity of DNA recovered—

or

7.1.4 Hybridization—Hybridization with probes to repetitive DNA specific to human/primate DNA. This procedure provides an estimate of the quantity of human DNA recovered.

7.2 Restriction Enzymes

7.2.1 Prior to its initial use, each lot of restriction enzyme should be tested against an appropriate viral, human or other DNA standard which produces an expected DNA fragment pattern under standard digestion conditions. The restriction enzyme should also be tested under conditions that will reveal contaminating nuclease activity.

7.3 Demonstration of Restriction Enzyme Digestion—Digestion of extracted DNA by the restriction enzyme should be demonstrated using a test gel which includes:

7.3.1 Size Marker—Determines approximate size range of digested DNA.

7.3.2 Human DNA Control—Measures the effectiveness of restriction enzyme digestion of genomic human DNA.

7.4 Analytical Gel—The analytical gel used to measure restriction fragments *must* include the following:

7.4.1 Visual Marker—Visual or fluorescent markers which are used to determine the end point of electrophoresis.

7.4.2 Molecular Weight Size Markers—Markers which span the RFLP size range and are used to determine the size of unknown restriction fragments. Case samples must be bracketed (as defined by the sample lanes) by molecular weight size markers.

7.4.3 Human DNA Control—Documented human DNA control of known phenotype/genotype which produces a known fragment pattern with each probe and serves as a systems check for the following functions:

a. electrophoresis quality and resolution
b. sizing process
c. probe identity
d. hybridization efficiency
e. stripping losses

7.4.4 A procedure should be available to identify and compensate for possible migrational differences in the DNA fragments.

7.5 Southern Blots/Hybridization—The efficiency of blotting, hybridizations and stringency washes is monitored by the human DNA control and size markers.

7.6 Autoradiography—The exposure intensity is monitored by the use of multiple X-ray films or by successive exposures in order to obtain films of the proper intensity for image analysis.

7.7 Image and Data Processing—The functioning of image and data processing is monitored by the human DNA control allelic values.

8. Data Analysis and Reporting

8.1 Autoradiographs and data *must* be reviewed independently by a second examiner/analyst. If a detection method other than autoradiography is used, the analytical results must be reviewed independently by a second examiner/analyst. Both examiners/analysts *must* agree on the interpretation of the data to be reported.

8.2 For proper data interpretation, the measured sizes of the restriction fragment bands of the human DNA control *must* fall within the established tolerance limits. Visual matching of the restriction fragment bands *must* be confirmed with statistical analysis based on tolerance limits.

8.3 The matching restriction fragment bands are assigned a frequency of occurrence calculated using a scientifically valid and accepted method from an established population data base.

8.4 The human DNA controls should be monitored by means of established statistical quality control methods (American National Standard ANSI/ASQC Z1.1-1985, ANSI/ASQC Z1.2-1985, ANSI/ASQC Z1.3-1985, ANSI/ASQC A1-1987; AT&T Technologies 1985; Bicking 1979; Gryna 1979; National Bureau of Standards 1966; Westgard *et al.* 1981).

8.5 Report Writing and Review—Laboratories should have policies, checks and balances in place which ensure the accuracy and completeness of reports. It is highly desirable that the laboratory reports be signed by the reporting analyst.

8.5.1 Contents—In addition to the presentation of the findings and conclusions of the examiner/analyst, the DNA locus (defined by the Nomenclature Committee of the International Gene Workshop), as identified by a particular probe, as well as the restriction enzyme used to digest the DNA, should be included.

8.5.2 All reports should be reviewed by appropriately designated personnel.

9. Proficiency Testing

9.1 Open Proficiency Test—Open proficiency tests *must* be submitted to the DNA testing laboratory on a semi-annual basis (minimum) such that each analyst is tested at least annually. These are labeled as proficiency test specimens and may be prepared internally and/or may be part of an external proficiency-testing program.

9.2 Blind Proficiency Test—Blind proficiency test specimens *must* be submitted to the DNA testing laboratory on a regular basis and are prepared in such a way as to appear as routine case specimens. These specimens may be prepared internally and/or may be part of an external proficiency-testing program. Each analyst should be tested at least annually.

9.3 A file will be maintained of all proficiency test results. Each laboratory will have a written policy for dealing with deficiencies. When deficiencies are noted, the file will identify the cause of the deficiency and the corrective action taken.

10. Audits

Audits are an important aspect of the quality assurance program. They are an independent review conducted to compare the various aspects of the DNA laboratory's performance with a standard for that performance. The audits are not punitive in nature, but are intended to provide management with an evaluation of the laboratory's performance in meeting its quality policies and objectives.

10.1 Audits or inspections should be conducted annually by individuals separate from and independent of the DNA testing laboratory. It is highly desirable that at least one auditor be from an outside agency.

10.2 Records of each inspection should be maintained and should include the date of the inspection, area inspected, name of the person conducting the inspection, findings and problems, remedial actions taken to resolve existing problems, and schedule of next inspection.

11. Safety

11.1 Policy—The DNA testing laboratory shall operate in strict accordance with the regulations of the pertinent federal, state and local health and safety authorities.

11.2 Written Manuals—Written general laboratory safety and radiation safety manuals shall be prepared by the laboratory and be made available to each member of the DNA analysis laboratory and/or other persons affected. (Bond 1987; Code of Federal Regulations 1988a, b; Gibbs and Kasprisin 1987; National Fire Protection Association 1986; National Research Council 1981, 1983; Sax and Lewis 1987; Steere 1971; Wang *et al.* 1975).

APPENDIX I

Participants—Technical Working Group on DNA Analysis Methods

Lucy A. Davis
Kentucky State Police
Forensic Laboratory
1250 Louisville Road
Frankfort, KY 40601

Michael A. DeGuglielmo
North Carolina State Bureau of
Investigation
3320 Old Garner Road
Raleigh, NC 27626

Jacqueline A. Emrich
Commonwealth of Virginia
Crime Laboratory
401-A Colley Avenue
Norfolk, VA 23507

Ronald M. Fourney
Royal Canadian Mounted Police
Central Forensic Laboratory
1200 Alta Vista Drive
Ottawa, Ontario, Canada K1G 3M8

Richard A. Guerrieri
Commonwealth of Virginia
Crime Laboratory
401-A Colley Avenue
Norfolk, VA 23507-1966

M. Roger Kahn
Metro-Dade Police Department
1320 N.W. 14th Street
Miami, FL 33125

Kenneth C. Konzak
California Department of Justice
140 Warren Hall
Berkeley, CA 94720

Terry L. Laber
Minnesota Forensic Science
Laboratory
1246 University Avenue
St. Paul, MN 55104

Henry C. Lee
Connecticut State Police
294 Colony Street
Meriden, CT 06450

Don MacLaren
Washington State Patrol
Crime Laboratory
Public Safety Building
Seattle, WA 98104

Susan D. Narveson
Arizona Department of Public Safety
Crime Laboratory
Post Office Box 6638
Phoenix, AZ 85005

Mark S. Nelson
North Carolina State Bureau of
Investigation Crime Laboratory
3320 Old Garner Road
Raleigh, NC 27626

Pamela J. Newall
Centre for Forensic Sciences
25 Grosvenor Street
Toronto, Canada M7A 2G8

James M. Pollock
Florida Department of Law
Enforcement
Post Office Box 4999
Jacksonville, FL 32211

Margaret C. Kuo
Orange County Sheriff's-
Coroner Office
601 Ross Street
Santa Ana, CA 92701

Willard C. Stuver
Metro-Dade Police Department
1320 N.W. 14th Street
Miami, FL 33125

Scott A. Wanlass
Nassau County Police Department
1490 Franklin Avenue
Mineola, NY 11501

John S. Waye
Royal Canadian Mounted Police
Central Forensic Laboratory
Post Office Box 8885
1200 Alta Vista Drive
Ottawa, Ontario, Canada K1G 3M8

Raymond White
Howard Hughes Medical Institute
University of Utah School
of Medicine
Salt Lake City, UT 84132

Dwight E. Adams
DNA Analysis Unit
FBI Laboratory

Stephen P. Allen Jr.
Forensic Science Research Unit
FBI Laboratory

F. Samuel Baechtel
Forensic Science Research Unit
FBI Laboratory

George F. Sensabaugh
University of California
School of Public Health
Berkeley, CA 94720

Bruce Budowle
Forensic Science Research Unit
FBI Laboratory

Catherine T. Comey
Forensic Science Research Unit
FBI Laboratory

William Eubanks
DNA Analysis Unit
FBI Laboratory

George E. Forsen
Technical Services Division
FBI

James J. Kearney, Chief
Forensic Science Research
and Training Center
FBI Laboratory

Keith L. Monson
Forensic Science Research Unit
FBI Laboratory

James L. Mudd
Forensic Science Research Unit
FBI Laboratory

Lawrence A. Presley
DNA Analysis Unit
FBI Laboratory

REFERENCES

AABB Standards Committee (1989). P7.000 DNA Polymorphism Testing. In: Standards for Parentage Testing Laboratories. American Association of Blood Banks, Arlington, Virginia.

Alwan, L. C. and Bissell, M.G. (1988). Time series modeling for quality control in clinical chemistry, Clin. Chem. 34:1396–1406.

American National Standard ANSI/ASQC A1-1987 (1987). Definitions, Symbols, Formulas, and Tables for Control Charts. American Society for Quality Control, Milwaukee, Wisconsin.

American National Standard ANSI/ASQC Q90-1987 (1987a). Quality Management and Quality Assurance Standards—Guidelines for Selection and Use. American Society for Quality Control, Milwaukee, Wisconsin.

American National Standard ANSI/ASQC Q90-1987 (1987b). Quality Management and Quality System Elements—Guidelines. American Society for Quality Control, Milwaukee, Wisconsin.

American National Standard ANSI/ASQC Z1.1-1985 (1985). Guide for Quality Control Charts. American Society for Quality Control, Milwaukee, Wisconsin.

American National Standard ANSI/ASQC Z1.2-1985 (1985). Control Chart Method of Analyzing Data. American Society for Quality Control, Milwaukee, Wisconsin.

American National Standard ANSI/ASQC Z1.3-1985 (1985). Control Chart Method of Controlling Quality During Production. American Society for Quality Control, Milwaukee, Wisconsin.

American National Standard ANSI/ASQC Z1.15-1979 (1979). Generic Guidelines for Quality Systems. American Society for Quality Control, Milwaukee, Wisconsin.

American National Standard ANSI/ASQC A3-1978 (1978). Quality Systems Terminology. American Society for Quality Control, Milwaukee, Wisconsin.

American National Standard ASQC Standard C1-1968 (1968). Specification of General Requirements of a Quality Program. American Society for Quality Control, Milwaukee, Wisconsin.

ASCLD (1986). Guidelines for Forensic Laboratory Management Practices. American Society of Crime Laboratory Directors, September.

ASCLD (1985). ASCLD Accreditation Manual. American Society of Crime Laboratory Directors, Laboratory Accreditation Board, February.

AT&T Technologies (1985). Statistical Quality Control Handbook. AT&T Technologies, Indianapolis, Indiana, May.

Baird, M. (1989). Quality Control and American Association of Blood Bank Standards. Presented at the American Association of Blood Banks National Conference, April 17–19, Leesburg, Virginia.

Bicking, C. A. (1979). Process Control by Statistical Methods. In: Quality Control Handbook. 3d ed., (Juran, J. M., ed.), McGraw Hill, New York.

Bond, W.W. (1987). Safety in the Forensic Immunology Laboratory. In: Proceedings of the International Symposium on Forensic Immunology, U.S. Government Printing Office, Washington, D.C.

Box, G.E.P. and Bisaard, S. (1987). The scientific context of quality improvement, Quality Progress 20(6)54–61.

Bradford, L. W. (1980). Barriers to quality achievement in crime laboratory operations, J. Forensic Sci. 25:902–907.

Brunelle, R. L., Garner, D. D. and Wineman, P. L. (1982). A quality assurance program for the laboratory examination of arson and explosive cases, J. Forensic Sci. 27:774–782.

Budowle, B., Deadman, H. A., Murch, R. S. and Baechtel, F. S. (1988). An introduction to the methods of DNA analysis under investigation in the FBI Laboratory, Crime Lab. Digest 15:8–21.

Bussolini, P. L., Davis, A. H. and Geoffrion, R. R. (1988). A new approach to quality for national research labs, Quality Progress 21(1):24–27. 2Code of Federal Regulations (1988a)1. Title 10, Part 19—Notice, Instructions, and Reports to Workers; Inspections. U.S. Government Printing Office, Washington, D.C.

Code of Federal Regulations (1988b). Title 10, Part 20—Standards for Protection Against Radiation. U.S. Government Printing Office, Washington, D.C.

Ford, D. J. (1988). Good laboratory practice, Lab. Practice, 37(9):29–33.

Gautier, M. A. and Gladney, E. S. (1987). A quality assurance program for health and environmental chemistry, Am. Lab. July, pp. 17–22.

Gibbs, F. L. and Kasprisin, C. A. (1987). Environmental Safety in the Blood Bank. American Association of Blood Banks, Arlington, Virginia.

Gryna, F. M. (1979). Basic Statistical Methods. In: Quality Control Handbook. 3d ed. (Juran, J. M., ed.), McGraw Hill, New York.

Hay, R. J. (1988). The seed stock concept and quality control for cell lines, Anal. Biochem. 171:225–237.

Juran, J. M. (1979). Quality Policies and Objectives. In: Quality Control Handbook, 3d ed. (Juran, J. M., ed.), McGraw Hill, New York.

Kenney, M. L. (1987). Quality assurance in changing times: proposals for reform and research in the clinical laboratory field, Clin. Chem. 33:328–336.

Kidd, G. J. (1987). What quality means to an R & D Organization. 41st Annual Quality Congress Transactions, May 4–6, American Society for Quality Control, Milwaukee, Wisconsin.

Kilshaw, D. (1987a). Quality assurance. 2. Internal quality control, Med. Lab. Sci. 44:73–83.

Kilshaw, D. (1987b). Quality assurance. 3. External quality assessment, Med. Lab. Sci. 44:178–186.

Kilshaw, D. (1986). Quality assurance. 1. Philosophy and basic principles, Med. Lab. Sci. 43:377:–381.

National Bureau of Standards (1966). The Place of Control Charts in Experimental Work. In: Experimental Statistics. National Bureau of Standards Handbook 91. U.S. Government Printing Office, Washington, D.C.

National Fire Protection Association (1986). Standard on Fire Protection for Laboratories Using Chemicals. National Fire Protection Association, Batterymarch Park, Quincy, Massachusetts.

National Research Council (1983). Prudent practices for disposal of chemicals from laboratories. National Research Council's Committee on Hazardous Substances in the Laboratory, National Academy Press, Washington, D.C.

National Research Council (1981). Prudent practices for handling hazardous chemicals in laboratories. National Research Council's Committee on Hazardous Substances in the Laboratory, National Academy Press, Washington, D.C.

Pereira, M. (1985). Quality assurance in forensic science, Forensic Sci. Int. 28:1–6.

Ruzicka R. K. (1979). Documentation: Configuration Management. In: Quality Control Handbook, 3d ed. (Juran, J. M., ed.), McGraw Hill, New York.

Sax, N. I. and Lewis, R. J. (1987). Hazardous Chemicals Desk Reference. Van Nostrand Reinhold, New York.

Simpson, J. (1983). National Bureau of Standards Approach to Quality, Test and Measurement World, December, p. 38.

Steere, N. V. (1971). CRC Handbook of Laboratory Safety. 2nd ed. (Steere, N.V., ed.), The Chemical Rubber Co., Cleveland, Ohio.

Taylor, J. K. (1987). Quality assurance of chemical measurements. Lewis Publishers, Chelsea, Michigan.

Taylor, J. K. (1985). The quest for quality assurance, Am. Lab., October, pp. 67–75.

Wang, C. H., Willis, D. L. and Loveland, W. D. (1975). Radiotracer Methodology in the Biological, Environmental, and Physical Sciences. Prentice-Hall, Englewood Cliffs, New Jersey.

Westgard, J. O., Barry, P. L., Hunt, M. R. and Groth, T. (1981). A multi-rule Shewart chart for quality control in clinical chemistry, Clin. Chem. 27:493–501.

Whitehead, T. P. and Woodford, F. P. (1981). External quality assessment of clinical laboratories in the United Kingdom, J. Clin. Pathol. 34:947–957.

APPENDIX II

The Combined DNA Index System (CODIS): A Theoretical Model

October 15, 1989

Technical Working Group on DNA Analysis Methods (TWGDAM)
James J. Kearney, TWGDAM Chairman
FBI Laboratory
Quantico, Virginia

Database Subcommittee

James J. Kearney, Subcommittee Chairman
FBI Laboratory
Quantico, Virginia

F. Samuel Baechtel
FBI Laboratory
Quantico, Virginia

Bruce Budowle
FBI Laboratory
Quantico, Virginia

George E. Forsen
FBI Laboratory
Quantico, Virginia

Richard A. Guerrieri
Commonwealth of Virginia Crime Lab
Norfolk, Virginia

M. Roger Kahn
Metro-Dade Police
Miami, Florida

Kenneth C. Konzak
California Department of Justice
Berkeley, California

Keith L. Monson
FBI Laboratory
Quantico, Virginia

Scott A. Wanlass
Nassau County Police Department
Mineola, New York

John S. Waye
Royal Canadian Mounted Police
Ottawa, Ontario
Canada

TABLE OF CONTENTS

APPENDICES

TABLES

I. INTRODUCTION

For many years, molecular biologists have recognized the uniqueness of DNA (deoxyribonucleic acid), the chemical substance which makes up the chromosomes of living cells. Using genetic analyses they showed that each individual is unique, except for identical twins, and hypothesized that this uniqueness was present in the primary structure or sequence of the DNA. It has been possible for several years to exclude large portions of the population as the source of a biological specimen using "conventional" antigen and protein genetic markers. However, it was not until late 1985 that English and American scientists, working independently, reported that specific, highly variable regions of human DNA could be used as a personal identifier. These scientists proposed that DNA extracted from blood or semen stains found at the scene of violent crimes would yield DNA profiles that could identify the donor of the stain, theoretically, to the exclusion of all others in the world.

Several methods of using DNA extracted from body fluid stains to build DNA profiles have been investigated. One method that has received substantial development efforts has already been put into practice by several commercial and law enforcement organizations, including the FBI. It is referred to as the Restriction Fragment Length Polymorphism (RFLP) method of DNA analysis. It examines restriction digested target DNA and uses a series of specific single locus DNA probes. The DNA fragments identified with these probes can be accurately sized and catalogued into specific DNA indexes. Other methods for building DNA profiles utilize the Polymerase Chain Reaction (PCR) to amplify the DNA. The amplified DNA of the DQ Alpha gene, for example, can then be profiled. The DNA indices produced by these analyses provide the potential for building statistical and investigative DNA profile databases which can assist law enforcement in identifying individuals who commit violent crimes.

Over the past two years, FBI laboratory scientists, in collaboration with scientists from Canada and other U.S. crime laboratories, have been engaged in an extensive research effort in order to develop and implement the RFLP DNA profiling method that uses a set of specific single locus DNA probes. This research effort represented an unprecedented bringing together of resources and the cooperative effort of many individuals in the forensic science community. As a result of these extraordinary efforts, a technically valid RFLP DNA profiling method has been developed which has sufficient precision and accuracy to be used in building DNA profiling databases. While this document focuses on the RFLP technique, the database model described is generally applicable and may be expanded to include other genetic analyses useful in profiling.

Database Development

It is proposed that the Combined DNA Index System (CODIS) be developed to support law enforcement operations only, and that the database system include initially one statistical and one investigative support database. The investigative support database would initially include open case, convicted violent offender and missing person/unidentified body DNA profiles.

Privacy Concerns and Data Security Issues

This model addresses the privacy concerns and data security issues which development of databases on individuals tend to raise. A DNA profile obtained using the RFLP method is simply a set of numbers that represent the sizes of several specific DNA fragments detected by a set of genetic markers, or "probes." It has been shown that although these sizes are the same for every nucleated cell in a single individual, they are different from one individual to the next. In addition there is no known relationship between these numbers and any physical or mental condition. Selection of the genetic markers used in the database will be made with an eye to eliminating the potential interest of the data to the private sector. Probes will not be used that are linked to disease conditions or personality traits.

Only information relevant to law enforcement requirements will be maintained in the national DNA database. The database will be limited to profiles of persons convicted of violent crimes or those involved in unsolved cases. No names of individuals or other identifying or characterizing data will be stored with the DNA profiles.

The national DNA database will only reference the sources of DNA profiles, and case data will be secured and controlled by local agencies. A unique coded identifier of the sample from which the DNA profile was obtained will be stored to allow the law enforcement agency to identify the source of the DNA accurately. Only the crime laboratory that submitted the DNA profile will be able to link this identifier with a known person.

Access to the national DNA database will be limited to government agencies on a "need to know" basis as described below. Access by private contractors and the military, in the service of the federal or state government (such as for personnel on foreign "hazardous duty" assignments), would be through the appropriate law enforcement agency only. Criteria will be developed for periodically examining the databases for the purpose of expunging data no longer relevant and for maintaining the currency and accuracy of the databases, and to ensure that there is no conflict with relevant laws of the states and of participating organizations.

Statistical DNA Population Database

This database would include DNA profiles produced from the examination of DNA from randomly selected unrelated individuals from a minimum of five populations, e.g., Blacks, Caucasians, Orientals, Hispanics and American Indians, and yet to be determined sub-populations. The statistical database would be built collaboratively and maintained by the FBI Laboratory for use by all crime laboratories in North America. This information would be used to establish the allele frequencies for the DNA probes being used by the forensic community. Statistical data on gene frequencies, devoid of any personal identifiers, could be accumulated and made available on compact discs (CD-ROMS) to qualified public and private organizations doing DNA analysis.

Open Case DNA Profiles

This portion of the CODIS investigative support database would contain DNA profiles obtained from body fluid stains (semen, blood, saliva, urine) found on evidentiary materials recovered from violent crime (rape, homicide, assault) cases having no suspects. With sufficient participation, these indices would provide immediate benefits allowing for the grouping of serial rape cases based on DNA profiles obtained from semen stains and the tracking of serial rapists in individual states as well as nationwide. Open case DNA profile indices would be maintained at both the state and Federal level. The CODIS would be used to facilitate nationwide coverage and networking.

Convicted Violent Offender DNA Profiles

This portion of the investigative database would contain the DNA profiles of individuals who have been convicted of violent offenses. The benefits from the inclusion of this set of indices would be longer range than the unknown subject (open case) indices. It may take from 2 to 5 years to build violent offender indices that can be used effectively for intra- and inter-state comparisons to assist law enforcement. An important example of the benefit of this type of index will occur when DNA profiles obtained from evidence in unknown subject sexual assault cases can be checked against a set of sex offender DNA indices to determine if the offense was committed by a repeat sex offender.

Several states (California, Washington, Virginia, Nevada, Colorado) are in the process of or have passed legislation which allows for the collection of blood samples from persons convicted of a sex offense to be used to develop DNA profiling databases. Complete convicted sex offender DNA profile databases would be developed and maintained at the state level. Only the DNA index and listing of the submitting agency of a specific profile would be maintained at the Federal level. When specific information on a match was required, the submitting agency would have to be contacted to provide the details. Using the convicted sex offender situation as a model, the DNA profiling database could also include individuals convicted of other felony crimes, such as homicide or aggravated assault, in accordance with state laws.

Missing Persons/Unidentified Bodies DNA Profiles
This portion of the database would contain DNA profiles from unidentified bodies, body parts and bone fragments. These indices would provide their greatest benefit when DNA profiles from immediate relatives (parents) could be used to reconstruct the DNA profile for a missing person which could be used to compare with the missing person/unidentified body DNA profile database. Although there would be immediate benefits from the development of this type of data the actual number of cases of this type would be small compared to unknown subject sexual assault cases.

Missing persons/unidentified body case details would be maintained at the state levels, and the DNA profiles would be used as an index at the federal level.

JUSTIFICATION

The justification for the development of these types of indices is based on the fact that individuals who commit violent crimes are often repeat offenders. According to a March 1989, Bureau of Justice Statistics report entitled, "Recidivism of Prisoners Released in 1983," of the 108,580 persons released from prisons in 11 states in 1983, an estimated 62.5% were rearrested for a felony or serious misdemeanor within 3 years. An estimated 67,898 of the 108,580 prisoners who were released in 1983 were rearrested and charged with 326,746 new offenses by year end 1986. More than 50,000 of the new charges were violent offenses, including 2,282 homicides, 1,451 kidnappings, 1,291 rapes, 2,626 other sexual assaults, 17,060 robberies, and 22,633 other assaults. In summary, of those individuals released for violent offenses, 59.6% were rearrested within 3 years for a similar offense. Recidivism rates were highest in the first year—1 of 4 released prisoners were rearrested in the first 6 months and 2 of 5 within the first year after their release. More than 1 of every 8 rearrests were made in states other than the state in which the person was released.

In addition to the above, it was reported in the FBI's Uniform Crime Report that in 1987 there were 91,111 forcible rapes reported and 20,096 murders and negligent manslaughters. It is estimated of all homicides and rapes committed only 70% of the homicides (6,000 unsolved) and 50% of the rapes (45,000 unsolved) are closed by arrest.

The above facts are strong justification for the development of a database of DNA profiles of unknown subjects (open case) and convicted violent crime offenders. For example, this database would provide law enforcement with a powerful tool in linking sexual assault cases together through DNA profiles and tracking the activities of the serial rapists. In light of the high recidivism rate (50%), the convicted sex offender data would provide the investigator with a logical first place to look for assistance in solving unknown subject sexual assault cases.

COST, COMMITTMENT and POLICY

The actual costs associated with the development of the statistical and investigative DNA profile databases will be described in detail later in this document. However, here it is sufficient to say that the cost can be divided into two main areas, equipment (hardware and software) and personnel. The equipment costs needed to support the development of the prototype DNA database system are less than may be expected. Not all crime laboratories will be producing DNA profiles during this period. In five years, there will likely be fewer than 200 persons qualified and actively involved in generating DNA profiles. It is estimated that CODIS would have less than 250,000 entries at the end of five years. The size of the database can be supported by a minicomputer. A state or regional index system would require a system with even less storage capacity. Many states already have sufficient computer power to support a statewide system. How the state and Federal systems would be linked has not been fully explored. It may be solved by using an existing nationwide communication system of dedicated phone lines. Dial-up lines can be used during prototype development.

The personnel costs will be the greater expense. It will take people to develop the systems as well as to develop the DNA data from probative evidence to be put into the databases. It is important that a commitment towards early growth of this system be made. Its utility is a function of its completeness and its future is a function of its utility. The use of open case DNA profiles in a centralized database will require that crime laboratories change their policies in regards to examination of evidence from unknown subject cases. At present, many laboratories simply cut and store evidence from unknown subject cases and the work is only begun when a subject is developed. The most effective use of the open case DNA data would be to examine the DNA in all cases immediately, develop the DNA profiles and then enter them into the database in order to make comparisons. An open case DNA profile would be left in the system as an unknown, to be compared with subsequent DNA profile submissions for a specified period of time, or until the suspect is apprehended and the case adjudicated. If convicted of a violent crime, the profile might be submitted as a known.

II. SYSTEM ORGANIZATION AND FUNCTIONAL RESPONSIBILITIES

INTRODUCTION

The recommended approach to a nationwide DNA profile database is to establish a centralized DNA index, with supporting records for each profile maintained by the submitting agency. This type of approach, of which the National Crime Information Center (NCIC) system is an example, provides many of the advantages of database administration and policy standardization inherent in a fully centralized approach, yet allows individual agencies to retain control of their case information.

Most design considerations for a centralized nationwide DNA database would also apply to a centralized statewide or regional database.

The principal benefit of an approach using a centralized index is that all participating agencies may share a larger pool of DNA profiles. By adhering to common methodology, reporting and quality standards, agencies submit DNA profiles that can be directly compared. Many advantages will result from wide compliance with standards:

—will demonstrate to the courts that DNA analysis is a mature technology, with wide acceptance and with consistent results between laboratories;

—it is the only way for different laboratories to make effective use of one another's results;

—training will be easier.

Individual law enforcement agencies would contribute DNA profiles to the centralized database, yet they would retain absolute control of their case records, both in terms of ownership and in the way records are maintained. The only concession to uniformity would be in those aspects of protocols, controls and results formats necessary to ensure inter-laboratory comparison via the centralized database. Individual jurisdictions would still be free to select the data management hardware and software they prefer, and which would be compatible with what they already have installed.

An advisory committee, made up of representatives from federal, state, and local agencies would be necessary to establish policies. Technical advisory and quality assurance sub-committees must also be established to deal with changes in technology, and to establish controls to insure the quality of submissions, guide individual agencies in their own quality control programs, and administer a proficiency testing program.

PROPOSED ORGANIZATION OF THE COMBINED DNA INDEX SYSTEM

Depending on the type of crime involved, DNA databases could logically be maintained at the national, state or regional level, and shared with other law enforcement agencies. Access by military and other organizations to the national database would be on a "need to know" basis and would be handled through the appropriate law enforcement agency. Table 1 lists several potential DNA database user communities that were considered.

After reviewing the costs involved in serving each of the database users and the anticipated benefits which might be provided, the technical working group recommends that a database be maintained of population statistics and that three sets of indices be included in the investigative DNA profile database:

(1) Open cases,
(2) convicted sex offenders, and
(3) missing persons/unidentified bodies

TABLE 1—OPERATIONAL CONSTRAINTS FOR POSSIBLE TYPES OF INDICES

Indices Maintained	Info Maintained	Who can Access	Who can Submit
National			
Statistical	Categorical data	Fed/State/Local Private companies	Fed/State/Local
Unknown subject case	Index	Fed/State/Local	Fed/State/Local
Missing persons/ Unidentified bodies	Index	Fed/State/Local	Fed/State/Local
Hazardous duty (foreign)	Index	Fed/State/Local	Federal Private Companies
Convicted sex offender	Index	Fed/State/Local	Fed/State/Local
Terrorist	Index	Federal	Federal
State/local Level			
Convicted sex offender	Index + case file	Fed/State/Local	Fed/State/local
Convicted felon	Index + case file	Fed/State/Local	Fed/State/Local
Missing persons/ Unidentified bodies	Index + case file	Fed/State/Local	Fed/State/Local
Regional Center/ Referee Labs			
Convicted sex offender	Index	Fed/State/Local	Fed/State/Local
Convicted felon	Index	Fed/State/Local	Fed/State/Local
Missing persons/ Unidentified bodies	Index	Fed/State/Local	Fed/State/Local
Military			
Personnel assigned foreign	Index + case file	Mil/Fed	Mil

These indices would comprise DNA profiles and ancillary data submitted by law enforcement agencies and/or their forensic laboratories from case files across North America. This system should be supported and maintained by the FBI, and be accessible only to qualified participants. It is essential that the three types of indices be mutually comparable. For example, an open case file containing both an evidence profile and a dead victim's profile might lead to a search of the missing persons indices and/or a search of the convicted sex offender indices.

The following sections outline the proposed operation and use of DNA indices in each of the above categories. Within the guidelines established by the previously mentioned quality assurance and technical advisory group, the FBI and participating agencies would have certain responsibilities.

Federal Responsibilities:

—Coordinate with the quality assurance and technical advisory group to implement the appropriate guidelines
—Coordinate with other agencies that have a law enforcement interest in the development and operation of the database
—Provide hardware and software for the database server and for access to the database by states
—Provide hardware to store and back up the database server
—Provide training to states on forensic DNA technology, quality control and database access
—Determine format for database input/output
—Update index with state and federal new submissions
—Assemble population data for all probes used; calculate and disseminate population frequencies
—Modify the system to accommodate new DNA profiling methods.

State/Local Agency Responsibilities:

—Perform DNA analysis of forensic material using the consensus methods, i.e., operate within the guidelines recommended by the technical and quality assurance advisory group
—Submit new information in specified format for incorporation into databases
—Guarantee quality of new submissions from that agency
—Provide hardware and software on state's image analysis workstation(s) for telephonic access to centralized index
—Maintain centrally indexed case files for as long as they remain on the index
—Provide relevant information from case files which are indexed centrally to other law enforcement agencies which subscribe when requested.

Open Case DNA Profiles:

The goal of using this type of DNA profile is to link cases having a common subject. DNA evidence from the case under investigation would be compared to all open case indices, as well as to other indices, such as convicted sex offenders. Any linkages produced would be useful in communicating the modus operandi and in tracking geographical and temporal movements of the suspect. Solution of any one case could lead to the solutions of all linked cases, and to multiple convictions. The open case indices would come from two types of cases:

(1) cases which include a *dead* victim's profile and one or more evidence profiles (e.g., homicide, rape/homicide), and
(2) cases which include a *live* victim's profile and one or more evidence profiles (e.g., serial rape, aggravated assault).

The open case DNA profiles produced in the local jurisdiction from each open case would be compared to the centralized open case indices for possible matches and the new profile would also be added to the open case indices for a specified period. Putative matches would be flagged on each of the cases, and coordinated with the agencies involved. When all putative matches were either confirmed or disproved, the cases involved would be either removed, or the "match flag" would be canceled, respectively. Storage of the detailed case records would be the responsibility of the submitting agencies. In addition, each submitting agency would periodically review their submissions to remove adjudicated cases and non-probative evidence profiles. All submissions would be subject to a quality control appraisal.

Convicted Violent Offender DNA Profiles:

The justification for the use of this type of DNA profiles is the demonstrated recidivism of sex offenders. Within three years of release from prison, more than half of convicted rapists are rearrested, and about 70% of these are re-convicted (U.S. Department of Justice Special Report NCJ-116261, "Recidivism of Prisoners Released in 1983," by A.J. Beck and B.E. Shipley, 1989). The goal is to apprehend serial sex offenders more quickly, even if their crimes are committed in different jurisdictions.

Where the appropriate statutes are in place, the state and local agencies will obtain blood samples for convicted sex offenders. The first time that a jurisdiction using the national system determines a DNA profile for that individual, it would immediately send the profile to the central facility for inclusion in the database. Storage of the detailed case record would be the responsibility of the submitting agency. Subsequent determinations of the DNA profile of the same individual, for example, by another jurisdiction, would also be added to the central index. Thus, an inquiry to the centralized index regarding an individual would reveal all jurisdictions in which he was involved in one or more DNA-related cases. Details on these cases could be made available by the jurisdictions in which they occurred.

The profile could also be submitted to determine if the person is a repeat offender, possibly using an alias. The submitted profile would, of course, also be compared automatically to open case indices. Each submission would be subject to a quality assurance appraisal.

Missing Persons and Unidentified Bodies DNA Profiles:
The goal of using this type of DNA profile is to link missing individuals with their close biological relatives or to determine the victim's name when body parts are found. Both index and case file information should be centralized for effective identification. The parents or siblings who reported the persons as missing would provide samples from which their own DNA profiles would be determined. Profiles from material such as hair or body fluids believed to be from the victim might also be correlated with the profiles of close blood relatives. These profiles could then be searched against profiles for unidentified body or body parts to determine if an association can be made. Profiles from stains taken from suspects involved in cases with unknown victims might also be stored in this index.

DNA STATISTICAL DATABASE

A DNA statistical database will be maintained at the federal level for several reasons:

—No laboratory can justify the expenditure of resources to generate a large database for many probes and racial groups
—Allows all laboratories to share results only available in certain areas due to endemic distribution of racial groups
—Acquisition of results shared by many laboratories allows accumulation of larger databases, so that the databases will more accurately represent the target populations
—When the database is administered and published at the federal level, it is equally accessible to all states
—A centralized frequency database can be updated regularly, so that the most current figures are available to all
—A centralized frequency database encourages uniform reporting of results by all agencies. Matters such as the validity of sampling, adequacy of sample size, and the methods of calculation population frequencies quickly become a matter of record, and after their acceptance in several courts, no longer will require detailed justification.

Submissions to the DNA statistical database would be handled in a similar manner as for the DNA profile database. State, local, and federal laboratories would submit their raw data to the database administrator, who would:
—Determine if the submission meets the quality control guidelines for inclusion in the database;

—Analyze the distributions of the existing database with groups of new submissions to determine if they both derive from the same population;
—If the first two conditions are true, (1) merge the two populations, (2) recalculate allele frequencies, and (3) post the new values for all participants' use.

The statistical database will be augmented by the continuing submission of profiles of known racial groups, both victims and suspects, resulting in more precise exclusionary probabilities and sub-population groupings. The submitted profiles will be batched periodically for comparison and possible inclusion in the statistical database, subject to certain statistical requirements. These files will contain no identifying links to specific individuals and would be used for law enforcement only. Furthermore, precautions are necessary to preclude multiple entries from the same individual or close relatives.

III. DATA ACCEPTANCE, DATA ENTRY, AND REPORTS

CONSIDERATIONS FOR DATA ENTRY AND INFORMATION RETRIEVAL

There are two levels at which data entry and information retrieval must be considered for the Combined DNA Index System. The first level concerns the analytical and quality control procedures used by an agency to generate DNA typing data and the second level relates to data entry and information retrieval.

Analytical and Quality Control Procedures

Genetic information in the form of restriction fragment length polymorphisms (RFLP) can be exchanged among laboratories through a computer database only if certain procedural criteria are strictly observed during the analysis. Failure to adhere to these criteria will jeopardize the accuracy of the RFLP results and render them useless for database comparisons.

All participating laboratories must accept the following standard practices as designated by the technical and quality assurance advisory group:
—DNA must be fragmented through use of the restriction endonuclease *Hae III*.
—The use of electrophoretic gels of a size and composition which can be shown to give comparable results to the data already in the database.
—The employment of an approved set of core DNA probes for which sufficient statistical and family data exist. Presently, D1S7, D2S44, D4S139, D14S13, D16S85, and D17S79 meet these criteria. Future experimentation with the prototype database will provide the data needed for the technical and quality assurance advisory group to identify the most useful set.
—The use of a standardized procedure for the correction of fragment migrational aberrations.

—The use of approved DNA sizing standards and computer imaging techniques for estimation of DNA fragment sizes.

Accuracy and reliability of RFLP data can be guaranteed only if specified quality control procedures are rigidly observed during the analytical and imaging phases of the test procedure. Quality control procedures applied to the RFLP analysis include:

—Inclusion of control human DNA on each analytical gel. This control is important to confirm the identity of the probe used and for insuring that the analytical phases of the RFLP technique have been carried out properly, but is also vital during the imaging steps to confirm that the fragment sizes are estimated correctly.

—Having two scientists in the laboratory independently determine the DNA fragment sizes on each autoradiogram.

—Analysis of proficiency test samples on a regular basis. Proficiency test samples will consist of body fluids stains for comprehensive analysis and pre-prepared autoradiograms that test imaging and fragment sizing competence.

DATA ENTRIES

NOTE: Some items mentioned in this section will be stored locally, and other items entered into the central index system.

Entries common to all types of indices:

—Submitting agency information. This information would include: location of agency, phone number, names of scientists who conducted the DNA typing analysis, name of individual entering the data into the database, and agency contact information. Upon acceptance of an agency into the database network, a code number will be issued that permits database access.

—Types of indices that will be received:

1. open cases
2. convicted sex offenders
3. missing persons/unidentified bodies
4. statistical data

—Specimen identification. This will be a series of entries that identifies the type of specimen (e.g., body fluid stain, solid tissue, exemplar blood specimen), and a unique specimen indentifier. Additional entries will indicate the condition of the sample, exceptional handling/storage, and other factors that might affect DNA quality, and the evaluation of partial patterns.

—RFLP profiling data. The fragment size data from each locus that has been successfully probed will be entered as the number of base pairs determined for each of the fragments. This also would include entry of the sizing data for the human DNA control specimen that is incumbent to each analysis. Similar data would be input for other methods.

Data entries unique to a particular set of indices:

—Open case indices. This set of indices will be designed to accept RFLP patterns from more than one category of specimen. The database must be able to receive RFLP data from victim exemplar blood specimens for subsequent comparison with biological materials associated with a suspect who has been arrested elsewhere. Also, any crime scene evidentiary body fluid stains that cannot be associated with the victim will be entered as possibly originating from the unknown subject.

—Convicted sex offender indices. Entries will include the name of the offender, dates of offenses/convictions, and the DNA profiling data. Only profile indexes will be centrally stored. Case data will be stored locally and its distribution will be under the control of its owners.

—Missing persons/unidentified bodies/unidentified body remains. This set of indices will include DNA typing entries developed from unidentified found bodies, and unidentified body parts. These DNA patterns will be entered into the database for comparison with RFLP data derived from biological materials remaining at the crime scene, or for comparison with RFLP data obtained from putative biological relatives. For the case of missing persons, the array of possible DNA profiles, derived from DNA analysis of biological parents and/or siblings, can be entered as reference data.

—Statistical database. Input will comprise RFLP profiles from blood specimens obtained from random donors of known populations. Similar profiles from victims and convicted sex offenders will also be added if the data meet the appropriate statistical criteria. The input format will include data on race/ethnic origin and sex, if known. Profiles of genetically related individuals cannot be entered into this database.

INFORMATION OUTPUTS

The information output will be a function of the types of indices being interrogated. A no match response should be communicated to the requesting agency as well as any match(es). The agency may leave an unmatched request in a pending status for a specified length of time. A match is an indicator for communications to occur between law enforcement agencies. The available evidence, the existence of a strong DNA profile match, and the additional information from the agency that previously submitted the matching profile may facilitate probable cause for obtaining blood samples that could confirm the original match. When the database produces a match between entered profiling data and reference data in the database, the following outputs would be expected:

Open case DNA profiles:

—Indication that one or more matches occurred. Multiple matches can occur if the subject's profile has been entered more than once, such as in serial rape and/or homicide cases.

—Names and contact points of agencies who entered the reference data into the database.

—Details of crimes whose evidence matched the entered profile.

Convicted sex offender DNA profiles:

—Indication that one or more matches occurred.

—Names (and aliases) of all individuals who, within defined statistical limits, match the entered profile.

—List of agencies (jurisdictions) and their contact points that entered the reference profile.

Missing persons/unidentified bodies:

—Indication that an association has been made.

—Identify/describe where the same DNA profile was found in association with other crime scenes.

—Identify associations between profiling data entered for biological relatives and found bodies/body parts.

Statistical database:

—Frequencies of occurrence of specified DNA fragments at each genetic locus in the general population, or racially/ethnically/geographically unique segments of the population.

—Calculated numeric value that indicates the likelihood of selecting a person at random that possesses a defined array of DNA fragments.

IV. PERFORMANCE REQUIREMENTS

WORKING ASSUMPTIONS

System Operational Lifetime:

—The operational lifetime of the Combined DNA Index System (CODIS) will be limited by its ability to continue to provide the desired service at an acceptable cost of operations and maintenance. Changes in forensic technology, in computer technology, and a reappraisal of the requirements will likely necessitate major revisions in order to extend the operational lifetime of the system past its original design lifetime.

System Design Lifetime:

—System design lifetime is the period of time during which the system should operate as originally designed without major revisions. The system design lifetime will be assumed to be five years. This means that the system must be designed to handle the expected workload occurring five years from now. Five years is a nominal value and is typical for properly designed systems that operate in a relatively stable environment.

Analytical Process:

—The FBI has developed what it believes to be a viable approach to the use of DNA for forensic purposes. While the national database model will concentrate on this RFLP technique, the database system could be expanded to accept other genetic marker system data, as population data, as the typing systems gain general acceptance in this role. The following characterizes the initial DNA profiling process:

—Use of the RFLP technique*

—Use of the *Hae III* restriction enzyme

—Use of a core set of single-locus probe sites, chosen, for example, D1S7, D2S44, D4S139, D14S13, D16S85, and D17S79.

—Use of accepted quality control guidelines.

—Searching on a partial probe set will be allowed.

Primary Files:

—While many uses of DNA profiles have been conjectured, the initial information stored in the system will be limited to:

—Unknown Subjects, e.g., in serial rape/homicide cases

—Known Sex Offenders, e.g., persons already convicted

—Unidentified Bodies, e.g., vis-a-vis missing persons

—Population Statistics, e.g., Caucasians, Blacks, Orientals, Hispanics, American Indians.

Limits to Growth:

—Several factors will limit the growth of the DNA Profile Database over the next five years. These include:

—Number of sex offenders required to give blood sample

—Number of violent crimes producing body fluid stains

—Acceptance of DNA results in courts

—Funding for new technology

—Training of personnel using RFLP techniques

*It is anticipated that other DNA analysis methods shown to be of use to the forensic community will be accommodated soon after the RFLP-based system has been satisfactorily developed.

—It is assumed herein that the best known limiting factor on the growth of the system will be the effective number of trained individuals capable of processing forensic DNA samples using common techniques, e.g., RFLP. The major training center is the FSRTC at Quantico. It is currently training about 75 persons a year. This may continue for several years. For reasons cited in Appendix A the effective number of full-time personnel processing DNA may be 200 or less in five years.

—At peak efficiency, a competent full-time scientist can process 1000 DNA samples in a 50-week year. This suggests that 200,000 samples could be processed per year by the end of the first five years. Not all are for known subjects.

—Further assumptions in Appendix A suggest that at the end of five years the system must be able to add 100,000 DNA profiles to its index per year. Assuming linear growth, at the end of the first five year period, system storage must have the capacity for a total of 250,000 DNA profiles. The actual number of profiles stored is likely to be less.

PERFORMANCE CRITERIA

Introduction:

Performance criteria that are acceptable to the end-users of the DNA database index system must be established prior to its design. The performance criteria for the system fall into at least six categories:

—Access and Access Control
—Primary On-Line Storage Capacity
—Secondary Storage Capacity
—Transaction Response Times
—Availability
—Quality Assurance Measures

Access and access control refers to who will access the system for storage and retrieval, modification and other database functions, and who will control the access privileges. Primary on-line storage capacity refers to the storage of files that are used most often and for which transaction response times should be low. Secondary storage capacity is concerned with reference files that are not used as often and with archival files. Transaction response times are most noticeable to the end-user, particularly if they are too long. A transaction usually involves several database operations concerning data about a particular item. An example would be the request to determine if a matching DNA profile exists and to determine where additional information may be obtained about the source of the matching profile. Availability refers to the extent to which the system is up and running and is accessible by a legitimate end-user. Data processing quality assurance measures are required to ensure availability, the integrity of the data, and the quality of the information available.

Access and Access Control:
Access to the system will be limited to federal, state, and local law enforcement agencies and crime labs. Crime laboratories are likely to be the most frequent users of the system. There are only 300 such laboratories in the United States, and not all of these have direct access to the National Crime Information Center.

No direct access will be provided to individuals, their lawyers, or to commercial and private organizations, or non-law enforcement government agencies. Statistical databases can be provided on CD-ROM ready-only media to universities and to commercial organizations involved in DNA forensic work for a fee.

Inter-agency access privileges shall be determined by the owners of the data, i.e., those agencies generating the data.

An overriding consideration is the quality of the data. No data can be part of the system if its accuracy is questionable. No data submitted will be included if it fails to pass certain quality assurance tests. These tests are discussed elsewhere.

Primary On-Line Storage Capacity:
As listed above, the investigative database includes known sex offenders, open cases and unidentified bodies/missing persons. In addition, a file on participating organizations is required to provide the means to obtain information regarding a previous convicted sex offender with a matching DNA profile.

The primary files must be maintained "on-line," i.e., continuously and quickly accessible. This requires the use of high performance disk storage devices with the capacity to store all of the "current" DNA profiles, e.g., data entered within the last five years.

Using the assumptions described above and in Appendix A, five years of operation may require storage of up to 250,000 DNA profiles for known sex offenders. This is estimated to require about 200 Mbytes* of disc storage. (See Table II for details.) This is well within the capacity of current mini-computer disc hardware.

The capacity requirements for unknown subjects might require as much as 160 Mbytes for 100,000 cases. Unidentified bodies data would add only 5 Mbytes more. Organization data would require an insignificant amount for the number of agencies and crime labs involved. The total on-line capacity, excluding imagery, for the first five years accumulation is not expected to exceed 400 Mbytes. This amount is typical of super-micro or mini computer installations. Mainframe storage capacities are not indicated for the prototype facility.

On-line storage for the FBI's DNA Analysis Unit and for those states for whose data the FBI acts as custodian is not included in these estimates. The FBI's DNA database will migrate to a separate computer system.

*A megabyte is one million bytes. A byte is eight binary digits, enough to uniquely specify one of 256 unique characters. Typically, a single-spaced type-written page document requires 2,000 bytes of storage, using one byte per character.

Secondary Storage Capacity:

Secondary storage is needed for sub-population sample data, archival storage, and autoradiograph images. Secondary storage requires higher capacity to hold data accumulated over several years. The response-time requirements of secondary storage are not as critical. Also, these data are not likely to be changed.

Each population sample represents a statistical grouping common to North America, e.g., Caucasians, Blacks, Hispanics, American Indians, and Orientals. It is estimated that after five years 500,000 DNA profiles will be available for the population statistics database, requiring about 100 Mbytes of storage.

Secondary storage of text and numeric data for the known sex offenders, open cases and unidentified bodies/missing persons indices is required for a longer period for archival purposes. It is estimated that ten years of such data will require approximately 1.1 Gbytes** (1,100 Mbytes) of storage capacity.

It is extremely important to be able to archive the data onto a reliable storage medium. WORM (write once ready many) disk technology is useful for archiving data because a large amount of data can be stored in a small space, it has a relatively long lifetime (greater than 10 years), and is not easily altered. One $400 (2.4 Gbyte, 12") optical disk can store an amount of data equivalent to single-spaced lines of text typed on sheets of paper and stored.

Transaction and Response Times:

A transaction comprises a series of communications, each of which should be timely, concise, and correct, allowing the end-user to make meaningful decisions during the transaction. Transaction time is defined herein as the time it takes to complete a transaction from the initiation of the transaction to the time when the desired results are obtained. Thus while a transaction may take several minutes to complete, each communication should only take several seconds to transmit.

Response time refers to the time it takes the system to notify the user that it is acting on an entry made using some input device such as a keyboard or mouse. The time it takes for a data entry screen to appear following its selection from a menu is an example of response time.

Transaction and response times should be specified relative to the desired improvement over alternative means of obtaining the necessary information. End-user impatience and attention span factors are also relevant. The transaction time for generating a hardcopy report resulting from a search of the system indices in response to the submission of an unknown profile should be specified in the context of how long it took to generate that profile. On the other hand, it is important that

**A gigabyte is 1,000 megabytes. in 88 five-drawer file cabinets. It is estimated that in the first five years as many as 500,000 autoradiograph images may be generated. If deemed necessary, these images could be stored in about 50 Gbytes, using image compression. See Section II of Appendix A for more details.

TABLE II—PERFORMANCE REQUIREMENTS

Access and Access Control:
Law Enforcement Agencies, e.g., Crime Labs

Primary On-Line Storage Capacity:

—Sex Offender Profiles for 250,000 Cases (5+ Years):	
Organization Pointers @0.2KB per case	50 MB
Case Data @0.2KB per case	50 MB
Processing Data @0.2 KB per case	50 MB
Autorad Analysis Data @0.2KB per case	50 MB
Total character/numeric data	200 MB
—Open Case Indices for 100,000 Cases (5+ Years):	
Organization Pointers @0.2KB per case	20 MB
Case Data @1.0KB per case	100 MB
Processing Data @0.2KB per case	20 MB
Autorad Analysis Data @0.2KB per case	20 MB
Total character/numeric data	160 MB
—Unidentified Bodies Indices for 5,000 Cases (5+ Years):	
Organization Pointers @0.2KB per case	1 MB
Case Data @0.4KB per case	2 MB
Processing Data @0.2 KB per case	1 MB
Autorad Analysis Data @0.2KB per case	1 MB
Total character/numeric data	5 MB

Organization data capacity is an insignificant amount compared to the other data requirements.

Secondary Storage Capacity:

—Population Samples for 500,000 Profiles @0.2KB:	100 MB
—Ten Year Archival Storage (text/numeric only):	1.1 GB
—Autorad Image Databases for 500,000 images (5+ years):	
Compressed Images (10X) @0.1Mb	50 GB
Total secondary (WORM) storage	51.2 GB

Transaction and Response Times:	
—Screen appearance	<1 sec.
—Confirm initiation of profile matching	10 sec.
—Generate and send match detection notice	1 min.
—Generate and send report of search results	15 min.
—Retrieve autorad image	30 min.
—Add/modify case data	10 min.
—Add to population sample	5 min.
—Request and receive capabilities information	30 sec.
—Request and receive help information	10 sec.

Peak Transaction Processing Rates Per Hour:*	
—Request/receive profile search results	500
—Retrieve autorad image	50
—Add/modify case data	500
—Add to population samples	1000
Total non-image transactions per hour	2000

Availability:	
—Nominal time periods and zones	7AM—9PM; All US Zones
—Percentage up-time in period	95%
—Effect of single CPU failure	None
—Effect of single mirrored-disk failure	None
—Effect of single point network failure	None
—Effect of system maintenance	None

Quality Assurance Measures:	
—Automatic journaling	REQUIRED
—Automated recovery procedures	REQUIRED
—Automatic (scheduled) back-ups	REQUIRED
—Centralized control of data integrity rules	REQUIRED

*Peak rates may be ten times as high as average and last for several hours during times of maximum utilization. Average rates based on 100,000 requests for profile match, 100,000 profile additions/modifications, and 200,000 additions to the population statistics over a 2000 hour year.

the end-user be notified of progress in accomplishing each step of the transaction. Generally, response times more than ten seconds are not tolerated. Ideally, response times should be less than one second. Response times between 1 and 5 seconds are considered reasonable in a wide-area network.

Several desired transaction and response times are specified in Table II. These values are subject to change as modeling and experience dictates. They include: the time to initiate a search for matching profile(s), the time to generate and send report(s) of search results, the time to retrieve an autoradiograph image, and the time to add new case data or modify existing data.

Availability:

Availability is defined herein to mean the condition of the system that allows all of its functions to be used by the end-user. Ideally the system should be available at any time the end-user needs it. The nominal up-time will be specified as regular working hours plus several hours on either side to provide early/late workers access and to account for time zone differences. During this time the system must be up at least 95% of the time, on average, and not be down for more than one hour.

Computer hardware and software require regular maintenance. This is usually performed during periods of minimal usage, e.g., overnight. However, it is common for computer systems to fail at the least opportune time. To maximize availability sufficient redundancy must be part of the computer hardware configuration and a sophisticated database management system must be used.

The availability of the system must be immune to the following events: (1) single processor failure, (2) single disk failure, (3) single network node failure, (4) standard database system maintenance, and (5) peak utilization.

Database Quality Assurance Measures:

The following quality assurance measures will enhance system availability and utility: (1) automatic journaling, (2) automatic recovery, (3) automated back-up, (4) automatic switching to back-up hardware, and (5) centralized control of data integrity rules.

V. RESOURCES REQUIRED

INTRODUCTION

The three types of resources that are required to implement the Combined DNA Index System (CODIS) are (1) Financial, (2) Technological, and (3) Human. This note will use previous planning documents to estimate (1), and make general statements about (2) and (3).

TABLE III—MAJOR DEVELOPMENT COSTS OF DNA DATABASE MANAGEMENT SYSTEM

Item	Cost
Computer hardware for database server, including magnetic disk storage and other peripheral devices, and VAX/VMS operating system and license for sixteen users	$85K
Archival disk storage and WORM media	$15K
Additional software, including system management, communications, and development software	$45K
Database management system software*	$50K
Terminals and software for the system users	$25K
Local area network equipment	$15K
Means to transfer DNA profile data to NCIC	TBD
Modifying NCIC system	TBD
Modifying end-user equipment/software	TBD
Means to transfer serial crime data to VICAP	TBD
Modifying the VICAP system	TBD
Development of database applications, including training of developers	TBD
User training and documentation	TBD
Initial testing period	TBD
Setting up operations	TBD
Initial database build	TBD
Performance tuning	TBD
Initial maintenance of applications	TBD
Miscellaneous travel, liaison, conferences, etc.	TBD

*Acquired

Working Assumptions:

Resource requirements vary according to design and implementation approaches. The requirements given below assume:

—The system will perform an indexing function. As currently conceived, the primary purpose of the system is to match an unknown DNA profile entered by a state or local law enforcement agency with a DNA profile belonging to a known individual, e.g., a previously convicted sex offender. The identity of the known individual, along with the associated case information, would be stored and controlled by the state or local agency owning the data. While the system would control which profiles can be included in the database (to maintain a high quality standard), access rights to case data would be controlled by individual agencies.

—The FBI will act as custodian for the data of state and local law enforcement agencies that have no other satisfactory means to manage their own DNA-related data. Because this is an unknown amount, and because it is not the primary purpose of the system, this function will be considered as a separate database system.

—The FBI's proposed prototype DNA computer system will be used for the following functions:

—Temporary support of FBI DNA Analysis Unit Operations, including, but not limited to, the DNA profiles database, quality assurance records, case management, processing status, probability calculations, population statistics, and correspondence.

—Functional prototype* for custodial support of DNA data for state and local law enforcement agencies without DNA database facilities.

—Functional prototype* for CODIS configuration.

FINANCIAL RESOURCES

The computer system to provide national access has not been specified. NCIC has been proposed as a vehicle to provide communications access to the various law enforcement agencies. Another approach is to publish DNA profiles on CD-ROM periodically. In either case, the proposed computer (MicroVAX 3400) for the FBI's DNA Analysis Unit operations will be the platform on which the functional prototype of the system will be developed.

Table III lists the items that have to be budgeted in order to develop the system. It is based on the use of the DNA Analysis Unit's system. The cost of many items can not be estimated until the final operation computer system is more completely specified.

*A functional prototype implements all of the functions of the operational system, but not at the capacity level required by the complete operational version.

The MicroVAX 3400 was chosen for the FBI's DNA Analysis Unit system for several reasons:

—The current computing facilities that support the Laboratory Division are from Digital Equipment Corporation (DEC). The operating system used is VAX/VMS.

—Systems support and applications programming personnel are in short supply. In it both impractical and uneconomical to introduce an additional facilities management requirement. Thus it is a DNA analysis Unit system requirement to stay within the VAX/VMS environment.

—The system is intended to support only DNA applications so that application development will not conflict with other system development operational schedules. Thus the capacity of the system need only support DNA requirements over the system design lifetime (five years).

—High system availability is mandatory. Thus a dual-processor, mirrored-disk configuration is required.

—DEC recently introduced the MicroVAX 3400 in a dual-processor configuration that is tailored to database applications. It has special disk I/O handling features that enhance server performance. The disks can be mirrored and will provide adequate initial capacity. Furthermore WORM (write once, read many) disk drives are available for MicroVAX.

—The database management system software of choice (Sybase) runs of VAX's and MicroVAX's.

The MicroVAX 3400 is intended to have some excess capacity beyond its initial needs. It will share this capacity initially with the DNA Analysis Unit's operational databases and possibly with another application that uses dial-up modem communications. Both of these systems will migrate to other platforms when operational. The Sybase DBMS is scalable and memory units can be added as requirements grow. Most important, it will serve as a high-availability system.

Included in the MicroVAX configuration is a WORM (write-once, read-many) disk drive. This is required for archiving case files that are no longer active. However, because of the need to quickly respond to inquiries about previous cases, the archival data must be on-line. WORM technology is well-suited for this purpose. It can also serve as the media for image storage, should the need arise.

Sybase is the DBMS of choice. The following are some of the reasons for selecting Sybase for the DNA application:

—Sybase is used for the National GeneBank DNA Database.

—Sybase can control the operation of a dual-processor, mirrored disk MicroVAX computer system.

—Sybase is oriented towards on-line transaction processing applications.

—Sybase requires less memory per additional user.

TABLE IV—HARDWARE AND SOFTWARE FOR STATE/LOCAL DNA DBMS SYSTEM

HARDWARE

IBM Corp.:			
(1)	1	8580311 PS/2 Microcomputer (Model 80) with a 311 Mbyte hard disk	$7,750
(2)	1	4869001 5-1/4" external floppy disk drive	300
(3)	2	6451013 PS/2 Dual Async. Adapter @ $200 ea.	400
(4)	1	6450378 80387 Math Co-Processor, 20 mhz.	750
(5)	1	6450605 Memory Option Board for mdl. 80	1,000
(6)	1	6450603 1 Mbyte Memory Kit for mdl. 80	450
(7)	3	6450604 2 Mbyte Memory Kit @ 900 ea	2,700
(8)	1	4202003 ProPrinter II XL	600
(9)	1	8514001 16" Very High Resolution Color Display	1,000
(10)	1	1887972 Display Adapter for 8514 Display	800
(11)	1	1887989 Memory Expansion for 8514 Adapter	200
Hayes:			
(12)	2	08-00246 V-Series Smartmodem 9600 External	$1,800
Mountain:			
(13)	1	01-30090-01 300 Mb. Filesafe Ext. Tape Backup	$2,300
(14)	1	01-09246-01 Ext. Tape Controller for PS/2	200
3Com:			
(15)	2	Etherlink Plus Cards for PS/2 @ $650 ea.	$1,300
(16)	2	Ethernet/IEEE 802.3 BNC Transceivers @ $200 ea.	400
(17)		Thin Ethernet Cables, Connectors and Terminators	250
		Total for PS/2 System Hardware and Peripherals	$22,300

SOFTWARE (Note: Most software listed below operates in OS/2 environment.)

IBM Corp:			
(18)	OS/2 Ext. Ed. v1.1	Operating System	N times $600*
Ashton-Tate, Inc.:			
(19)	dBase IV v1.1 for OS/2	Database Front End	N times $600*
(20)	SQL Server	Sybase SQL Server	2,500
3Com Corp:			
(21)	3+ Open Lan Manager	Entry Network Operating System	$1,000**
(22)	3+ Open Internet	Internetwork Communication	$1,500
(23)	3+ Open Mail	Electronic Mail Utility	$1,200

Total OS/2 DBMS Network Software Cost (N=5) $13,200

*Need as many copies as workstations on the network.

**Unlimited user license is available for $2,000 more.

—Sybase permits data integrity controls to be associated with a centralized data dictionary, thus reducing the coding effort normally associated with each application.
—Sybase has a client/server architecture, thus partitioning the computer load between the central server and the distributed clients.
—Sybase is oriented towards the use of the Structured Query Language (SQL), which is recommended for federal use and which is becoming generally available for all computers.
—Sybase has a comprehensive set of tools for forms specification, communications with other programs, communications with other database systems and writing transaction-oriented applications.

Table IV lists the major cost items of a functionally equivalent database system based on the use of the IBM PS/2 Model 80 microcomputer and the version of Sybase that runs on the PS/2 under the OS/2 operating system (SQL Server). This system assumes a network configuration, and the network operating system of choice is 3Com's 3+ Open Lan Manager. This configuration is likely to give the highest performance and flexibility in a network of file server and workstations and will be compatible with the Combined DNA Index System. The total cost for the hardware and software of the file server and network components is estimated to be approximately $35,500 for five users. This does not include user workstations.

TECHNOLOGY RESOURCES

The objectives of this project can be met with the use of existing technology, although an approved encryption technique is pending. Adequate computer power is available. Database management software is available. Most items require straightforward application of known technology.

As with financial resources, technology resource requirements vary with the total system configuration specifications. Several such configurations have been suggested:

—Use of the NCIC System for Nationwide Access: This approach has attracted some support because the NCIC model is similar to that proposed for the system, i.e., a centralized index. NCIC management has expressed an interest, but the advisory board has yet to act on this matter, as of this date. The main benefit that NCIC can offer is the use of the existing set of leased lines and a communications control computer that could easily handle DNA communication data. In addition, NCIC is a known entity and has an existing policy advisory board that is quite aware of state/local needs.
—NCIC currently may not be the optimum system for CODIS: Extensive redesign plans exist. This raises the question of what would be the best time to make use of NCIC networking capabilities, before or after the new design is implemented. The prototype DNA database system will have server-client architecture. Screen

operations will be performed locally. Certain immediate-response functions will be performed by the client portion. Data integrity operations will be performed centrally by stored procedures in the database server. This architecture gives both flexibility and control. Local operations, including the automated retrieval of DNA profile data from image processing computer systems, may pose some difficulties for existing terminal equipment. The translation of locally generated screen entries into the Structured Query Language is not a simple task.

—While using NCIC's communications network is an alternative worth evaluating, incorporating the DNA databases into the current NCIC database structure would require more than adding a few fields to those existing in order to have a viable system. Inherent in the task of preserving the utility of the database system are stringent data and process integrity considerations. DNA methods are likely to change. In the interest of obtaining low system lifetime cost it is essential that modern database and programming technology be used. This suggests that no DNA database functions should be incorporated into NCIC until it is implemented using a relational database management system. In terms of numbers of transactions handling daily and response times, NCIC performs at a level far beyond the needs of the DNA database. But sheer raw power is not to be the answer to handling all of the DNA database functions in a manner consistent with today's database management technology. Dialog is continuing with NCIC staff to determine its suitability for meeting the full-scale CODIS requirements.

—Use of a Separate Stand-Alone System: This approach has the advantage of being unencumbered with competing requirements and expectations. Independent development can proceed at a pace determined mainly by the capabilities of available resources. While this approach may lead to an earlier implementation, it requires that a separate access method be established. It is conceivable that many state or local law enforcement agencies performing DNA analyses will have adequate computing facilities that can communicate with a stand-alone system over standard telecommunications lines. However, this condition has yet to be ascertained. The prototype system will use dial-up modems initially. This approach will be unsatisfactory if images must be transmitted.

—Publishing DNA Profiles on Read-Only Media: This approach is attractive because it can provide for a small investment in a CD-ROM disk drive the ability to perform a DNA profile database search locally. As a minimum, the statistical database could be published periodically, or upon major updates. The information published on CD-ROM would be entirely under centralized quality control. CD-ROM distribution might be used in conjunction with other solutions.

Computer Security Matters at FBIHQ:

Computer system security requirements complicate the data communications situation. Current FBI ADP security policy excludes any computer that has any form of external non-secure access from being attached to the FBI Headquarter's local area network (HQLAN).

TABLE V—HUMAN RESOURCE REQUIREMENTS

Task	*Resource*	*Experience*
Modify NCIC Software	Programmer/Analyst(TSD)	2 yrs direct on NCIC programs
Design Sybase Applications	Computer Scientist (FSRTC/TSD)	6 mos direct on DNA database plus Sybase training and knowledge of DBMS principles
Implement Sybase Applications	Programmer/Analyst (FSRTC/Consultant	2 yrs experience with Sybase DMBS; one year on VAX/MicroVAX
Supervise Sybase Application Implementation	Computer Scientist (FSRTC/TSD)	See above
End-User Training (Sybase)	Instructor (Sybase)	Instruction experience plus knowledge of Sybase
End-User Training (Applications)	Computer Scientist (FSRTC)	Familiarity with DNA DBMS applications
Documentation	Publications Staff and Comp. Scientist (FSRTC/SPS)	Previous documentation
Initial Installation	Systems Analyst (FSRTC or Sybase)	One year hands-on with Sybase plus one year hands-on with VMS or MicroVMS
Performance Optimization	Systems Analyst (FSRTC/TSD)	Same as for initial installation
Database maintenance	Database Administrator (TSD)	Six months experience with Sybase or equivalent courses plus one year experience in database administration/design
Database Policy	Data Administrator (Lab Div)	One year direct on DBMS systems, preferably Sybase, knowledge of principles of data administration, good working relationship with Division 7 policy makers

TABLE VI—PROPOSED IMPLEMENTATION SCHEDULE

Task	Description	Completion	Status
1.	Acquire database management system to support a DNA profiling and statistical database	May 1989	Complete
2.	Acquire hardware for the DNA database prototype	May 1989	In Place
3.	Train FSRTC scientists on DBMS	May 1989	Initial
4.	Develop draft of theoretical model	June 1989	Complete
5.	Finalize theoretical model (This document)	October 1989	Complete
6.	Develop working DNA database prototype	January 1990	Pending
7.	Acquire hardware to support the operational DNA profiling and statistical database	January 1990	Proposed/ Approved
8.	Implement DNA profiling and statistical databases in FBI Laboratory	March 1990	Pending
9.	Implement investigative DNA profiling database in selected states	June 1990	Pending
10.	Recommend network system to link state DNA profiling databases	June 1990	Pending
11.	Implement prototype network to link state DNA profiling databases	October 1990	Pending

Currently the HQLAN can not handle data with multiple levels of security. Handling secure data on the HQLAN will require the use of an encryption technique that is awaiting testing and approval.

The decision has been made to operate the prototype as a stand-alone system and to allow the DNA Analysis Unit application to migrate to another computer when operational. This will allow the CODIS prototype to have dial-up communications and/or direct links to NCIC.

HUMAN RESOURCES

On the list of financial resources above are several items that require persons having certain skills. (See Table V.) Modifying NCIC software requires the use of programmer/analysts familiar with the NCIC system program code. These exist in the Technical Services Division. Modifying end-user systems will be required, but not enough is known to make a statement at this time as to what is required and who should do it. FSRTC/TSD computer scientist staff will manage the development of the Sybase applications by an outside consultant familiar with Sybase and MicroVAX's. User training and documentation will be the responsibility of the FSRTC publications staff working with the Sybase application developer(s). Initial testing, setting up operations, building the database, performance and performance tuning will require the expertise of a systems analyst and/or a database administrator familiar with MicroVAX operating systems and with Sybase. A data administrator should set policy and control the data and its integrity over the life of the system.

APPENDIX A

PERFORMANCE REQUIREMENTS

I. DETAILED WORKING ASSUMPTIONS

Combined DNA Index System Operating Policy:

—Only indexing information will be stored centrally for sex offender data. By itself the DNA profile is such an index. No names of convicted sex offenders will be stored centrally. The state and local agencies who have their own storage and retrieval systems will store case details and have control over release of the data.

—The FBI DNA Analysis Unit's case data will be handled in the same manner as that belonging to state and local agencies. The FBI will provide storage and retrieval facilities for those states not having their own.

Basic Growth Assumptions:

—Several factors will limit the growth of DNA Profile Database over the next five years. These include:
  —Number of sex offenders required to give blood sample
  —Number of violent crimes producing body fluid stains
  —Acceptance of DNA results in courts
  —Funding for new technology
  —Training of personnel using RFLP techniques.

—The number of sex offenders being released in states requiring a blood sample prior to parole is likely to be less than the number of violent crimes in this country producing body fluid stains. (Over 90,000 forcible rapes were reported to local police in 1987.) State and local laboratory personnel are likely to be understaffed to handle the potential case load. While it may take several years before the use of DNA profiling is accepted in most state and local courts, this is not anticipated to be the most limiting factor. Because of the promise of the technique, funding is also not expected to be the main limiting factor. The primary limiting factor of the growth of the Combined DNA Index System will be the number of qualified individuals generating DNA profiles.

—It is assumed herein that the limiting factor on the growth of the system will be the effective number of trained scientists capable of processing forensic DNA samples using common techniques, e.g., RFLP. The major training center is the FSRTC at Quantico. It is currently training about 75 persons per year. This may continue for several years.

—Assuming the FBI trains three-quarters of the total, and continues to maintain its current training schedule for five years, a total of 500 qualified individuals will be available in five years. It is also expected that these scientists will likely perform other forensic laboratory duties. Most state and local laboratory personnel will likely perform all related functions. Combined with their other laboratory duties, only a fraction of full time is available for lab work on DNA samples. A gross assumption is that on the average a maximum of 40 percent of their time will be used for processing samples of DNA. Thus the effective number of scientists producing DNA results in five years is assumed to be 200.

—Based on the current FBI DNA Analysis Unit workload, three well-trained and hard-working scientists can handle forty samples per week using an average of (slightly less than) four probes per sample. To be conservative in the design, let us assume that productivity can be improved by 50 percent. This suggests a rate of twenty samples per week, or 1,000 samples per 50-week year, per scientist. At the end of the first five years the limit placed on the number of DNA samples processed annually nationwide by trained personnel is approximately 200,000.

II. DETAILS OF PRIMARY STORAGE REQUIREMENTS

Combined DNA Index System Volume:

—State and local lab work will be split between samples from convicted sex offenders about to be released on parole and from on-going case work for violent crimes. Only one sample is generated for each parolee, and there is usually one suspect convicted per criminal case. An average of five samples, including known victim and questioned samples, are processed per case while only one sample is processed per parolee. Furthermore, not all suspects are convicted. Thus for each DNA profile submitted many more are generated. A five-to-one ratio of processed to submitted profiles will be used. However, to be conservative in design, this figure will be multiplied by 2.5. This produces a system design parameter of 100,000 new DNA profiles submitted for indexing per year at the end of the first five year period.

Autoradiograph Volume:

—The average number of single-locus probes applied to each sample is four. An average of five samples are processed per case and all can be placed on a single gel. One autoradiograph is generated per probe per case. However, many samples of known parolees could be placed on a single autoradiograph. Thus the number of autoradiographs generated depends on the mix of parolee and case work. Assuming an average of two autoradiograms per submitted profile and 100,000 profiles annually, 200,000 autoradiographs could be generated each year at the end of the first five years.

Cumulative DNA Profile Capacity:

—It is further assumed that growth in the generation of DNA profiles will be linear. Starting at zero and ending in five years at an effective conservative rate of 100,000 per year produces a cumulative number of 250,000 DNA profiles submitted for indexing by the end of the first five year period. About twice that many could have been submitted for use in the population samples.

FBI's DNA Analysis Unit Capacity:

—At current complement (six technicians, three examiners and one unit chief) and hypothesizing a 50 percent higher efficiency than is current, the FBI Data Analysis Unit can expect to handle 140 samples per week, or 7,000 per year. The number of FBI technicians processing DNA will increase to 12 and the number of examiners to 8 in the near future. The new staff will come from other FBI laboratories. This represents a relatively large laboratory devoted exclusively to DNA. Although it is currently doing most of the forensic DNA analysis at the moment, in five years the FBI would be processing less than ten percent of the total.

III. DETAILS OF SECONDARY STORAGE REQUIREMENTS

Introduction:

Secondary storage is for sub-population sample data, archival storage and autoradiograph images. Secondary storage requires higher capacity because it must hold the accumulation over many years of operation. However, the response-time requirements of secondary storage are not as critical as for primary storage. Secondary storage data is also not likely to be changed.

Population Samples:

For every offender there is a victim. Furthermore, DNA profiles will be obtained frequently for known suspects who do not match the profiles found in the questioned samples. Thus at least twice as many DNA profiles are available for inclusion in the population samples as there are for convicted sex offenders. No linkage with an individual is stored with his or her DNA profile in the population sample database. Only broad categorical data that specify which population sample to which the profile belongs are kept.

Each population sample represents a commonly recognized statistical grouping of the nation. Five such groups are envisioned: Caucasians, Afro-Americans, Hispanics, Orientals, and American Indians. If the ethnic origins of the suspect are known, the appropriate population statistics can be applied to determine the likelihood of two unrelated individuals of the same ethnic origins having the same DNA profile. The use of a very small likelihood value can have a strong influence on a jury's decision. Thus it is important that the population statistics be accurate. The population sample must be sufficiently large to remove the chance of small-sample bias and be sufficiently representative of the population as a whole. As more DNA samples are processed, the population sample size will grow and the purported power of the technique to uniquely identify an individual will be enhanced.

It is estimated that after five years 500,000 DNA profiles will be available for the population statistics database, requiring about 100 Mbytes of storage. These data need only be in secondary storage because they will only be used periodically to update the frequency estimates (histograms) of DNA fragment sizes found by each DNA probe for a given population sample.

The collection of DNA profiles for such a large population sample sizes represents a unique opportunity for continued research into such factors as population stability, mixing, and for the detection of the possible influence of geographic or regional factors. This may lead eventually to a larger number of better-defined population samples with the potential for obtaining better exclusionary likelihood values. Thus additional information about an individual's parentage, origins and/or background may be requested and recorded on a voluntary basis.

Archival Storage:
Secondary storage of text and numeric data for known sex offenders, open case and unidentified bodies/missing persons indices is required for a longer period for archival purposes. It is estimated that ten years of such data will require approximately 1.1 Gbytes of storage capacity.

Archival data should never be altered. Currently WORM (write once, read many) optical disk storage media provides an intuitively obvious solution to archival storage. One $400 optical disk can store well over two gigabytes of data. Savings in floor space alone is worth the cost of an optical disk drive and storage media.

Autoradiograph Image Storage:
If WORM technology is useful for archiving text and numeric data, its utility is even more apparent when considering the storage of autoradiograph images. If such storage becomes a reality, much larger amounts of secondary storage capacity will be required. Although slower than magnetic media disks, optical disks are fast and inexpensive enough to maintain image data on-line, thus eliminating the need for physically handling the media when retrieving such data. It is estimated that in the first five years as many as 500,000 autoradiograph images may be generated. With image compression techniques, these images could be stores on about 50 Gbytes of optical disk media. It is possible that the physical location of these data would be distributed. They could also be stored on a set of CD-ROM disks that are similar to those used for music recordings, that can be played back on inexpensive disk drives. These could be periodically published and distributed.

IV. TRANSACTION AND RESPONSE TIMES

Transaction time is defined herein as the time it takes to complete a transaction from the initiation of the transaction to the time when the desired results are obtained. A transaction usually involves several operations on related data. An example of a transaction would be the generation of a report in response to a request to search the system for a matching DNA profile. This would involve a series of steps including forwarding the necessary access privilege information, entering the search profile data, obtaining the location of the information relating to any match, requesting permission to obtain such information from the custodial law enforcement agency, and obtaining that information in report form. Response time refers to the time it takes the system to notify the user that it is acting on an entry made using some device such as a keyboard or a mouse. The time it takes for a data entry screen to appear following its selection from a menu is an example of response time.

Transaction and response times should be specified relative to the desired improvement over alternative means of obtaining the necessary information. End-user impatience and attention span factors are also relevant. Generally, response times more than ten seconds are not tolerated. Ideally, response times should be less than one second. Response times between 1 and 5 seconds are considered reasonable in an wide-area network. A transaction comprises a series of communications, each of which should be timely, concise and correct, allowing the end-user to make meaningful decisions during the transaction. Thus while a transaction may take several minutes to complete, each communication should only take several seconds to transmit.

If the end-user understands the amount of processing that is required to generate the results and how busy the system is at certain periods of the day, realistic transaction times will be tolerated. The need to obtain "instantaneous" response from the database search is not consistent with the fact that the DNA profiles requires from two to six weeks to develop for one to four probes obtained from a single membrane. One the other hand, it is important that the end-user be notified of the progress in accomplishing each step of the transaction so that he/she can be reassured that the system is responding. It is also important to notify the user of the magnitude of the response and provide an opportunity to cancel the transaction if, for example, no report is possible, or if too many reports will be generated. In the latter case, the information given to the system must be further refined, e.g., by the use of results from an additional probe, in order to narrow the search to a manageable response.

APPENDIX III

SUPPLIERS OF DNA ANALYSIS EQUIPMENT, PRODUCTS, OR SERVICES

The following companies were listed as suppliers of DNA analysis equipment, products, or services, in the United States, at a recent International Symposium on the Forensic Aspects of DNA Analysis sponsored by the FBI.

*AMERICAN TYPE CULTURE COLLECTION
12301 Park Lawn Drive
Rockville, MD 20852
301-231-5585

APPLIED BIOSYSTEMS
850 Lincoln Centre Drive
Foster City, CA 94404
414-570-6667

BETHESDA RESEARCH LABORATORIES
Life Technologies Incorporated
8717 Grovemont Circle
Gaithersburg, MD 20877
301-670-8343

BIOS CORPORATION
291 Whitney Avenue
New Haven, CT 06511
203-733-1450

BIO IMAGE
Ann Arbor, MI 48108
313-971-7500

BIO-RAD LABORATORIES
Research Products Division
1414 Harbour Way South
Richmond, CA 94804
415-232-7000

CELLMARK DIAGNOSTICS
P.O. Box 1000
Germantown, MD 20874
301-428-4980

CETUS CORPORATION
1400 - 53rd Street
Emeryville, CA 94608
415-420-4220

COLLABORATIVE RESEARCH INCORPORATED
2 Oak Park
Bedford, MA 01730
617-275-0004

*Suppliers of tissue culture cell lines (not listed at the International Symposium)

GENELEX CORPORATION
1000 Seneca Street
Seattle, WA 98101
206-285-3132

GENESCOPE CORPORATION
1380 Blountstown Highway
Tallahassee, FL 32315
904-562-3261

GENE SCREEN
2600 Stemmons Freeway, Suite 133
Dallas, TX 75207
214-631-8152

GENETIC TECHNOLOGIES
30 Whiting Street - Top Floor
Artarmon, Australia

GENETICS & IVF INSTITUTE
3020 Javier Road
Fairfax, VA 22031
703-698-3950

GENMARK
417 Wakara Way
Salt Lake City, UT 84108
801-582-2600

ICI DIAGNOSTICS
20271 Goldenrod Lane
Germantown, MD 20874
301-353-9625

I.I.T. RESEARCH INSTITUTE
6800 Versar Center, Suite 106B
Springfield, VA 22132
703-658-0086

INTERMOUNTAIN FORENSIC LABORATORIES, INC.
11715 NE Glisan Street
Portland, OR 97220
503-257-7177

LIFECODES CORPORATION
Sawmill River Road
Valhalla, NY 10595
914-592-4122

MOLECULAR DYNAMICS
240 Santa Ana Court
Sunnyvale, CA 94086
408-773-1222

*NIGMS HUMAN GENETIC MUTANT CELL REPOSITORY, AND THE NIA AGING CELL REPOSITORY
Coriell Institute for Medical Research
401 Haddon Avenue
Camden, NJ 08103
800-752-3805
609-966-7377

PERKIN-ELMER CORPORATION
761 Main Avenue
Norwalk, CT 06859
203-834-6737

PHARMACIA LKB BIOTECHNOLOGY INCORPORATED
800 Centennial Avenue
Piscataway, NJ 08854
201-457-8000

PROMEGA
2800 S. Fish Hatchery Road
Madison, WI 53711
608-274-4330

PUCKETT LABORATORY
4200 Mamie Street
Hattiesburg, MS 39402
601-264-3856

SEROLOGICAL RESEARCH INSTITUTE
3053 Research Drive
Richmond, CA 94806
414-223-7374

SIGMA CHEMICAL COMPANY
Box 14508
St. Louis, MO 63178
800-346-6405

S.W. & C. LOUISIANA LABORATORIES OF FORENSIC SCIENCE
4808 Antonini Drive
Metairie, LA 70006
504-568-5867

Some of the above also have branches or distributors in Canada and the U.K. Three additional companies in Canada are

BIO/CAN SCIENTIFIC INC
2170 Dunwin Drive, Unit 5
Mississauga, Ontario L5L 1C7
800-387-8125

FISHER SCIENTIFIC
112 Colonnade Road
Nepean, Ontario K2E 7L6
800-267-7424

HELIX BIOTECH CORPORATION
215-7080 River Road
Richmond, British Columbia V6X 1X5
604-270-7468

Two companies in the U.K. are

CAMBRIDGE RESEARCH BIOCHEMICALS
Button End, Harston
Cambridge, England CB2 5NX
44-223-871674
(0800) 585396

ICI DIAGNOSTICS
Gadbrook Park Rudheath
Northwich Cheshire
England CW9 7RA
(0606) 41100

GLOSSARY

Accuracy. A measure of the closeness of a test result to the absolute value.

Adenosine triphosphate (ATP). A nucleoside triphosphate that upon hydrolysis results in energy availability for processes such as muscle contraction and synthesis of macromolecules, including proteins and carbohydrates.

Agar. A polysaccharide extracted from certain seaweeds.

Agarose. The neutral gelling fraction of agar commonly used in gel electrophoresis.

Algorithm. A step by step process for solving a problem.

Allele. One of a series of alternative forms of a gene (or VNTR) at a specific locus in a genome.

Alu. A family of repeat DNA sequences, cleaved by the restriction enzyme Alu I, dispersed throughout the genomes of many animal species. The family consists of about 500,000 copies, at 300 bp each, per human genome.

Amino acids. The building blocks of proteins coded by triplets of bases in the DNA blueprint. For example, the mRNA transcript code AUG codes for the amino acid methionine; whereas, CGU, CGC, CGA, CGG, AGA and AGG code for arginine.

Amniotic fluid. The fluid, containing fetal cells, surrounding the developing fetus.

AMP-FLP. Polymerase chain reaction amplified restriction fragment lengths consisting of variable number of tandem repeats.

Antenatal diagnosis. In utero diagnosis to determine fetal defects.

Antibody. A protein produced for body defense in response to an antigen.

Anticoagulant. An agent, such as EDTA, used to prevent the clotting of whole blood.

Antigen. A foreign substance, usually a protein, capable of stimulating an antibody response for body defense.

Archive. To file data in a record bank, such as computer disks.

Artiodactyl. Any order of hoofed mammals with an even number of toes on each foot.

Assortment. See Mendel's laws.

ATP. See adenosine triphosphate.

Autopsy. A post-mortem examination.

Autoradiogram (autoradiograph). The resultant X-ray film after having been exposed to a radioactive source. A DNA probe tagged with a radioactive isotope such as ^{32}P (radioactive phosphorus) will expose an X-ray film where the probe hybridizes to complementary sequences on the blot in contact with the film.

Autosomal dominant. See dominant.

Autosomal recessive. See recessive.

Autosome. Non-sex chromosome. There are 22 autosomes in the human genome.

Background radiation. Radiation from a source other than the test sample specifically being analyzed.

Bacterial colony. A clone of bacterial cells.

Bacterial stab. A small container of agar with bacterial colonies suitable for storage or shipment.

Bacteriophage (phage). A bacterial virus.

Band shifting. The phenomenon where DNA fragments in one lane of an electrophoresis gel migrate across the gel more rapidly than identical fragments in a second lane.

Base pairs (bp). The bases A=T or C≡G, linked by hydrogen bonds, binding DNA complementary strands. See DNA.

Base sequence. Order of bases in a DNA molecule.

Bayes' theorem. A statistical procedure to assess the relative probability of two alternative possibilities.

Beta counter. An instrument used to measure radioactive decay.

ß-globin. A 146-amino-acid protein molecule of hemoglobin.

Biopsy. A small piece of tissue excised for the purpose of analysis.

Blood group. A classification of red blood cell surface antigens, for example, ABO.

Blot. See DNA blot and Southern blotting.

Buccal cells. Cells derived from the inner cheek lining. These cells are present in the saliva or can be gently scraped from the inner cheek surface.

Buffy coat. The whitish layer of cells (white blood cells plus platelets) overlaying the red cell pellet after centrifugation of whole blood.

Carrier. An individual with only one of two possible copies of an abnormal gene. The heterozygous state.

Cassette film holder. A light-proof holder, used in autoradiography for exposing X-ray film to radioactive blots.

cDNA. See complementary DNA. DNA produced using mRNA as a template. The gene introns are, thus, not included.

Cell division. The process whereby a mother cell gives rise to two identical daughter cells (see mitosis) or four gametes (see meiosis).

Central dogma. The theory of DNA — RNA — protein transfer of genetic information.

CF. See cystic fibrosis.

Chi-square test. (χ^2). A statistical test to determine how closely an observed set of data values correspond to the values expected.

Chimera. An individual composed of two different cell lines originally derived from the union of different zygotes.

Chloroplast. A chlorophyll-containing plastid that functions as a plant photosynthetic site.

Chorionic villi sample (CVS). A specimen of fetal tissue aspirated from the placenta at about 10 weeks gestation in the human.

Chromatid. One of the two attached daughter strands or "chromosomes" of a mother chromosome observable during mitosis or meiosis.

Chromosome. A nuclear structure in eukaryotes that carries a portion of the genome. The human has 46 chromosomes per nucleus; 22 homologous pairs of autosomes and 2 sex chromosomes. Prokaryotes have only a single circular chromosome.

Clone. A group of identical cells descended from a single cell. See cloned DNA and bacterial colony.

Cloned DNA. A DNA sequence ligated into a vector and replicated as the vector replicates in its host.

Cloning vector. See vector.

Code. See amino acids and degenerate genetic code.

Codon. A nucleotide triplet coding for the amino acid to be inserted at a specific position in a protein.

Coefficient of inbreeding (F). The proportion of homozygous loci in an individual; or, the probability that both alleles at corresponding loci were inherited from the same ancestor. Siblings or parent and child: $F = 0.25$; grandparent and grandchild: $F = 0.125$; first cousins: $F = 0.0625$.

Coefficient of relationship (r). The proportion of genes that two individuals have in common. It is the proportion of the genomes inherited from a common ancestor; or, the probability that two individuals have inherited a specific gene or DNA fragment from a common ancestor. Siblings or parent and child: $r = 0.5$; grandparent and grandchild: $r = 0.25$; first cousins: $r = 0.125$.

Coefficient of variation (CV). A statistical measure of dispersion relating the measure of dispersion *(SD)* to the average or mean $(\overline{X})$. $CV = SD \div \overline{X} \times 100$.

Colony. See bacterial colony.

Complementary strands. See deoxyribonucleic acid.

Complementary DNA (cDNA). DNA formed by the action of the enzyme reverse transcriptase on mRNA.

Confidence interval. A statistical measure of confidence in a calculated value. A 95% confidence interval equates to the expectation that the value in question will lie within the range stated 95% of the time and outside the range 5%. A certain allele in a population may have a calculated frequency of 1 in 500 people with confidence limits of 1 in 400 to 1 in 600 at the 99% confidence interval.

Confidence limits. Limits attached to a confidence interval. The 95% confidence limits, for example, when measuring a 10-kb allele on a gel in a certain laboratory may be 9.9 to 10.1-kb.

Controls. Samples of predetermined concentration (known or unknown to the analyst) treated as unknowns in an assay. Controls are included as part of quality control for each test run.

Cosmid. A cloning vector consisting of a plasmid and λ-phage cos nucleotide sequence. The cos sequences facilitate protein capsule formation for the DNA.

Covalent. A strong chemical bond formed between atoms, for example, oxygen 0=0, water H—0—H.

Crime specimen. A specimen left at the scene of a crime by the perpetrator.

Crossing-over. Exchange of homologous chromosome segments during meiosis.

Cross-hybridization. The binding of a probe to a DNA sequence other than the intended target sequence. This occurs because of homology between the probe and the sequence and because low stringency hybridization wash conditions are followed.

Cultivar. A variety of plant developed through selective breeding programs.

CVS. See chorionic villi sample.

Daughter cell. See cell division.

Decant. The process of pouring off the supernatant during separation from a pellet after a mixture has been centrifuged or left to settle.

Degenerate genetic code. The phenomenon of more than one DNA nucleotide triplet of bases coding for the same amino acid. See amino acids.

Deletion. Loss of one or more nucleotides from a DNA strand. This may result in a gene mutation.

Deoxyribonucleic acid (DNA). The molecule of heredity. DNA is composed of deoxyribonucleotide building blocks, each containing a base: adenine (A), thymine (T), cytosine (C), or guanine (G); a deoxyribose sugar (S); and a phosphate group (P). The DNA molecule forms a double helix with the nucleotides of each strand held together at the 3' and 5' carbons of the sugar, and the antiparallel complimentary strands held by hydrogen (H) bonds between the base pairs: A=T and C≡G. During replication, the hydrogen bonds break, the complementary strands separate and each acts as a template for production of its complement. DNA does not contain the base uricil; this is found only in RNA in place of thymine of DNA.

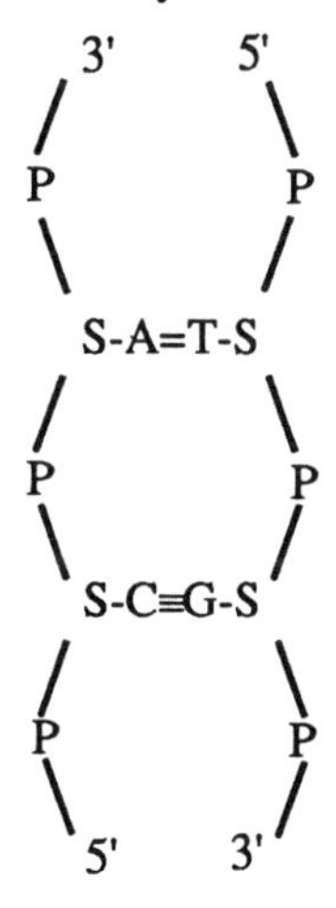

Deoxyribonucleotides. The building blocks of DNA. See deoxyribonucleic acid.

Depurination. The process of partial DNA hydrolysis by acid at purine (adenine and guanine) sites, resulting in the cleavage of large DNA fragments into smaller pieces. This process improves Southern transfer.

Dialysis. The process of separating different size molecules in solution by means of their differential transfer across a porous membrane.

Digested DNA. DNA cleaved by the action of restriction enzymes.

Digital. Expression directly in the form of digits.

Diploid (2N). The nuclear state in which both members of each homologous chromosome pair are present, for example, human somatic cells.

Dizygotic. Twins produced from two separate zygotes. Also called fraternal twins. On average, one-half of their genomes are shared.

DMD. See Duchenne muscular dystrophy.

DNA. See Deoxyribonucleic acid.

DNAase (deoxyribonuclease). An enzyme capable of cleaving DNA into small fragments.

DNA band. A DNA fragment or allele on a Southern blot autoradiogram. With reference to an identity profile, a band is a tandem repeat DNA sequence (allele) produced by cleaving a genome into fragments with a restriction enzyme having recognition sites *(a)* flanking the allele and *(b)* usually at millions of other genome locations.

DNA blot. A membrane (usually nylon) with covalently bound single-stranded DNA.

DNA fingerprinting. See DNA identification analysis.

DNA haplotype. A pattern of DNA polymorphisms.

DNA identification analysis. The characterization of one or more features of an individual's genome by developing a DNA fragment band (allele) pattern If a sufficient number of different size bands are analyzed, the resultant bar code profile will be unique for each individual except identical twins.

DNA library (genomic or cDNA). A collection of vectors containing fragments (a maximum of one per vector) of the total genome or cDNA of an individual.

DNA library screening. The process of testing for vector colonies containing a cloned DNA sequence of interest.

DNA ligase. An enzyme that catalyzes nucleotide phosphodiester bond formation.

DNA polymerase. An enzyme that catalyzes the linking of deoxyribonucleotide triphosphates using complementary DNA as a template.

DNA profiling. See DNA identification analysis.

DNA typing. See DNA identification analysis. DNA typing is the term that will be used in North America as set forth by TWGDAM.

Dominant. A trait phenotypically expressed when an abnormal gene is present in the heterozygous or homozygous state.

Dot-blot. A DNA analysis system where sample DNA is directly pipetted onto a membrane, as opposed to the Southern blot procedure of enzymatic digestion, electrophoresis, and Southern transfer.

Double helix. See deoxyribonucleic acid.

Duchenne muscular dystrophy (DMD). An X-linked muscular disease.

Duplex DNA. DNA consisting of complementary strands.

Electrophoresis. The process of separating charged molecules, for example, negatively charged DNA fragments, in a porous medium such as agarose, by the application of an electric field. DNA separates according to size with the small fragments moving most rapidly.

Enterotoxin. A toxic substance, produced by certain microorganisms, that causes gastrointestinal disorders.

Enzyme. A recycling protein molecule that catalyzes a specific chemical reaction.

Eukaryote. A multicellular organism having true membrane-bound nuclei containing chromosomes that undergo mitosis.

Evidentiary specimen. See crime specimen.

Evolution. A stable genotype change resulting in new taxa or classification.

Excision. Enzymatic removal of nucleotides from nucleic acid molecules.

Exclusion. A crime suspect or putative father DNA identity profile not matching that of a crime sample or offspring paternally-derived alleles.

Exons. Portions of a gene coding for spliced mRNA. Exons consist of the code signals for *(a)* the initiation of RNA transcription and ribosomal accommodation, *(b)* termination of translation and addition of poly-A tail, *(c)* specific protein formation.

Exonuclease. An enzyme that digests DNA strands starting at their termini.

Exponential increase. An increase at a rate defined by raising a number to a power, for example, 10, 10^2, 10^3.

Expression vector. A vector designed to express a cloned gene.

Extra chromosomal inheritance (extra nuclear inheritance). Cytoplasmic inheritance of DNA via organelles, such as mitochondria (or plasmids), in female gametes. The human egg, for example, transmits approximately 10,000 mitochondria.

Extranuclear DNA. DNA located in organelles such as mitochondria and plastids. This material is also referred to as cytoplasmic DNA and its inheritance as maternal or cytoplasmic since the organelles are transmitted only from the female via gamete cytoplasm.

Fingerprinting. See DNA identification analysis.

Fixed tissues. Tissues preserved with an agent such as formalin.

Fluorometric assay. The process of molecules absorbing light of one wavelength and emitting at a different (higher) wavelength.

Forensic. Analysis information suitable for use in a court of law.

Forensic sciences. The application of scientific facts to legal problems.

Formalin. A solution of 40 grams formaldehyde/100 ml of solution.

Fragment band. See DNA band.

Fraternal twins. See dizygotic.

Frequency. Specifically refers to the number of individuals or measurements in a subgroup of the total group under consideration. The term is often more loosely equated to proportion, that is, to define a fraction or percent.

Frye standard. A set of standards established by the Court of Appeals of the District of Columbia in 1923 for Frye vs. the United States. The standards in general define when a new scientific test should be admissible as evidence in the court system.

Galactosemia. An autosomal recessive inborn error of metabolism resulting from a block in the conversion of galactose 1-phosphate to glucose 1-phosphate.

Gamete. A reproductive cell (egg or sperm).

Gamma rays (γ rays). A type of radiation from certain radioactive sources, for example, ^{125}I.

Geiger-Mueller counter. A radiation monitor capable of detecting medium and high intensity α, ß and to a lesser degree γ rays.

Gel. Agarose gels used in electrophoresis. See electrophoresis.

Genes. Units of heredity. Regions of DNA containing the blueprint for specific RNA formation or regulation of the formation.

Genetics. The scientific study of heredity.

Genome. The genetic material in a cell.

Genomic DNA. DNA sequence as it appears in the genome.

Genomic library. See DNA library.

Genotype. The genetic make-up or hereditary blueprint. See DNA identification analysis.

Genotyping. See DNA identification analysis.

Germ cell. Sex cell.

Gestation. The time period from conception to birth.

Glucose 6-phosphatase. The enzyme that catalyses the conversion of glucose 6-phosphate to glucose plus inorganic phosphate.

Half-life ($t_{1/2}$). The time required for a radioisotope to decrease to one-half of its original quantity, (the $t_{1/2}$ for ^{32}P is 14 days).

Haploid (N). The nuclear state in which only one member of each homologous chromosome pair is present, for example, human gametes (egg or sperm).

Haplotype. See DNA haplotype.

Hardy-Weinberg law. In a large random intrabreeding population, not subjected to excessive selection or mutation, the gene and genotype frequencies will remain constant over time. The sum $p^2 + 2pq + q^2$ applies at equilibrium for a single allele pair where p is the frequency of allele A, q the frequency of a, p^2 the frequency of genotype AA, q^2 the frequency of aa and $2pq$ the frequency of Aa.

Hemoglobin. The oxygen carrying pigment of red blood cells composed of two α-chains, two ß-chains, and heme groups.

Hemoglobinopathy. A disease resulting from a defect in the ß- or α-globin genes.

Hemophilia. An X-linked disorder which results in reduced blood clotting ability.

Hereditary material. See deoxyribonucleic acid.

Heterochromatin. Nontranscribed, late replicating, repetitious chromosomal DNA.

Heterogametic sex. The sex that produces gametes each with either an X or Y chromosome. In mammals, the Y-bearing sperm are male determinant and the X female.

Heterologous. Refers to segments of DNA derived from different sources.

Heterozygous. The presence of different alleles at corresponding homologous chromosome loci.

Histones. Chromosomal DNA-binding proteins.

HLA (human leukocyte antigen). Antigens located on the surface of most cells, excluding red blood cells and sperm. These antigens are closely associated with transplant rejection.

Homogametic sex. The sex that produces gametes with only one type of sex chromosome. In mammals, each egg carries one X chromosome. Sperm carry an X or a Y chromosome.

Homologous. Refers to the chromosome pairs found in diploid organisms. The human has 22 homologous pairs of autosomes (non-sex chromosomes) plus two sex chromosomes per nucleus The members of each pair have an identical sequence of genes; however, the alleles at corresponding loci may be identical (homozygous) or different (heterozygous).

Homozygous. The presence of identical alleles at corresponding homologous chromosome loci.

Huntington disease. A severely debilitating, autosomal dominant, neurological disorder.

Hybridization. The pairing of complementary strands of DNA, or RNA and DNA, derived from different sources.

Hydrogen bond. A relatively weak bond between a hydrogen (H) atom, covalently bound to a nitrogen (N) or oxygen (O) atom, and another atom. These bonds bind complimentary DNA strands. They can be broken by increasing temperature.

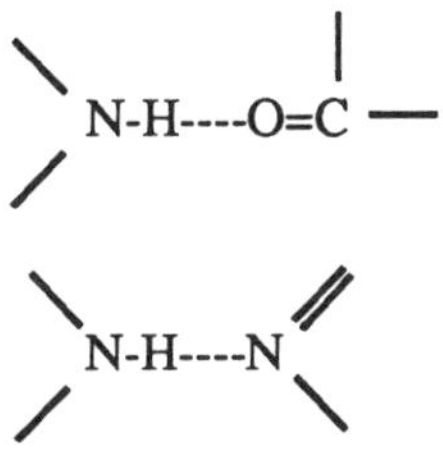

N-H----O=C —

N-H----N

Hypervariable region. A segment of a chromosome characterized by considerable variation in the number of tandem repeats at one or more loci.

Hypothesis. A reasonable suggestion that remains to be proven.

Hypothyroid. A state of thyroxin deficiency in humans. If untreated, severe mental retardation can develop in an affected individual.

Identification analysis. See DNA identification analysis.

Identity testing. See DNA identification analysis.

Identity profile. See DNA identification analysis.

In vitro. Means “in glass” and refers to a biological process carried out in the laboratory separate from an organism.

In vivo. Refers to a biological process within a living organism.

Inborn error. A gene defect resulting in disruption of a metabolic pathway.

Inbreeding. Reproduction between related individuals.

Inclusion. A crime suspect or putative father DNA identity profile matching that of a crime sample or offspring paternally-derived alleles.

Incriminating value (IV). The ratio of the probability *(x)* of a match for the characteristic measured if the suspect and crime specimens are from the same source to the probability *(y)* if they are from different sources. $IV = x \div y$.

Insertion. Addition of one or more nucleotides into a DNA strand. This may result in a gene mutation.

Intergenic. Nucleotide sequences located between genes.

Intervening sequence (IVS). See intron.

Intron. A segment of DNA within a gene of a multicellular organism that is transcribed into mRNA but excised and degraded prior to translation. A number of introns of variable length may be present in a gene separating the exons.

Inversion. A reversed chromosome segment.

Isozyme. One of several forms of a specific enzyme. Different isozymes may be most efficient under different environmental conditions such as temperature and pH.

IVS. See intron.

Klenow fragment. The portion of bacterial DNA polymerase I lacking 5' to 3' exonuclease activity.

λgt 11. A lambda phage cloning vector.

Lambda phage (λ phage). A double stranded *E. coli* bacteriophage.

Likelihood. A statistical measure of the correctness of a hypothesis given certain observations.

Linkage disequilibrium. The phenomenon of a specific allele of one locus being associated or linked to a specific allele or marker of another locus, on the same chromosome, with a greater frequency than expected by chance.

Linkage. A measure of association between loci. Loci on different chromosomes are nonlinked. Those close together on the same chromosome are closely linked and are usually inherited together.

Litigation. To contest in law.

Locus (plural, loci). A specific position on a chromosome.

Lymphoblast. A cell type derived from antigen, for example, phytohemagglutin (PHA), stimulation of T lymphocytes.

Marker. See size marker.

Mean (arithmetic). A statistical measure of central tendency equating to an arithmetic average of a group of values. The mean of $10 + 20 + 50 + 5 + 15 = 100 \div 5 = 20$.

Median. A statistical measure of central tendency equating to the mid value in a ranked series. The median of the series 5,10,15,20,50 is 15.

Meiosis. The process whereby a sex cell nucleus, after chromosomal replication, divides twice to form four nuclei each with one-half the original chromosome number.

Melt. The process of disrupting the hydrogen bonds linking complementary DNA strands.

Mendel's laws. (1) Segregation: during meiosis only one member of each homologous chromosome pair is transferred to a specific gamete. (2) Independent assortment: during meiosis the members of the different homologous chromosome pairs assort independently when transferred to a specific gamete, for example, AA' and BB' homologous chromosome pairs could give rise to AB, AB', A'B or A'B' possible gametes.

Mer. See 17-mer.

Methylation (me). One form of methylation, the most common in mammals, involves the conversion of cytosine to 5-methyl cytosine. Methylation can prevent cleavage of DNA at a restriction enzyme recognition site, for example, Hpa II cleaves at CCGG but not at $C^{me}CGG$

Microfuge. A high speed (usually 10,000+ rpm) centrifuge for centrifugation of small (usually < 2 ml) specimens.

Minisatellites. Regions of tandem repeat sequence DNA scattered throughout animal (and probably plant) genomes.

Mismatch. Bases that do not match in "complementary" DNA strands. Depending on the blot wash stringency conditions, some mismatch can be tolerated between hybridized sample and probe DNA complementary regions, for example,

```
             G
   A C C T     T T G
≈                      ≈
   T G G A     A A C
             T
```

Missense mutation. A codon change resulting in a code for a different amino acid in a protein.

Mitochondrion. A DNA-containing cytoplasmic organelle of eukaryotes. Mitochrondia are referred to as the powerhouses of the cell because they are sites of ATP production.

Mitosis. The process whereby a somatic cell nucleus, after chromosomal replication, divides to form two identical nuclei.

Monochromatic light. Radiation consisting of a single wavelength.

Monomorphic bands. DNA fragments of specific sizes found in most individuals. Each different size monomorphic fragment is detected by cleaving genomic DNA with a specific restriction enzyme and hybridizing with a specific monomorphic probe. These fragments provide excellent markers for use in quality control especially as related to band shifts.

Monozygotic. Twins produced from a single zygote which later splits. The genomes are, therefore, identical

Mosaic. An individual composed of two genetically different cell lines originally derived from the same zygote.

mRNA. See Spliced mRNA and exons.

Multiallele. Refers to a number of different possible alleles at a specific locus. See allele.

Multilocus. Refers to a number of different loci or positions in the genome. See locus.

Multifactorial trait. A trait characterized by genetic as well as nongenetic factors.

Mutation. Any change in the sequence of genomic DNA. This may result from one or many base pair changes.

Myoglobin. The heme protein of vertebrate muscle consisting of 152 amino acids in the human.

Neutral mutation. Any change in the sequence of genomic DNA that does not affect the physical make-up of the individual.

Nonsense codon. A codon that has arisen due to a mutation in a normal codon and that causes termination of translation.

Nucleic acid. See DNA and RNA.

Nucleotide. A building block unit of DNA or RNA. See deoxyribonucleic acid.

Nucleus. The genome-containing (membrane-bound) structure of eukaryotic organisms.

Oligo-dT. A polymer of deoxyribothymidylate used in the purification of mRNA.

Oligonucleotide. A polymer composed of a few, usually less than 100, nucleotides. Oligonucleotides are usually synthesized by automated machinery and used as primers in the PCR and as probes.

Organelle. See mitochondrion and plasmid.

Ovum. Female gamete or egg cell.

Palindrome. A DNA site where the base order in one strand is the reverse of that in the complementary strand, for example, 5'GAATTC 3', 3'CTTAAG 5'.

Patent. Documented exclusive legal right to a process.

Paternity. Refers to the biological father of an offspring.

Paternity index (PI). The ratio of the probability *(x)* that the putative father is the biological father to the probability *(y)* that a random man is the biological father. PI = $x \div y$.

Pathogen. A disease causative agent such as a virus.

PCR. See Polymerase chain reaction.

Pedigree. A chart outlining an ancestral history.

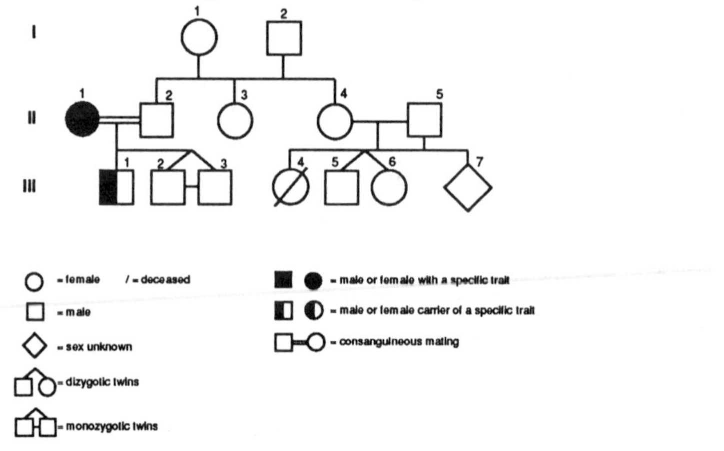

Pellet. The button of particulate material formed after a suspension has been centrifuged.

Percentile. A division of the members of a group into 100 parts, for example, a value at the 95th percentile indicates that 95 percent of the values are below that value.

pH meter. An instrument used to measure the degree of acidity or alkalinity of a solution. pH 7 (neutral), pH < 7 (acidic), pH > 7 (alkaline).

Phenol (carbolic acid). A poisonous and caustic organic compound used in the isolation of DNA from cellular proteins.

Phenotype. The physical make-up of an individual as defined by genetic and nongenetic factors.

Phenylalanine hydroxylase. The enzyme that catalyzes the conversion of L-phenylalanine to L-tyrosine.

^{32}P. Radioactive phosphorous used to label DNA probes.

PKU (phenylketonuria. An inborn error of phenylalanine metabolism. See phenylalanine hydroxylase. If untreated, severe mental retardation can develop in an affected infant.

Plaintiff. The complaining party in a litigation.

Plaque. A clear circular region formed by the viral lysis of bacterial colonies growing on a solid such as an agar support.

Plasmid. An extrachromosomal, circular, double stranded DNA element native to certain bacteria. Plasmids are used as vehicles to replicate cloned (recombinant) DNA sequences.

Plastid. A plant cytoplasmic DNA-containing organelle such as a chloroplast. Each, as with mitochondria, contains chromosomes whose genes together with those of nuclear DNA control the organelle function.

Poach. Illegal taking of wildlife.

Poly A-tail. The adenine nucleotide polymer attached to the 3' end of pre-mRNA.

Polymerase chain reaction (PCR). An in vitro process whereby specific regions of double-stranded DNA can be amplified more than a millionfold using thermal stable Taq polymerase. Oligonucleotide primers must be annealed to the target DNA sequence 5' flanking regions. The PCR may be likened to a molecular xeroxing machine.

Polymorphism. More than one form. See RFLP.

Precision. A measure of reproducibility of a test result.

Prehybridization. The process of incubating a DNA blot with a hybridization solution to, in part, block cross-reactive sites. This precedes the addition of labeled probe.

Primer DNA. A short, perhaps 20-mer, oligonucleotide annealed to the 5' end of a DNA template. The primer provides an initiation point for addition of deoxyribonucleotides in DNA replication.

Print profile. See DNA identification analysis.

Probability. A statistical means of describing uncertainty.

Probe. A single-stranded segment of DNA, or mRNA, capable of being tagged with a tracer, such as ^{32}P, and hybridized to its complementary sequence.
Profiling. See DNA identification analysis.
Prokaryote. A unicellular organism lacking a membrane-bound nucleus, for example, bacteria. Prokaryotes do not undergo mitosis.
Proportion. The fraction or percent of individuals or measurements in a specific subgroup of the total group under consideration.
Protein. A biological molecule composed of amino acids.
Proteinase K. A hydrolytic enzyme used in the digestion of proteins to amino acids.
Putative father. A man accused but not proven to be the biological father of an offspring.
Quality control (QC). A monitoring system to ensure assay results adhere to appropriate precision and accuracy levels.
Race. As applied to the human species (*Homo sapiens*), a population subgroup with a gene pool characteristic only of that group.
Radiation. As related to radiation protection, is energy emission from a radioactive source.
Radionuclide. A radioactive element. = ^{32}P (radioactive phosphorus) is a high energy ß-nuclide.
Random "man". Any individual in the population whose DNA profile matches a crime specimen profile or an offspring's paternally-derived profile.
Rape. Sexual intercourse between a man and a woman without the woman's consent. This may occur by deception or by force.
Reading frame. The DNA codon sequence transcribed into pre-mRNA.
Recessive. A trait phenotypically expressed only when an abnormal gene is present in the homozygous state.
Recombinant DNA. DNA formed by the union of two heterologous DNA molecules, for example, the ligation of a human growth hormone gene into a plasmid.
Recombinant DNA technology. Refers specifically to in vitro ligation of two heterologous DNA molecules, for example, the splicing of a human DNA tandem repeat sequence into plasmid pBR322. As a generalization, the technology refers to the new approach of direct DNA analysis.
Recombination. Combinations of genes in offspring different from those in the parents due to independent assortment and crossing-over.
Repetitious DNA. DNA consisting of repeated chromosomal nucleotide sequences.
Repetitive sequence. A repeated series of bases in a DNA molecule.
Restriction endonucleases. Enzymes (molecular scissors) that cleave double-stranded DNA at specific palindromic base recognition sequences. The sequences are usually different for each enzyme. Restriction enzymes are named according to the bacterial species of origin. Eco RI is derived from *E. coli* RY 13. Bam H I, for example, cleaves as follows:

↓
5'-G—G-A-T-C-C-3'
3'-C-C-T-A-G—G-5'
↑

to form staggered sticky ends at ↑↓; other enzymes such as-Hind II cleave to form blunt ends ↕.

Restriction enzymes. See restriction endonucleases.

Restriction fragment. A piece of DNA resulting from digestion with a restriction endonuclease.

Restriction fragment length polymorphism (RFLP). The different length fragments of DNA produced by the action of a restriction enzyme at a specific polymorphic site. If the restriction endonuclease recognizes the variable site, then two fragments are produced; if not, only one is formed.

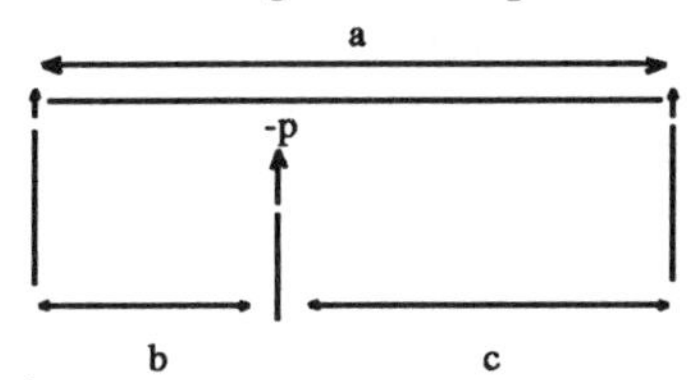

↑ = constant recognition site
↑-p = variable (polymorphic recognition site)
"a" = fragment length if ↑-p site is absent
"b" and "c" = fragment lengths if ↑-p site is present

An individual may be homozygous or heterozygous, that is, the site may be present or absent on both chromosomes of the homologous pair, or it may be present on one and absent on the other. It should also be noted that different fragment lengths are produced if the distance between the fixed recognition sites changes. This can occur because of the insertion or deletion of nucleotides, or because of a variable number of tandem repeats. See VNTR.

Reverse transcriptase (RNA-dependent DNA polymerase). An enzyme that catalyzes DNA synthesis using RNA as a template.

RFLP. See restriction fragment length polymorphism.

Ribonucleic acid (RNA). Single-stranded nucleic acid molecules composed of the ribonucleosides adenosine, guanosine, cytidine, and uridine. The three types of RNA include mRNA, rRNA, and tRNA.

Ribosomal RNA (rRNA). RNA molecules that, together with specific proteins, form the subunits of the ribosomes.

Ribosome. A ribonucleoprotein particle consisting of two subunits, each with one rRNA and approximately 25 protein molecules. The ribosomes function as sites for mRNA translation.

Ribosomes. The ribonucleoprotein units in which translation occurs.

RNAase. An enzyme capable of degrading RNA.

rRNA. See ribosomal RNA.

Satellite DNA. See repetitious DNA, and Alu.

Scheduled quantities. Quantities of radioisotopes as defined in atomic energy control.

Screening. See library screening.

Segregation. See Mendel's laws.

Self-replication. A duplication process whereby each strand of the DNA double helix acts as a template for replication of its complementary strand.

Semen. Sperm plus seminal fluid.

17-mer. An oligonucleotide consisting of 17 nucleotides.

Sex chromosome. A chromosome containing sex factor determinants.

Sex-linked. See X-linked, X-chromosome.

Sheared DNA. DNA broken into fragments by mechanical action.

Shewhart rules. A recommended set of actions to be taken in the event that one or more controls in an assay deviate significantly from their respective means. The rules were originally designed for manual radioimmunoassays; however, they provide a good base for establishing quality control systems for many other types of testing.

Size marker. DNA fragments of known molecular weight and base pair length, such as λ-phage digested with the restriction enzyme Hind III, run on electrophoresis gels for the determination of DNA sample fragment sizes.

Somatic cells. All cells of eukaryotes excluding gametes.

Satellite DNA. DNA located in telomeres and centromeres.

Southern blotting. A technique developed by E. Southern for the direct transfer of DNA fragments from an agarose gel onto a solid support such as a nylon membrane The transfer occurs by salt solution capillary action.

Specific activity (SA). A measure of the incorporation of radiolabeled nucleotide into a probe, for example, counts per minute (cpm) per microgram of probe.

Spectrophotometer. An instrument used to measure the intensity of a specific wavelength of light entering and leaving a solution.

Spliced mRNA. mRNA after removal of the intron coding regions from the premessenger and linking of the exon coding portions.

Splicing. See spliced mRNA.

Standard deviation (SD). A statistical measure of the dispersion of values around a mean.

Standard error (SE). A statistical measure of the dispersion of a series of sample means around the population mean.

Standards. Test reagents, of known concentration used to establish standard curves for the determination of unknown sample concentrations.

Stop codon. One of UGA, UAG, or UAA signalling the end of mRNA translation into protein.

Stringency. The buffer salt concentration and temperature used in the DNA blot wash posthybridization process. As these parameters are changed, the degree of binding of probe to target DNA changes.

Strip. Removal, by melting of hydrogen bonds, of hybridized probe from DNA blots.

Structural gene. A DNA sequence coding for a specific message.

Subpoena. A written command summoning a specific individual to appear in court under penalty for failure to do so.

Supernatant. The liquid portion separated after a suspension is centrifuged.

Suspect specimen. A specimen obtained from a suspect for analysis.

T_{25}. A tissue culture flask with a culture surface area of 25 square centimeters.

Taq polymerase. A DNA polymerase isolated from the bacterium *Thermus aquaticus* that lives in hot springs. This enzyme is capable of withstanding high temperatures and is, therefore, very useful in the polymerase chain reaction (PCR).

Tag. A tracer linked into a molecule, for example, ^{32}P labeling of a DNA probe.

Tandem repeat. The end to end duplication of a series of identical or almost identical stretches of DNA. See VNTR.

Target DNA. The DNA sequence to be hybridized to a specific probe.

Template. The single-stranded DNA blueprint for complementary strand assembly or the production of pre-mRNA.

Tetrads. Combinations of replicated homologous pairs (four chromatids) of chromosomes.

Thalassemias. Anemias resultant from α- and ß-globin chain imbalances.

Tissue culture. Growth of cells in vitro.

TLD (thermoluminescent dosimeter). A film-badge type monitor used to record radiation exposure from a radioactive source.

Transcription. Production of pre-mRNA using DNA as a template and RNA polymerase to catalyze linking of the nucleotide building blocks.

Transfer RNA (tRNA). RNA molecules that transfer amino acids to ribosomes for protein production during mRNA translation.

Transformation. Genome modification.

Translation. The process of linking amino acids to form proteins, using mature mRNA as a template.

Translocation. Change in the position of a chromosome segment within a genome.

Triplet code. A series of three successive DNA or RNA nucleotide bases that codes for a specific amino acid, for example, AUG codes for methionine.

tRNA. See transfer RNA.

Trypsinize. The process of cleaving protein molecules, at the sites of the amino acids arginine and lysine, with the enzyme trypsin. This process is used to dislodge growing fibroblast monolayers from the surface of tissue culture vessels.

Typing (DNA typing). See DNA identification analysis and DNA typing.

Uracil. The pyrimidine base in RNA that appears in place of the thymine found in DNA.

UV transilluminator. A source of ultraviolet (UV) light used to detect ethidium bromide stained DNA.

Variable number of tandem repeats (VNTR). The variable number of repeat core base pair sequences at specific loci in the genome. See tandem repeat. The variation in length of the alleles formed from the repeats provides the basis for unique individual identification.

Vector. A self-replicating DNA molecule capable of transferring foreign DNA into a cell. For example, the human insulin gene can be cloned into the plasmid vector pBR 322 which in turn will replicate in *E. coli* cultures.

VNTR. See variable number of tandem repeats.

Wash. The process of removing nonbound or loosely bound probe from blots after hybridization. This process reduces background interference.

Watson-Crick model. Refers to the DNA molecule that forms a double helix ladder with the complementary strands held by hydrogen bonds between specific base pairs. See deoxyribonucleic acid.

Wipe test. A radioactivity contamination test whereby an ethanol wetted filter paper is wiped over the test area and the radioactivity determined.

X-chromosome. A chromosome responsible for sex determination. Two copies are present in the genome of the homogametic sex and one copy in the heterogametic sex. The human female has two X-chromosomes and the male one X.

X-linked. Genes on the X-chromosome.

Y-chromosome. A chromosome responsible for sex determination in the heterogametic sex. This occurs in male (XY) mammals and female (ZW) birds.

Zygosity. Twin development from one or two zygotes. If one, the twins are identical (monozygotic); if two, they are fraternal (dizygotic).

Zygote. The diploid cell resulting from the union of a haploid egg and sperm.

GENERAL REFERENCES

Black HC. 1978. *Black's Law Dictionary*. 5th ed. West Publishing Company, St. Paul, MN.

Dorland's Illustrated Medical Dictionary. 1988. 27th ed. WB Saunders, Philadelphia.

King RC and Stanfield WD. 1985. *A Dictionary of Genetics*. 3rd ed. Oxford University Press, New York.

Lapedes DN, ed. 1989. *McGraw-Hill Dictionary of Scientific and Technical Terms* McGraw-Hill, New York.

Windholz M, ed. 1983. *The Merck Index: An Encyclopedia of Chemicals and Drugs*. 10th ed. Merck and Company, Rahway, NJ.

ABBREVIATIONS AND SYMBOLS

A	-	adenine
α	-	alpha
AIDS	-	acquired immune deficiency syndrome
~	-	approximately
$\simeq$	-	approximately equal
Alu I	-	*Arthobacter luteus*
AMP-FLP	-	amplified fragment length polymorphism
ASO	-	allele-specific oligonucleotide
ATP	-	adenosine triphosphate

ß - beta
bp - base pair(s)
BSA - bovine serum albumin
C - carbon
°C - centigrade
C - cytosine
^{14}C - radioactive carbon
Ca - calcium
cDNA - complementary DNA
CF - cystic fibrosis
Ci - curie
cm - centimeter(s)
CMV - cytomegalovirus
CO_2 - carbon dioxide
cpm - counts per minute
CV - coefficient of variation
CVS - chorionic villus sample (specimen)
dATP - deoxyadenosine triphosphate
dCTP - deoxycytosine triphosphate
dGTP - deoxyguanosine triphosphate
DMSO - dimethyl sulfoxide
DMD - Duchenne muscular dystrophy
DNA - deoxyribonucleic acid
DNAase - deoxyribonuclease
dNTP - deoxynucleotide triphosphate
dT - deoxythymidine
DTT - dithiothreitol, Cleland's reagent
dTTP - deoxythymidine triphosphate
dUTP - deoxyuridine triphosphate
DZ - dizygotic twin
E. coli - *Escherichia coli*
EBV - Epstein-Barr virus
Eco RI - *Escherichia coli RY 13*
EDTA - ethylenediaminetetraacetic acid
EtBr - ethidium bromide
EtOH - ethanol
F - coefficient of inbreeding
FBI - Federal Bureau of Investigation
FCS - fetal calf serum
5' - carbon atom 5 of deoxyribose
G - guanine
λ - gamma
g - gram(s)

g - gravity
GIBCO - Grand Island Biological Company
> - greater than
h - hour(s)
^3H - tritium, radioactive hydrogen
Hae III - *Haemophilus aegyptius*
HBSS - Hanks' balanced salt solution
HCl - hydrochloric acid
HEPES - (N-[2-hydroxyethyl] piperazine-N'[2-ethanesulfonic acid])
Hind II - *Haemophilusi influenzae rd*
Hinf I - *Haemophilus influenzae* R_f
HIV - Human Immunodeficiency Virus
HLA - human leukocyte antigen
HPR - horseradish peroxidase
HPLC - high pressure liquid chromatography
HTE - high Tris-EDTA buffer
H_20 - water
H_20_2 - hydrogen peroxide
HVR - hypervariable region
ID - identity
IQ - intelligence quotient
IV - incriminating value
IVS - intervening sequence, intron
kb - kilobase(s)
kBq - kilobaquerel
kg - kilogram(s)
λ - lambda
l - liter(s)
< - less than
LB - Luria-Bertani
LiCl - lithium chloride
LR - likelihood ratio
LTE - low Tris-EDTA buffer
μ - micro
μg - microgram(s)
μl - microliter(s)
μSv - microsievert
m - meter(s)
m - mobility
M - molar
MEM - minimum essential medium
Mg - magnesium
mg - milligram(s)

$MgCl_2$ - magnesium chloride
$MgSO_4$ - magnesium sulfate
μg - microgram(s) (10^{-6})
μl - microliter(s) (10^{-6})
min - minute(s)
ml - milliliter(s)
mm - millimeter(s)
mM - millimolar
mmol - millimole
mo - month
mRNA - messenger RNA
mSv - millisievert(s)
mtDNA - mitochondrial DNA
MW - molecular weight
MZ - monozygotic twin
n - number
NaCl - sodium chloride
Na_2EDTA - disodium - EDTA
NH_40Ac - ammonium acetate
ng - nanogram(s) (10^{-9})
NH_4Cl - ammonium chloride
nm - nanometer(s)
OAc - acetate
OD - optical density
OH - hydroxyl
oligo - oligonucleotide(s)
P - phosphorus
p - probability
^{32}P - radioactive phosphorus
PAGE - polyacrylamide gel electrophoresis
PBS - phosphate buffered saline
PCR - polymerase chain reaction
PEG - polyethylene glycol
pg - picogram(s)(10^{-12})
pH - logarithm of reciprocal of hydrogen (H) ion concentration
PHA - phytohemagglutinin
PI - paternity index
PKU - phenylketonuria
pp - page(s)
PP - probability of paternity
PVP - polyvinyl pyrolidone
QA - quality assurance
QC - quality control

r - coefficient of relationship
RBC - red blood cell(s)
RCMP - Royal Canadian Mounted Police
rem - roentgen equivalent man
RFL - restriction fragment length
RIA - radioimmunoassay
RFLP(s) - restriction fragment length polymorphism(s)
RNA - ribonucleic acid
RNAase - ribonuclease
rpm - revolutions per minute
rRNA - ribosomal RNA
RT - room temperature
^{35}S - radioactive sulfur
SA - specific activity
Sau 3A - *Straphlococcus aureus 3A*
SD - standard deviation
SDS - sodium dodecyl sulfate, lauryl sulfate
SE(SEM) - standard error (standard error of the mean)
SF - statistical frequency
sec - second(s)
17-mer - 17 nucleotide oligonucleotide
Sma I - *Serratia marciscens Sb*
SSC - sodium chloride sodium citrate
SSO - sequence-specific oligonucleotide
SSP - sequence-specific probe
SSPE - sodium chloride-sodium phosphate-EDTA
Σ - sum of
T - thymine
T_4 - thyroxine
T_{25} - 25 cm^2 growth surface area
$T_{1/2}$ - half-life
T_{75} - 75 cm^2 growth surface area
TAE - Tris-acetate-EDTA
Taq I - *Thermus aquaticus YT1*
TBE - Tris-borate-EDTA
TE - Tris EDTA buffer
3' - carbon atom 3 of deoxyribose
TLD - thermoluminescent dosimeter
Tris - tris (hydroxymethyl) methylamine
tRNA - transfer RNA
TSH - thyroid stimulating hormone
TWGDAM - Technical Working Group on DNA Analysis Methods

U	-	unit(s)
U	-	uracil
UV	-	ultraviolet
V	-	voltage, volt(s)
wk	-	week
VNTR	-	variable number of tandem repeats
vol	-	volume
WBC	-	white blood cell(s)
$\overline{X}$	-	mean
χ^2	-	chi squared
yr	-	year(s)

Name Index

C

D

E

F

G

H

I

J

K

L

M

N

O

P

Q

R

S

T

Subject Index

A

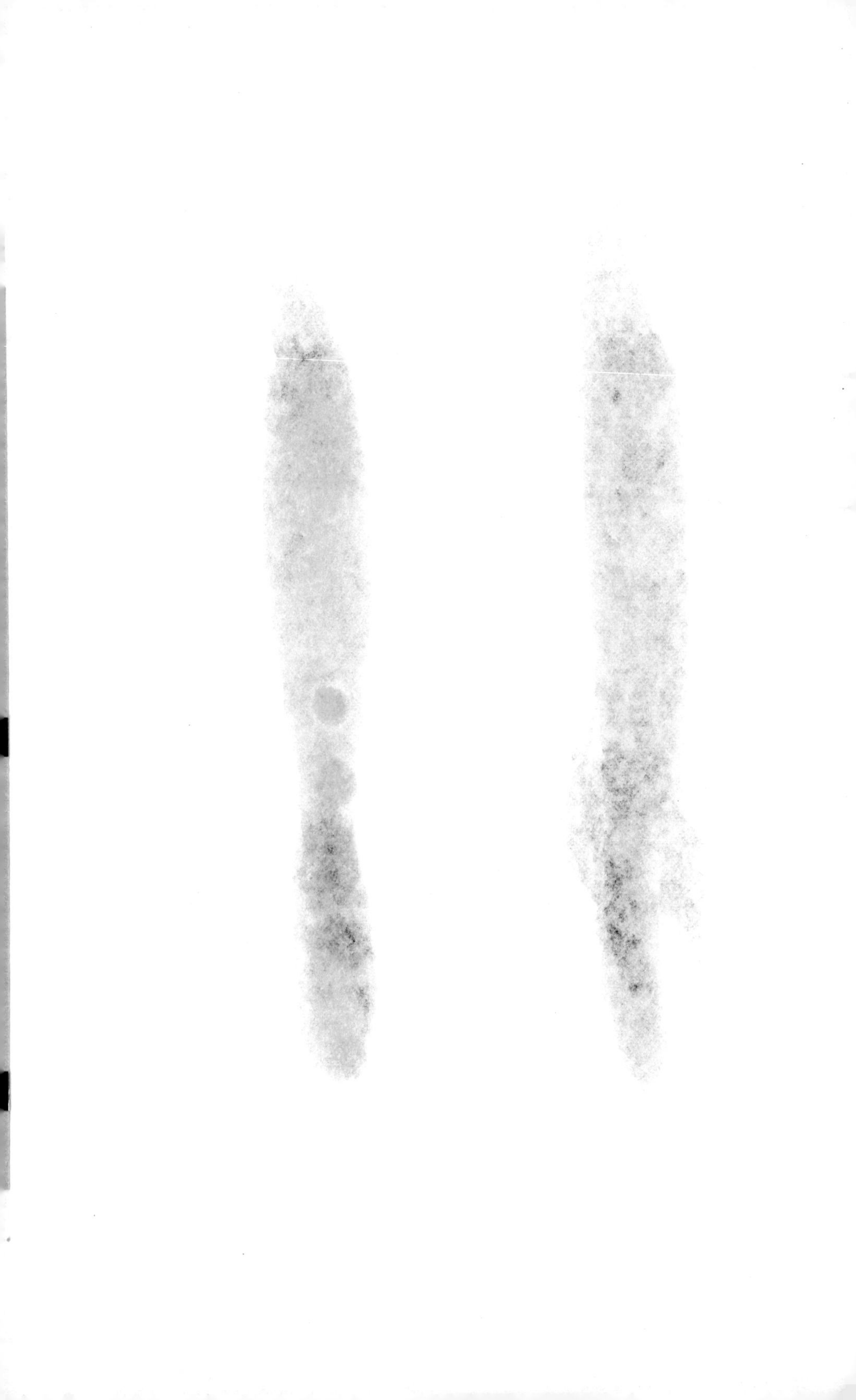

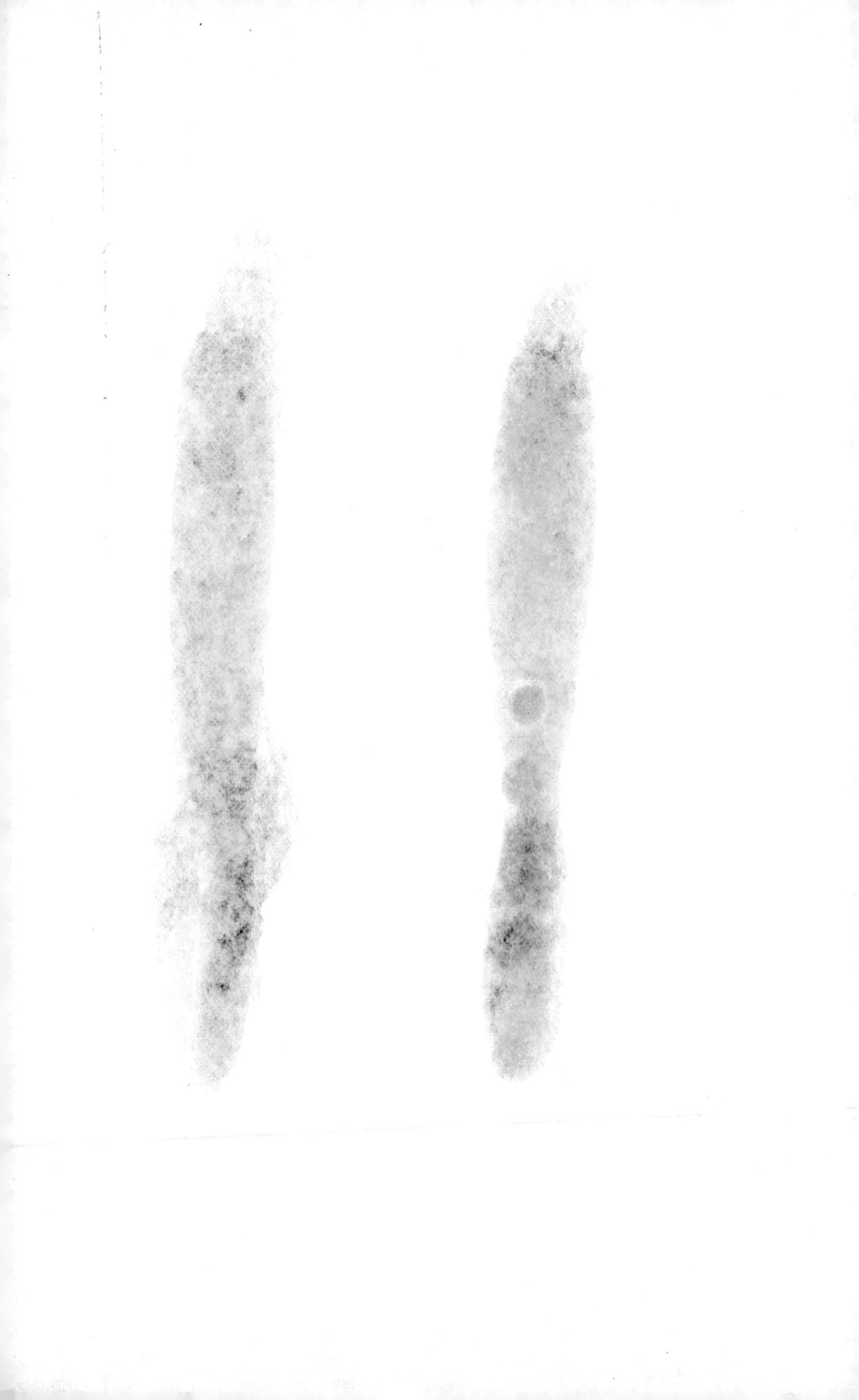